Control of Human Voluntary Movement

Control of Human Voluntary Movement

Second edition

John Rothwell

Head of Movement Disorder Section, MRC Human Movement and Balance Unit, Institute of Neurology, London, UK

CHAPMAN & HALL

London · Glasgow · New York · Tokyo · Melbourne · Madras

Published by Chapman & Hall, 2–6 Boundary Row, London SE1 8HN, UK

Chapman & Hall, 2–6 Boundary Row, London SE1 8HN, UK

Blackie Academic & Professional, Wester Cleddens Road, Bishopbriggs, Glasgow G64 2NZ, UK

Chapman & Hall Inc., One Penn Plaza, 41st Floor, New York NY 10119, USA

Chapman & Hall Japan, Thomson Publishing Japan, Hirakawacho Nemoto Building, 6F, 1–7–11 Hirakawa-cho, Chiyoda-ku, Tokyo 102, Japan

Chapman & Hall Australia, Thomas Nelson Australia, 102 Dodds Street, South Melbourne, Victoria 3205, Australia

Chapman & Hall India, R. Seshadri, 32 Second Main Road, CIT East, Madras 600 035, India

First edition 1986 Published by Croom Helm

Second edition 1994

© 1986, 1994 John Rothwell

Typeset in 10½ Palatino by ROM-Data Corporation, Falmouth, Cornwall

Printed in Great Britain at the University Press, Cambridge

ISBN 0 412 47700 9

A catalogue record for this book is available from the British Library

Library of Congress Cataloging-in-Publication data available

∞ Printed on permanent acid-free text paper, manufactured in accordance with ANSI/NISO Z39.48–1992 and ANSI/NISO Z39.48–1984 (Permanence of Paper).

Cover photograph supplied courtesy of Dr J. G. Colebatch

Contents

Introduction: Plans, strategies and actions $\quad$ 1

The most remarkable thing about moving is how easy it is. It is only when we watch someone who cannot move, perhaps after a stroke, or someone whose movements continually go wrong, like the contortions of a child with athetoid cerebral palsy, that we are reminded of the problems of movement control with which our nervous system copes so uncomplainingly. This book will deal with voluntary movements of the limbs and their disorders, which are particularly well illustrated by movements of the human hand and arm.

It is sometimes fashionable to admire the exquisite anatomical perfection of our bodies, and to imagine that the complex arrangement of muscles, tendons and joints is responsible for the vast repertoire of movements which we can perform. Yet this is not entirely true. For example, the **anatomical** difference between the human hand and that of the Old World monkeys is small, but the abilities to write, type and play musical instruments remain characteristically human. The reason for this is that the nervous control of movement has been improved rather than there being any change in the design of the mechanical parts.

Unfortunately, little is known about the overall strategies which the brain uses to plan and execute a movement, but there are a number of observations which have been made which point towards some general principles of movement control. For example, during the entire act of writing my signature I need only take one conscious thought, that is to take up my pen and write my name. The entire sequence of muscle contractions and movements which follow can occur entirely automatically, although I am, of course, free to intervene at will at any time. Even more remarkable is the fact that no matter which muscles I use to write, the writing turns out to be recognizably mine each time. The idea of the movement has been transformed into the appropriate action (Figure 1.1).

Thus, it is possible to distinguish the **idea** of the movement from the **plan** which is used to carry it out. The idea which I have of the form of

Figure 1.1 Two specimens of the author's handwriting. The signature at the top was written with a small fine pen and measured only 2 cm from start to finish; the signature at the bottom was written with a large felt-tip pen on a full sheet of A4 and measured 20 cm from start to finish. The large writing was performed mainly using the wrist, elbow and shoulder whereas the small writing mainly used the small muscles of the hand itself. Despite the differences in the muscles used, the character of the handwriting remains unmistakably the same.

my signature remains unchanged, while the motor plan is updated to suit the requirements of whatever muscles I use. But formation of the plan is only the first stage in movement. The nervous system has a good deal more work to do in timing the contraction of some muscles or the relaxation of others, and in checking the progress of the evolving movement before an action can be said to be successfully underway. The **execution** of the motor plan forms the last stage in the performance of a movement.

The essential stages in the production of any voluntary movement may be envisaged as following the sequence: (i) idea; (ii) motor plan; (iii) execution of programme commands; (iv) move. The stages in this sequence form an hierarchical chain of command, from the abstract concept of moving, to the practicalities of contracting the right muscles at the right time.

1.1 *Categories of muscles*

We all know that movements are complicated: that there are many tens of muscles involved in almost any simple movement. We owe to two pioneers of muscle physiology, Duchenne and Beevor, the terms used to categorize muscle action and which help to describe a movement in detail.

The French neurologist, Duchenne, was perhaps the first to point out that a simple antagonistic relationship can only occur between two muscles at a hinge joint.

However, most muscles have more than one action, either because they span more than one joint or because they can produce rotation as well as flexion/extension at the same joint. Thus, the vastus medialis flexes the hip as well as extending the knee, the finger flexors flex both the wrist and the fingers, and biceps supinates the forearm as well as flexing the elbow, and hence is best displayed in the 'body-builder' pose. Although the combined action of the muscle may sometimes be useful, such as the role of biceps in picking up a piece of food and raising it to the mouth, it is often necessary to activate other muscles as **fixators** in order to limit the extent of the desired movement. A common example is the use of the wrist extensors to fixate the wrist while the finger flexors are used as prime movers during exploration and manipulation of an object.

Finally, it is important to remember that, as the song has it, the toe bone is connected to the foot bone, which is connected to the ankle bone, and so on up to the head bone. Movement is possible at almost all these joints, and it is a physical fact that forces exerted at any one joint are communicated through other intervening joints to the point of body contact with the supporting surface. Because of this, lifting a weight with the hand requires that the force be taken at the wrist, elbow, shoulder and back, down to the buttocks if seated, and down to the feet if standing. **Postural** muscle contractions occur which not only stabilize the individual joints, but also may produce subtle movements of the trunk to counterbalance movements of the arms and maintain the body's centre of gravity within the postural base.

In fact, it turns out that this categorization of muscle types may have more profound implications than serving as a descriptive tool of movement types. Most muscles do not have one stereotyped function as agonist or fixator in all types of movement. They shift their function as the movement progresses. Thus, in reaching and grasping an object, the extensors of the wrist and fingers will be activated first as agonist to open the palm, and then as fixators to maintain wrist extension, as the fingers flex onto the object in a power grip.

The importance of this distinction between the use of the same muscle for different categories of movement is shown particularly well in some patients suffering from stroke. For example, a patient may be unable to use triceps to extend the elbow, but be perfectly capable of using the same muscle as an elbow fixator during supination of the wrist by biceps. It appears that there may be different anatomical pathways serving different categories of movement, or that different parts of the central nervous system (CNS) may be used to calculate muscle activity under different conditions.

1.2 *Problems of moving*

The major problem facing the motor control system, even in the simplest movements, is not only to contract the agonist, or prime moving muscle by the correct amount and at the appropriate time, but also to time and organize the pattern of antagonist, fixator and postural muscle contractions which are necessary to accompany its action.

As in the act of writing one's name, it is clear that there are some aspects of every action that never consciously enter into one's original **idea** of movement. But that is as far as introspection can take us. From there on it is a thorny problem of motor control as to what details are specified where and when in the chain of command producing a movement.

A working hypothesis, which will be used here and in the following chapters, is that the motor **plan**, like the initial idea of the movement, does not itself contain a complete description of the motor task. During execution, different parts of the brain and spinal cord are called upon to compute, for example, the exact postural/fixator activity needed to stabilize the primary movement. A computational analogy would be that the motor plan contains a general controlling program and that various subroutines are handed over to other brain areas when necessary. These other areas may call up further subroutines *ad infinitum*.

From the point of view of our conscious voluntary control of movement, we tend to concentrate on the action of the prime moving muscle. Thus, when I stand to write on a blackboard, I concentrate mostly on the motion of my fingers and hand. The delicate and equally important adjustments of my trunk, legs and shoulders all pass relatively unnoticed. Such details of movement are thought to be organized at low levels of the movement hierarchy, rather in the way that the Civil Service is expected to formulate the details of a minister's policy statement.

Nevertheless, convenient as this chain of command may be, it cannot be the only mechanism whereby voluntary control is exerted over the muscles. With practice, we can learn to use almost any muscle in our bodies as a prime mover. Indeed, we can even exert voluntary control over postural muscle activity. When we learn to lean outwards down the mountainside when skiing, against all our better instincts, we realize how important this more direct muscle control can be. In terms of the hierarchical model of motor control, it is necessary to allow for parallel pathways through which the conscious idea of movement may interact at all levels of movement control.

In the second half of the nineteenth century, John Hughlings Jackson devised a classification of types of movement that fits neatly into this description of movement control. He proposed that movements should be graded along a scale from the most to the least automatic. Those which could be least easily controlled by voluntary effort of will, such as the simple tendon jerk, were classified as the most automatic, whereas those most directly under conscious control of the individual were put into the least automatic category. The implication is that those movements which lie at low levels of the chain of command are the most difficult to activate via direct, parallel pathways.

1.3 *A legacy*

Most of these ideas of movement control were developed intuitively by the great clinical neurologists of the last century. They remain essentially the same today even when dressed up in the latest computer jargon. In fact, the idea of a chain of command is the only justification for treating the various areas of the CNS separately in discussion and investigation of movement control. It is assumed that there is some particular task or aspect of a movement that each area is best adapted to control, whether by virtue of its anatomical connections with other areas, or because of peculiarities of its internal circuitry. It is essentially a reductionist approach and follows very naturally from the hierarchical control of movement described earlier. However, useful as this approach may be, one must beware of pushing it to extremes. There is no simple answer to the question 'What does the cerebellum/basal ganglia **do**?' simply because neither structure ever does anything on its own.

The following chapters will describe each section of the CNS concerned with movement and what is known about the operation of each during different kinds of movements in man and animals. Unfortunately, it is easier to take the nervous system apart in this way than it is to reconstruct it later.

Mechanical properties *2* of muscles

2.1 *Review of muscle anatomy*

Skeletal muscle is made up of long fibres, terminated at each end by tendinous material attached to the bone. These fibres are formed from a syncitium of cells whose walls fuse during development, and hence have many nuclei spread throughout their length. Groups of individual muscle fibres are gathered together into bundles called fascicles which are surrounded by a connective tissue sheath (Figure 2.1). The internal structure of the muscle fibre is quite complex. The main elements visible under the light microscope are the myofibrils. These run longitudinally throughout the fibre and constitute the contractile machinery of the muscle. Each myofibril is traversed by striations. Usually, the myofibrils are aligned so that the striations appear to be continuous right across the muscle fibre.

The myofibrils can be seen in more detail under a polarizing light microscope. The basic structure consists of an alternating pattern of light and dark bands known as I (isotropic) and A (anisotropic) bands, respectively. In the centre of the I band is a dark Z line. The structures which repeat between adjacent Z lines are called sarcomeres. A paler line is sometimes seen to bisect the A band, and is known as the H zone (*Hell* is the German word for light) (see Figure 2.1).

The finest details of muscle structure only become clear under the electron microscope. Each myofibril consists of longitudinally orientated fine filaments called myofilaments. The bulk of these filaments is composed of two proteins, actin and myosin. Actin is present in the smaller diameter filaments, and myosin is found in the thicker filaments. Figure 2.1 shows the relationship between the filaments and the bands seen under the polarizing light microscope. The actin filaments are attached at one end to the Z line, and are free to interdigitate at the other end with the myosin filaments. The A band represents the extent of the myosin filaments. The I and H bands represent the regions where there is no overlap between myosin and the actin filaments; the I band contains only thin filaments, and the H band only thick filaments. The M line is composed of thickenings which connect adjacent myosin filaments.

The sliding filament hypothesis of muscle contraction

This hypothesis was formulated in the early 1950s after a single striking observation with the improved electron microscope (by H.E. Huxley and Hanson, 1954), and with the new polarizing light microscope (by A.F. Huxley and Niedergerke, 1954). When a muscle contracts, the length of the thick and thin myofilaments remains constant, and they appear to slide over one another. This observation solved the problem

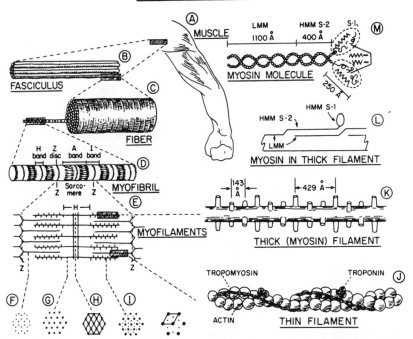

Figure 2.1 Structure of skeletal muscle. Panels A to M show gradually increasing details of structure from whole muscle (A) to myofibrils (E); F, G, H, I show cross-sections of myofibrils at different points throughout their length; G shows a cross-section where only thick filaments are present; F is a section where only thin filaments are present. The thin filaments (J) are made up of a core of actin molecules, together with troponin and tropomyosin. Thick filaments are made up of a myosin unit, as shown in L and M. Each myosin molecule consists of two parts, light meromyosin (LMM) and heavy meromyosin (HMM). The latter part includes the myosin heads, or S1 components. (From Gordon, 1982, Figure 7.8; after Bloom and Fawcett (1970), *A Textbook of Histology*, 9th edn, W. B. Saunders, Philadelphia.)

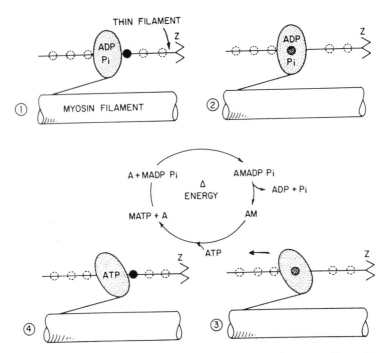

Figure 2.2 Postulated mechanism of cross-bridge cycling by actin and myosin filaments. The head of the myosin filament is stippled and forms the cross-bridge, or actin–myosin bond between the two filaments. Active sites on the actin filament are outlined by broken circles. (1) No bonds between filaments; (2) Initial attachment of myosin head to one of the active sites on the actin filament. The attachment takes place only in the presence of Ca^{2+} ions; (3) Formation of strong actin–myosin bond. This process causes a conformational change in the angle of the myosin head, which produces a relative movement of actin and myosin filaments across one another. ADP and phosphate ions are lost from the myosin; (4) If ATP is available, the actin–myosin bond can be broken. Subsequent hydrolysis of the ATP by the myosin ATPase returns the cycle to stage 1. If no ATP is available, the strong actin–myosin bond remains intact, as in muscular rigor. (From Gordon, 1982, Figure 7; with permission.)

of how muscle changes length. However, the problem of exactly how the filaments interact with each other and generate force is still a matter for debate.

The key force-generating elements appear to be cross-bridges between the actin and myosin filaments which can be seen in some preparations under the electron microscope. These consist of projections of the myosin filaments (called myosin heads) which extend towards adjacent actin filaments. In the presence of ATP, the myosin heads are thought to interact cyclically with special sites on the actin filaments. One possible mechanism is shown in Figure 2.2. In stage 1,

ADP and P_i are bound to the myosin head. In this stage the heads are free to bind to sites on the actin filament, and form an actin–myosin–ADP–P_i complex (stage 2). Shortly after binding, this complex undergoes a conformational change. The myosin head rotates through 45 degrees, and relative force is exerted between the two filaments. This process leads to release of P_i and then ADP, and the formation of an actin–myosin complex (stage 3). The cycle is completed by binding of ATP to the complex. Myosin dissociates from the actin and at the same time hydrolyses ATP, although the reaction products (ADP and P_i are not released but remain bound to the myosin head as a myosin–ADP–P_i complex (stage 1 and 4)

The critical stage of force generation is rotation of the myosin head. This causes the filaments to slide over one another if they are free to move (isotonic contraction). Otherwise an isometric force is generated.

In normal muscle, ATP is freely available, and the cycling process is controlled by the availability of actin binding sites. These are normally covered by a protein called tropomyosin, linked to a smaller protein known as troponin. The actin sites are revealed by changing the concentration of Ca^{2+} in the sarcoplasm. Troponin can bind Ca^{2+} when available, and in the process cause movement of the tropomyosin molecule to reveal active sites on the actin. At rest, the concentration of Ca^{2+} in the sarcoplasm is very low so that the majority of actin–myosin bonds remain unformed (state 1 in Figure 2.2). The Ca^{2+} ions are stored in special membranous structures called the sarcoplasmic reticulum, separate from the external fluid which surrounds the muscle.

Control of Ca^{2+} concentration occurs in the following way. Throughout its length, the cell membrane of the muscle fibre (sarcolemma) has numerous infoldings which form a system of membranes within the muscle fibre called the transverse tubular system. This forms regions of close opposition with the sarcoplasmic reticulum known as triads. Release of neurotransmitter at the neuromuscular junction initiates an action potential along the sarcolemma. This depolarizes the membrane of the transverse tubular system, which acts as a device to transmit the depolarization to the interior of the fibres. Depolarization of the transverse tubular system causes release of Ca^{2+} from the sarcoplasmic reticulum into the sarcoplasm and allows formation of actin–myosin bonds. Ca^{2+} is then actively pumped back into the sarcoplasmic reticulum and contraction ceases.

After death, Ca^{2+} is no longer pumped back into the sarcoplasmic reticulum and ATP synthesis ceases. The actin–myosin molecules in muscle become bonded together in rigor mortis (Huxley and Hanson, 1954; Huxley and Niedergerke, 1954; Huxley, 1974; Eisenberg and Greene, 1980; Pollack, 1983).

2.2 *Mechanical properties of muscle*

The reason that the detailed mechanical properties of muscle need to be known is to be able to predict how much force a muscle will produce under given conditions. The same command, in terms of action potentials along the muscle nerve, can generate varying levels of force which depend on the muscle length, the velocity of shortening and the state of activation of the muscle before receiving the command. Thus the CNS cannot assume that the same command always will produce the same movement. All these factors must be taken into account.

Studies on tetanically-activated isolated muscle

In these, now classical, studies, single muscles were isolated from an animal. No reflex arcs remained intact, and to remove the complicating factor of activation history, the experiments were performed either with the muscle completely inactive, or during maximum activation of the muscle with tetanic stimulation.

A useful model used to describe the mechanical behaviour of such a preparation is shown in Figure 2.3. It consists of three elements: the series and parallel elastic elements are purely passive components acting like mechanical springs, whereas the contractile element actively generates muscle force. The contractile element has its own mechanical properties, namely viscosity and stiffness, but these properties change with the level of muscle contraction, unlike those of the two passive components.

It is not as simple as this simplistic model suggests to divide the actual structure of muscle into three components. In fact, as seen below, the model only helps to describe the mechanical properties of muscle, and is not intended to imply specific relationships between hypothetical springs and muscle structure. This is because the model was formulated well before the sliding filament theory of muscle contraction, and before

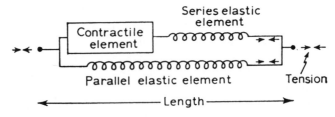

Figure 2.3 Mechanical model of muscle. Muscle consists of a contractile element and two (series and parallel) elastic elements. The properties of the elastic elements are constant, whereas the mechanical properties (that is, the stiffness and viscosity) of the contractile component vary with the state of muscle contraction. (From Roberts, 1978, Figure 2.4; with permission.)

detailed electron microscope pictures of muscle were available. Nevertheless, there are some relationships that can be drawn which help in understanding how the muscle works. Thus, the contractile mechanism can be said to reside in the interaction between the actin and myosin filaments; the series elastic element in the tendinous insertions of muscle, and in the actin–myosin cross-bridges themselves; and the parallel elastic element in the sarcolemma of the muscle cells and the surrounding connective tissue.

The length–tension relationship

To determine the effect of length on the force exerted by a muscle, the factors of velocity and time are held constant. The muscle is held at different lengths, either shorter or longer than its physiological 'resting' length, with one tendon attached to a stiff force transducer. It is then stimulated tetanically via the muscle nerve, and the force exerted on the transducer is recorded. The upper dotted curve in Figure 2.4 describes this behaviour for the soleus muscle of the cat. As the muscle length is

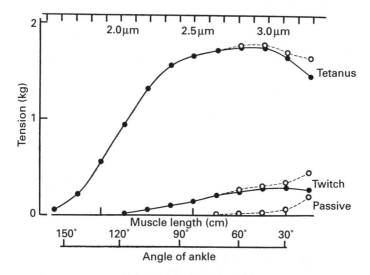

Figure 2.4 The effect of muscle length on the isometric tension exerted by isolated cat soleus muscle. The upper dotted line shows the total tension measured after 1.5 s tetanic stimulation of the muscle nerve at 50 Hz. The lowest dotted line shows the passive tension recorded in inactive muscle. Subtraction of these values from the upper dotted curve gives the true action tension generated by tetanic stimulation of the muscle (continuous line). The middle curves show similar data for peak twitch tensions produced by single supramaximal nerve stimuli. Changes in muscle length are shown in centimetres with the corresponding angle of the ankle joint in the intact animal below. Mean sarcomere lengths equivalent to the ankle positions have been added above the figure. (From Rack and Westbury, 1969, Figure 3; with permission.)

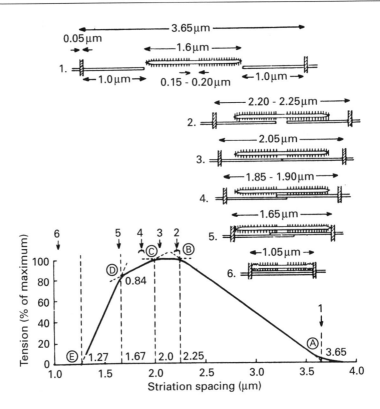

Figure 2.5 Relationship between striation spacing and the tension generated in a single frog muscle fibre. The upper diagrams (1–6) show the degrees of overlap of thin and thick filaments within each sarcomere, and the spacing between the Z lines. The bottom curve shows the length–tension relationship for the fibre. Maximum tension is generated with maximum overlap between the filaments (2 and 3). Tension drops if the overlap is less or if the thin filaments collide with each other. (From Gordon, Huxley and Julian, 1966; with permission.)

increased, the force exerted rises and often reaches a peak before falling slightly at the end of its normal length range.

Figure 2.3 shows that the tension recorded in these experiments is due not only to the activity of the contractile element, but also to passive stretch of the parallel elastic element. An estimate of the force produced by the parallel elastic element can be obtained by applying stretch to a relaxed muscle (the lowest dotted line in Figure 2.4). Under these conditions, the contractile element exerts no force, and all the tension is exerted across the parallel elastic element. Subtraction of this passive tension from the total tension recorded in active muscle gives the true force produced by the contractile element (upper continuous curve in Figure 2.4).

The reason for the influence of length on the force produced by the contractile mechanism lies in the way actin and myosin filaments interact in single sarcomeres. The length–tension relationship for a single muscle fibre is shown in Figure 2.5, with the overlap between the thick and thin myofilaments in a single sarcomere drawn above. As the overlap between the filaments increases, more force is produced because more cross-bridges can form. However, as the sarcomere spacing decreases further, the actin filaments begin to 'overlap'. It is believed that this interferes with cross-bridge formation and causes the force to decline. In whole muscle, the transition points between each part of the curve occur at slightly different lengths for each sarcomere and each fibre, so that the length–tension diagram is more rounded than for a single fibre (Gordon, Huxley and Julian, 1966; Huxley, 1974).

Twitch contractions at different muscle lengths

The classical length–tension diagram of Figure 2.4 shows how the maximum isometric force of muscle varies at different muscle lengths. The peak tension attained in a single maximal twitch follows a similar curve (Figure 2.4, middle curves). However, measurements of peak tension do not give the complete picture of how muscle twitches are affected by muscle length. Figure 2.6 gives a typical example of raw data from the isolated cat soleus muscle. With increasing muscle length, the time course of the twitch changes: both time taken to reach peak tension and duration of the relaxation phase are prolonged. The reason for this is thought to be that at short muscle lengths, there is a less efficient activation of the contractile machinery. A single nerve impulse may give a smaller and perhaps shorter period of activation (for example, by a

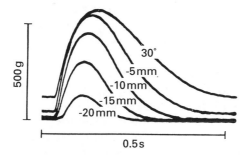

Figure 2.6 Records of isometric twitches from isolated cat soleus muscle held at different lengths. The longest duration twitch was at a muscle length equivalent to an angle of 30 degrees. The others were recorded after shortening the muscle by successive 5 mm steps. Note how the initial tension in the relaxed muscle changes with length (baseline of the twitch). The twitches reach peak tension earlier and relax faster when the muscle is short. (From Rack and Westbury, 1969, Figure 4; with permission.)

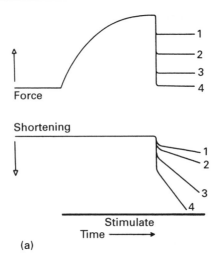

Force

Shortening

1
2
3
4

Stimulate

Time ———▶

(a)

Figure 2.7 Determination of force–velocity curve for a contracting muscle. (a) How the curve is built up. The muscle is held fixed at a constant length and stimulated tetanically via its nerve. When maximum isometric tension is achieved, the muscle suddenly is allowed to shorten against a load. The smaller the load (that is, trace 4), the faster the final velocity of shortening.

decrease in the amount of Ca^{2+} released from the sarcoplasmic reticulum). The implication is that the minimal stimulation rates needed to achieve a fused contraction will vary with muscle length.

Force–velocity relationship

The rate at which a muscle can shorten depends on the force it is exerting. The isometric force exerted by a muscle is always greater than the force exerted during shortening. The method used to measure force–velocity relationships is shown in Figure 2.7. The muscle is stimulated tetanically, and when tension has built up, a catch is released and the muscle shortens against a preset load. Some records of the shortening profile are shown in Figure 2.7a. After releasing the catch, the muscle shortens very rapidly and then settles down to a slower, steady rate of contraction. The initial rapid phase of shortening can be explained by reference to the mechanical analogue of muscle in Figure 2.3. In the isometric state, the contractile element exerts its maximum tension through the series elastic element. When the catch is released, the contractile element can only shorten at a finite speed, whereas the series elastic element shortens immediately to take up a length appropriate to the new load. It is only later that the slower contraction of the contractile element becomes evident in the position records as the ramp phase of shortening. The smaller the load, the faster the muscle shortens.

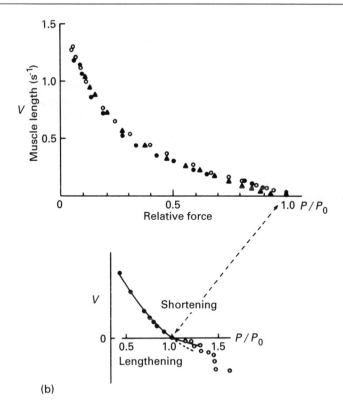

(b)

Figure 2.7 Continued. (b) The force–velocity relationship for a single muscle fibre (upper graph) and for a whole muscle (lower graph). Force is plotted as a fraction of the maximum isometric tension produced by the muscle at its initial length. Speed of shortening is in muscle lengths per second. In the upper graph, the different symbols refer to experiments carried out with the muscle fibres held at different starting lengths of (open circles: sarcomere length of 2.03 μm, closed circles: 2.23 μm and triangles: 2.43 μm). The starting length defines P_0, the maximum isometric tension (c.f. tension–length relationship). The lower graph, from a whole muscle, shows what happens when a load greater than P_0 is applied to the muscle. Rather than shortening, the muscle lengthens. However, the velocity at which it is stretched is lower than expected. The dotted line is extrapolated from the relationship of force and velocity at loads less than P_0. The solid line shows the true (experimental) relationship. It lies above the dotted line. The findings of a lower lengthening velocity than predicted indicates that the muscle is stiffer when lengthening than it is when shortening. The muscle gives way at forces above about 1.5 P_0. (From Gordon, 1982, Figures 7.35 and 7.36; with permission.)

The force–velocity curve also can be extended to lengthening contractions, that is, attempted shortening against a load greater than isometric. In this case, the muscle is extended by the load with a speed related to the excess tension. These points are plotted in Figure 2.7b as negative shortening velocities.

The shortening region of the force–velocity curve has been fitted by many equations. The best-known is that of Hill:

$$(T + a)(V + b) = (T_0 + a)b = a \text{ constant}$$

where a and b are constants, and T is the tension exerted by the muscle. T_0 is the maximum isometric tension of the muscle and V the rate of shortening. The curve is of interest when one considers the amount of work that a muscle can perform on a load. Work is defined as 'force × distance' moved by the load perpendicular to the direction of the force. Thus, in an isometric state, although force is a maximum, no work is done since the load does not move. Similarly, at the maximum velocity of shortening, the muscle cannot produce any force against an external load, and again the work done is zero. Maximum work is performed at intermediate velocities of shortening.

Neither Hill's equation (nor any of the other postulated functions) describes the lengthening part of the force–velocity curve. This is usually taken to mean that the process of force generation is different during lengthening than it is during shortening. During lengthening, a tetanically stimulated muscle can exert a force greater than isometric. As described later, this extra force is often used during normal movement.

There have been several attempts to explain the force–velocity curve on the basis of the cross-bridge theory of muscle contraction. The best known is that developed by Huxley in 1957. There were two important assumptions in this model: (i) that the cross-bridges themselves were elastic (that is, they represent part of the series elastic element of muscle); and (ii) that myosin heads could interact with a given actin site over a relatively wide range of distances (about ± 10 nm in the model).

If myosin heads interacted with a relatively distant actin site and underwent a conformational change as shown in Figure 2.3, the elastic element would be stretched, and force developed between the filaments. If a myosin head interacted with a relatively close actin site, then the conformational change might produce no stretch of the elastic element, and no force would be developed. Finally, if a myosin head remained attached to an actin site beyond the point of zero force, then by compressing the elastic element in the cross-bridge, it could exert a negative force between the filaments. Thus the force developed depends on both the number of cross-bridges and the distance over which each one interacts.

To fit the force–velocity curves of tetanically activated muscle, Huxley assumed that in the isometric state, the rate of cross-bridge detachment was lower than the rate of attachment. Thus, in the steady state, all the cross-bridges were formed and the force maximum. However, it was suggested that during shortening, there was an increase in the rate at which the cross-bridges detached, and also an increase in the number of attached cross-bridges which exerted negative force. The result would be a decrease in the total force exerted by the muscle. At maximum shortening velocities, very few bonds remained attached, and those that did remain were equally distributed between states exerting positive and negative force. No external force could be developed.

The sliding filament theory of muscle contraction also accounts for the difference between muscle behaviour during lengthening and shortening contractions. During lengthening contractions, the rate of detachment of actin–myosin bonds is slower than during shortening at the same velocity. The effect is for cross-bridges to remain attached and to be stretched forcibly until detachment finally occurs. Such stretching is beyond the normal range of myosin attachment, and lengthens the cross-bridge elasticity to increase the force exerted by the muscle.

The effect of the rate of muscle stimulation

The response of skeletal muscle to a single nervous impulse is a twitch contraction. The features of this contraction differ from muscle to muscle, and will be dealt with in Chapter 3. If more than one impulse is given at an interval which is less than the contraction time of the muscle, the forces produced by each impulse summate. A fused tetanus is produced when the force fluctuations to each individual impulse can no longer be distinguished.

The force developed depends also on the pattern of stimulation. An interesting example of this is the so-called 'catch' enhancement of muscle tension which is produced by the insertion of a single short interval into a low frequency stimulus train (Figure 2.8). As expected, the extra stimulus results in an immediate increase in muscle tension. Unexpectedly, however, the effect outlasts the stimulus for some time: the force produced by a low frequency train is elevated above normal for many seconds. The mechanism of this effect is unknown, although it may affect Ca^{2+} sequestration following activation.

2.3 *Behaviour of isolated muscle stimulated at subtetanic rates*

The previous section described the classical experiments on mechanical properties of fully activated isolated muscle. However, under normal conditions, muscle is never activated in a pattern resembling that of

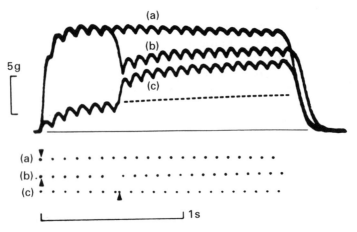

Figure 2.8 Catch property in a slow motor unit from a cat medial gastrocnemius muscle. Upper traces are isometric force records, lower traces (dots) represent timing of muscle nerve stimulation. Arrows show points at which extra impulses have been inserted into the trains. In (a) and (b), an extra impulse was inserted into the train at the arrow point, some 10 ms after the first. This produces the rapid rise in initial force, far greater than that seen in trace (c). The extra force level is then 'caught' throughout the rest of the train. In (c), an extra impulse was inserted a short time after the seventh impulse of the train, resulting in extra force production, that was again 'caught' at the new level. In (b), an impulse interval was purposely lengthened, with the opposite effect. (From Burke, Rudomin and Zajac, 1970; copyright American Association for the Advancement of Science, 1970, Figure 3; with permission.)

synchronous tetanic stimulation. First, the individual motor units tend to be activated asynchronously and, second, the stimulation frequency is seldom as high as that needed to produce a full tetanic contraction. When muscle is stimulated in a more physiological fashion, its mechanical properties are rather different from those seen during synchronous tetanic contractions.

Asynchronous activation of muscle

Rather than stimulating the whole muscle nerve tetanically, it is possible to produce a more 'physiological' contraction by subdividing the nerve into several filaments, so that each filament can be stimulated separately and asynchronously. Figure 2.9 shows the difference in tension produced by asynchronous stimulation of several nerve filaments, and synchronous stimulation of the whole muscle nerve. Synchronous stimulation at 5 Hz produces an unfused contraction, in which the variation in tension between each stimulus amounts to about 50% of the mean. Asynchronous stimulation of separate filaments at the same rate produces

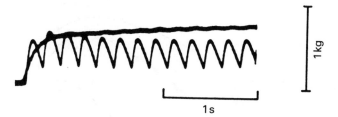

Figure 2.9 Isometric tension produced by isolated cat soleus muscle stimulated at 5 Hz. The grossly unfused contraction was recorded during synchronous stimulation of the muscle nerve. The smooth contraction was recorded when five filaments of the motor nerve were stimulated asynchronously at the same frequency. (From Rack and Westbury, 1969, Figure 5; with permission.)

a contraction which has less ripple on the tension record (in Figure 2.9 it is only of the order of 5% mean tension). More surprisingly, however, asynchronous stimulation produces a more forceful contraction in which the mean level of tension is some 50% greater than with synchronous stimulation. This difference between the force produced by the two types of stimulation is seen only at low rates of stimulation. When the rate of synchronous stimulation is rapid enough to produce a fused contraction, then there is no difference in the amount of ripple or the mean level of the tension record (see Figure 2.10).

The reason for this difference is as follows. During an isometric contraction, the muscle fibres are not stationary; any increase in tension stretches the tendon and allows the fibres to shorten. The tension fluctuations in an unfused tetanus are accompanied by visible movement of the muscle fibres. This is much smaller during asynchronous stimulation of several nerve filaments. It is thought that at low stimulus rates, muscle is best able to generate tension when the fibres are stationary. If this is true whenever a muscle contraction is made smoother at low stimulus rates by distributing the stimulation pulses, then more force is developed.

Length–tension relationship at different levels of muscle activation

The length–tension curve of partially activated muscle is very similar in form to those obtained during maximal tetanic stimulation. In Figure 2.10, the continuous lines show the length–tension relationship measured during distributed asynchronous stimulation of five separate filaments of a muscle nerve. Vertical bars show the limit of tension fluctuation during synchronous stimulation of the whole nerve. The graphs illustrate two important points:

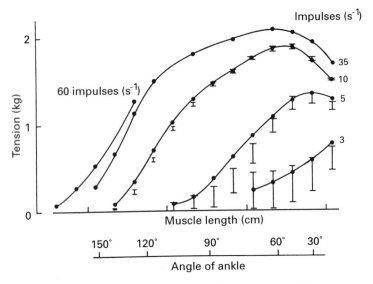

Figure 2.10 The effect of muscle length on the active isometric tension produced by isolated cat soleus muscle stimulated at different rates. Asynchronous stimulation of five motor nerve filaments was used to plot points joined by the continuous line. The vertical lines show the limits of tension fluctuations during synchronous stimulation of all five nerve roots. Changes in muscle length are shown in centimetres; the corresponding angle of the ankle in the intact animal is drawn below. (From Rack and Westbury, 1969, Figure 8; with permission.)

1. The previous section described how the differences in mean tension produced by the two methods was most pronounced at low stimulus rates (for example, curve at 5 Hz). In addition, Figure 2.10 shows that this difference also depends on muscle length. At short muscle lengths, twitch contraction time is short (see above), so that synchronous stimulation of the whole muscle nerve at low rates produces unfused contractions. Under these conditions there is much more movement within the muscle than with asynchronous stimulation, and the mean tension is correspondingly low. If the muscle is lengthened, the twitch contraction time increases, and the contractions produced by synchronous nerve stimulation become fused. There is less internal movement in the muscle, so that the tensions generated by synchronous and asynchronous stimulation become similar.

2. Although the shapes of the length–tension curves are similar at all levels of activity, they are displaced from each other along the x axis. That is, the steep part of the curve lies at different muscle lengths, depending on the level of activation. The probable explanation for this is that at short muscle lengths, the process of activation may function less effectively (see also section above on

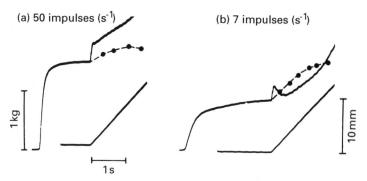

Figure 2.11 Tension recorded in isolated cat soleus muscle during forcible lengthening from a steady isometric contraction. Length records are below and tension records above. In (a), five separate motor nerve filaments were stimulated asynchronously at 50 Hz; in (b), the filaments were stimulated asynchronously at 7 Hz. The dashed lines indicate the isometric tension obtained during sustained contractions at some of the lengths that the muscle shortened through. Note the fall in tension during lengthening in (b), to a level below isometric (from Joyce, Rack and Westbury, 1969; Figure 2, with permission)

twitch contractions at different muscle lengths). Thus, a high stimulation rate is needed to produce maximum tension at short muscle lengths, while at longer lengths, relatively low stimulation rates produce high tensions.

Forced lengthening of partially activated muscle

Force–velocity curves for submaximally activated shortening muscle can be constructed during asynchronous stimulation of several nerve filaments. In many respects, they are similar to those seen in maximally activated muscle. However, the behaviour during lengthening contractions can be strikingly different. When maximally activated muscles are forcibly stretched, they exert a tension greater than isometric (see Figure 2.7). When submaximally active muscles are stretched, tension may fall below isometric. Figure 2.11 shows tension changes recorded in isolated cat soleus muscle during forced lengthening at constant velocity. On the left of the picture, the muscle was stimulated maximally at 50 Hz; on the right, several muscle nerve filaments were stimulated asynchronously at 7 Hz to produce a smooth submaximal contraction. In each case, tension rose steeply during the first 0.5 mm of movement, and in each case it rose gradually during the later part of the movement. However, the initial rise in tension at 7 Hz stimulation was followed by a fall which took the tension during lengthening below the isometric tension corresponding to the new muscle length. Thus, under these conditions, the muscle was being extended by a force smaller than it could have withstood during a sustained isometric contraction.

The explanation for this behaviour is thought to be as follows. In the initial isometric state, the rate at which myofibrillar cross-bridges break down is relatively slow. At the onset of forced lengthening, they remain attached over a short distance before they begin to dissociate at a rate faster than occurs during isometric contractions. The rapid rise of tension at the start of forced lengthening is due to stretch of attached cross-bridges. While the cross-bridges remain attached, the stiffness of the muscle is very high. Over this range (up to about 1 mm in a muscle 20 cm long) the muscle is said to show **short range stiffness**. The behaviour after the point at which cross-bridges begin to dissociate depends on the level of muscle activity. When the muscle is maximally activated, the rate at which cross-bridges form may be so high that the increased rate of destruction during lengthening might not seriously diminish the number of cross-bridges remaining intact. When the level of muscle activation is lower, then the cross-bridges may form more slowly so that during lengthening, the number of attached cross-bridges might diminish and the force falls below isometric.

The implication is that submaximally activated muscle may resist lengthening with a very high initial short range stiffness. However, if the muscle is stretched beyond the point at which cross-bridges become dissociated, then the stiffness falls dramatically. Such effects are important not only in describing the resistance of extrafusal muscle to external disturbances, but also affect the response of muscle spindles to applied stretch (see Chapter 4). (Rack and colleagues give details of the work covered in this section.)

2.4 *Muscle mechanics in intact humans*

It is not so easy to study the mechanical properties of single muscles in humans as it is in animal experiments. There are two main problems. First, it is usually very difficult to stimulate a single muscle tetanically without causing severe discomfort to the subject and without the stimulating current spreading to activate other muscles in the vicinity. Second, it is not possible to measure directly the force exerted by a muscle. Because muscles act across one or more joints, it is only possible to measure the torque exerted about the joint or the force exerted by the limb against a force transducer. If the characteristics of a single muscle are to be studied, then corrections must be applied to account for the changing mechanical advantage of the muscle at different joint angles.

Despite this, some useful studies have been made in humans which

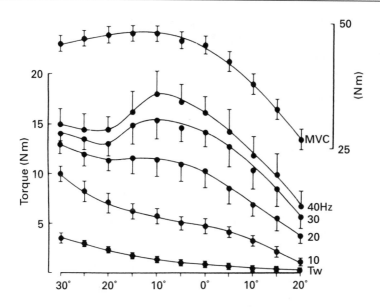

Figure 2.12 Effect of joint position on torque developed during twitch (Tw) and tetanic contractions of tibialis anterior, and during maximum voluntary contraction of the ankle dorsiflexors (MVC). Stimulation applied to the motor point of tibialis anterior at frequencies indicated by the numbers at the end of each curve. Mean values are plotted with standard error limits. (From Marsh *et al.*, 1981, Figure 3; with permission.)

confirm the basic animal data described above. For example, it has proved possible to measure the length–tension characteristics of the biceps brachii muscle. Tetanic stimulation can be applied to the motor point of biceps without danger of excessive spread of activity to other muscles, and being superficial it is relatively easy to measure the anatomical parameters of the muscle in intact humans. After corrections are made for the angle of insertion of the tendon at different elbow angles, it is found that the biceps changes length by about 6 cm over the full range of elbow angles, and has a peaked length–tension curve with maximum tension being exerted at a length corresponding to an elbow angle of 100–120 degrees.

However, it can be argued that it is not necessary to extract the length–tension characteristics by trigonometric corrections, since it is the force produced by the limb, or the torque exerted by the muscle around the joint that is the important mechanical characteristic for the CNS. Thus, other studies do not correct for the angle of insertion of the muscle or calculate the actual muscle length. They simply measure the angle–torque relationship around a joint. A typical curve from the tibialis anterior muscle is shown in Figure 2.12. In these experiments,

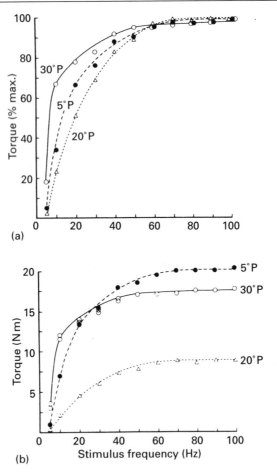

Figure 2.13 Torque developed at the ankle during stimulation of the motor point of tibialis anterior muscle at different frequencies. Curves plotted with the ankle held isometrically at mid and extremes of joint position. (a) The relative torques achieved at each position; (b) The actual torques achieved. P, plantarflexion; D, dorsiflexion. (From Marsh *et al.*, 1981, Figure 5; with permission.)

the tibialis anterior was stimulated at its motor point at several different frequencies, and the ankle torque measured at several different angles. At higher stimulation frequencies, the angle–torque relationship is peaked like that of the length–tension curves described above. It is similar to the angle–torque relationship measured during maximal voluntary dorsiflexion of the foot.

Peak voluntary torque is much higher than that seen during electrical activation because, during voluntary activation, force is produced by the long extensors of the toes in addition to tibialis anterior.

These experiments on tibialis anterior also illustrate the effect of changing muscle length on the twitch contraction time. When the ankle is dorsiflexed and the muscle short, the duration of a single twitch contraction is short; when the ankle is plantarflexed and the muscle extended, the twitch contraction time is longer. The result is that the twitches summate more readily when the muscle is long. Thus, for a given frequency of activation, the muscle exerts a larger proportion of its maximum torque when the ankle is plantarflexed (see Figure 2.13).

Force–velocity curves for a single human muscle have not been obtained. However, curves for all the flexor muscles acting at the elbow have been calculated by releasing the elbow during maximal voluntary activation. The force–velocity curves obtained in this way can be fitted, like those from isolated animal preparations, by Hill's equation (Wilkie, 1950; Ismail and Ranatunga, 1978; Marsh *et al.*, 1981).

2.5 *Effects of muscle properties on control of movement*

Effects arising from the length–tension relationship

1. *Force production at different muscle lengths.* A striking example of the effect of muscle length on the force developed is the position which the hand and fingers adopt during a power grip. The wrist is held extended by contraction of the extensor muscles in the forearm, while the fingers flex powerfully onto an object in the palm of the hand. A more flexed wrist allows the flexors to shorten further, but at this length they are unable to produce as much force. As every hero in a gangster movie knows, the way to make the villain release his grip on the gun, is to flex the wrist forcibly to reduce the power of the finger flexors. The gun will thereupon fall dramatically to the floor, to be kicked neatly away by the hero's foot. The importance of this example is that it provides evidence that the nervous system must somehow take into account the length–tension relationship of muscle during normal movement, in this instance by extending the wrist when a maximal flexor force is required.

2. *Stiffness of muscle.* Relaxed muscles present little or no resistance to changes in muscle length. When active, however, stiffness can be surprisingly high, particularly the short-range stiffness seen during small changes in muscle length, or at the start of larger muscle stretches. Stiffness increases with the force of muscle contraction. This stiffness plays the initial part in the response of a limb to a disturbing force. Thus if muscles co-contract at a joint, the mechanical stiffness of the limb is increased, helping to resist forces that

might be expected to be encountered, for example, in carrying a full glass across a crowded party room.

Effects arising from the force–velocity relationship

1. *Efficiency of muscle contraction.* As shown above, maximum work will be performed by a muscle shortening at intermediate muscle lengths. Power, which is the rate at which work is done, also is maximum at these lengths. Thus, in order to achieve maximum efficiency of muscle, it is necessary to arrange for the contraction to take place at about these velocities and lengths. A good example of the importance of these effects in normal movement is given by the gearing on a bicycle. Load and speed are matched to the properties of the muscle so that power output can be maintained regardless of the steepness of the incline.

2. *Lengthening contractions.* Lengthening contractions also are a common feature of normal movement. They are sometimes known as eccentric contractions as opposed to the more usual shortening or concentric contractions. In fact, lengthening contractions occur regularly in activities such as walking, jumping or hopping. Hopping is a particularly instructive example since the nervous system seems to make purposeful use of a lengthening contraction to provide extra force output from the muscle.

Hopping can be initiated voluntarily or reflexly. If the force under the foot is measured during a voluntary hop, the first thing to happen is that there is a momentary decrease in the force, followed by a rapid

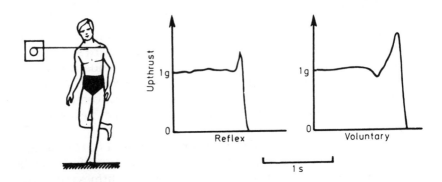

Figure 2.14 Time course of force changes on platform during reflex and voluntary hopping. Reflex hopping was induced by sudden pulls to the shoulder of the subject standing on one leg, which made him off-balance. Voluntary hops were self-initiated and were always preceded in the force record by a momentary decrease in force. (From Roberts, 1978, Figures 10.22 and 10.23; with permission.)

increase which produces the upthrust of the hop (see Figure 2.14). This preceding relaxation allows the body to fall a little, stretching the extensor muscles at a time when they are being reactivated for take-off. A voluntary hop always is initiated in this way, and subjects find it almost impossible to hop without first bending the knee.

Reflex hopping is produced in a subject standing on one leg by pulling him briskly off-balance. In this instance, the hop never shows the initial dip in the force record; there is only an uncomplicated contraction of the extensor muscles. Reflex hopping, unlike voluntary hopping, does not make use of the extra force available from the muscle in a lengthening contraction. Perhaps the nervous system, in attempting to restore balance as fast as possible, 'trades off' the loss of contractile force against the reduction in delay in starting the hop.

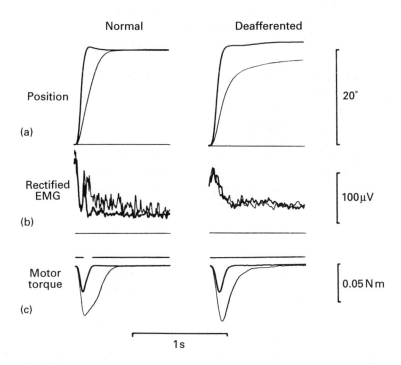

Figure 2.15 Response of a normal subject and a deafferented man to an unpredictable five-fold increase in the viscosity of the load opposing a thumb flexion movement. Control trials shown by thick lines, trials with unexpected increase in viscosity by thin lines. The traces are (a) thumb position, (b) rectified electromyogram from the long flexor of the thumb and (c) torque generated by motor against which subject is pressing, which approximates the torque exerted by the thumb on the lever. Note the large increase in torque during the addition of increased viscosity. (From Rothwell *et al.*, 1982, Figure 14; with permission.)

Intrinsic feedback control of muscle contraction

Both the length–tension and force–velocity relationships can contribute to compensation for unexpected disturbances of movement. For example, a sudden increase in load can produce increased muscle tension not only via reflex pathways (see Chapter 6), but also by the nature of the length–tension characteristics of the muscle.

The force–velocity relationship of active muscle also can produce surprisingly large effects on the force exerted by a muscle. An example is shown in Figure 2.15. On the left are the responses from a normal subject, and on the right are responses from a patient who was deafferented by a severe peripheral neuropathy of unknown origin. The responses from this individual are uncontaminated by any reflex contractions of the muscle. Both subjects were trained to perform fairly rapid thumb flexion movements to a given end position against a small force without visual feedback. On random trials, the viscosity of the load was suddenly increased at the start of the movement. To a normal person this feels as if the load has to be pushed through a jar of treacle, and slows the movement considerably. In the normal subject, the sudden introduction of this unexpected viscosity produced extra reflex and voluntary electromyogram activity in the thumb flexor muscle, and the thumb finally reached the intended end position. The deafferented man had no clues that the load had changed and his electromyogram was the same as in the control trials. However, the force output from his thumb still changed considerably, although not as much as normal. Slowing the movement increased the amount of force that the muscle could exert, and the area under the curve describing the force output of the muscle increased by a factor of five. The same effects are seen if the inertia of the load is changed.

2.6 *A theory of movement control which makes use of the mechanical properties of muscle*

Emilio Bizzi has argued that the length–tension relationships of agonist and antagonist muscles acting around a single joint could be used by the nervous system to specify particular joint angles. If, for example, both muscles are activated equally, then the tension in the agonist will increase if the joint is passively rotated so as to stretch that muscle. By the same reasoning, such a movement will shorten the antagonist and reduce its tension. The point of equilibrium will be given by the intersection of the respective length–tension curves, where the forces generated in each muscle are equal and opposite (see Figure 2.16). Activation of a muscle can be regarded as changing the length–tension

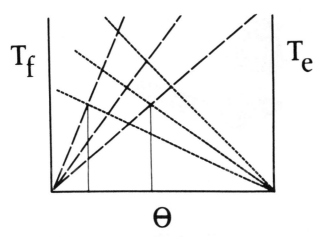

Figure 2.16 Diagram of how length–tension relationships of extensor and flexor muscles about a single joint might interact to specify a particular joint angle. Length–tension curves for flexor (T_f) and extensor (T_e) muscle are plotted against the angle of the joint (θ), rather than the length of the muscle. The point of intersection of the length–tension curves is the equilibrium position of the joint (vertical lines). The slope of the length–tension curves can be shifted by voluntary activity. For example, increasing flexor activity increases flexor muscle stiffness and shifts the flexor curve upwards. In normal circumstances this might be accompanied by decreased activity and hence decreased stiffness of extensor muscles. Thus, the two new length–tension curves then intersect at a different joint angle, that is, the equilibrium angle has been shifted. (From Bizzi *et al.*, 1982, Figure 1; with permission.)

relationship of the muscle, since it will increase the stiffness. Thus, if the agonist muscle is activated, its length–tension curve will be shifted upwards, and the intersection point of equilibrium with the antagonist curve will be shifted likewise. The joint will move automatically to take up the new position of equilibrium. Movement to any joint angle can be viewed in the same way, that is by changing the balance between the length–tension curves of each muscle.

The attraction of looking at muscle activation in terms of shifting length–tension relationships is that this provides a mechanism whereby the nervous system can specify the end (equilibrium) position of a movement without calculating the separate steps to get there. In other words, this is a way of specifying a movement to a particular place without necessarily calculating a specific trajectory. As long as the length–tension relationships show no prominent peaks, there can be only one point of joint equilibrium. (If there were peaks in the curves, then more than one position would be stable, and the initial starting position would determine the end point assumed by the joint.)

Bizzi and his colleagues (1982) have carried out some elegant

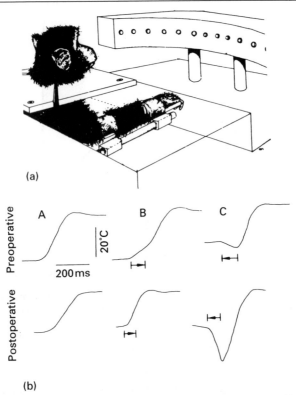

(a)

(b)

Figure 2.17 Movement of the elbow to a target position indicated by an illuminated lamp. (a) The monkey cannot see its own arm. (From Polit and Bizzi, 1979, Figure 1; with permission.) (b) Records of arm position before and after surgical deafferentation of the arm. Records in A are control trials in which no disturbance was given to the movement. In B, the movement was assisted briefly (arrows) and in C, it was reversed (arrows). Despite these disturbances, the monkey reaches the final target without substantial loss of accuracy even after deafferentation. (From Polit and Bizzi, 1978; copyright American Association for the Advancement of Science, 1978, Figure 1; with permission.)

experiments on monkeys which indicate that this may be the way in which some movements are organized. The experiment involved training monkeys to flex their elbows and point to a target light when illuminated in order to achieve a food reward (Figure 2.17). In random trials, the arm movement was perturbed by the experimenter either loading or assisting the movement for a short period. The monkeys could not see their arm, but they could feel the perturbation. They achieved the final end position with reasonable accuracy despite the disturbance. Of course, in these monkeys, both reflex and voluntary muscle activity might have helped them to compensate for the disturbance.

To determine to what extent these corrections were necessary, the animals then were deafferented by surgical section of the dorsal spinal roots in order to remove any sensory feedback from the arm, and allowed to relearn the task. After retraining, the monkeys were deprived of vision of the arm, and the accuracy of pointing in the absence of any sensation was assessed. Again in a number of trials, the movement of the forearm was interrupted briefly by passive manipulation by the experimenter. Since the monkey was deafferented, it could not feel the interruption of movement and no reflex events intervened to correct for them. Despite this, the monkeys reached the final end position with little loss of accuracy (Figure 2.17), suggesting that the final end position of the movement was planned rather than the trajectory necessary to achieve it.

These elegant experiments were consistent with the 'equilibrium point' hypothesis of movement control. Since that time there has been one further modification of the hypothesis. It is clear that in some instances, the trajectory of movement, as well as its final end position, is of importance. We need to move slowly on some occasions and rapidly on others, and in such circumstances, muscle activity will have to be modulated during the movement according to the velocity at which we wish to move. In order to allow for this under the equilibrium point hypothesis, it may be that the central nervous system can specify a **virtual** trajectory of the movement. For example, if we wish to move a limb faster than it would travel under the influence of the length–tension relationships that are normally set up, we could specify, initially, an endpoint beyond the target. The arm would move rapidly at the beginning of the movement, and then the endpoint could be adjusted to the actual position of the target whilst the movement progressed. Such virtual trajectory planning could be extended to many other types of complex movements. Although such a mechanism may occur, however, the idea of specifying a trajectory as well as an endpoint begins to undermine the simplicity of the original theory which was its attraction.

Another problem with the simple equilibrium point hypothesis is the assumption that the point of equilibrium between the length–tension relationships of different muscles will be specified with enough accuracy to allow precise movement to take place. Unfortunately, this is not always true. For example, in Figure 2.15, it is clear that the deafferented man did not reach the final intended end position when the viscosity of the load was increased. In fact, if the limb had moved according to the spring hypothesis, introduction of extra viscosity would have had no effect on the final end position of the movement. The fact that it had implies that the general applicability of the theory is still in question.

Despite the questions surrounding the theory, there is good evidence that, at least in some circumstances, the wiring of the spinal cord circuits may control shifts of limb position via a true equilibrium point mechanism. The experiments were performed on spinal frogs, with stimulating electrodes placed in the intermediate zone of the lumbar region of the spinal cord. Trains of stimuli lasting about 300 ms produced EMG activity in many leg muscles and, when the leg was free to move, produced a change in the position of the limb. If the leg was placed in different initial positions, then stimulation at the same site evoked movement in a different direction. However, the leg always tended to move towards the same final end position. If the stimulation site was changed, then the apparent end position also changed. In order to quantify the effect more completely, movement of the leg was not usually allowed. Instead, the leg was placed at different points on a two-dimensional surface parallel with the body, and the force exerted at the ankle together with its direction was measured during the periods of stimulation. Stimulation at any one point always produced a pattern of force vectors which pointed towards one location on the two-dimensional surface. This location is the two-dimensional equilibrium point of the contraction produced by the spinal stimulation. Stimulation of different parts of the intermediate zone produced different equilibrium positions.

There are two reasons why the leg would tend to move in different directions when the initial position was changed: there would be differences in both the starting length of various muscles, and in the pattern of afferent feedback from the limb. However, these factors cannot account completely for the profile of the force vectors described above. If the stimulating electrode was moved into the ventral horn so that motoneurones were stimulated directly, then the force fields rarely showed any equilibrium point. Instead, the force vectors at each position on the plane could be divergent or parallel rather than converging. The conclusion is that the input to the motoneurones from the intermediate zone imposes a structure on the pattern of motoneurone activation which is consistent with the form of the equilibrium point hypothesis. (Articles by Feldman, 1974; Day and Marsden, 1982; Bizzi *et al.*, 1982; Giszter *et al.*, 1991 give further details of the equilibrium point theory.)

Bibliography

Review articles

Eisenberg, E. and Greene, L. E. (1980) The relation of muscle biochemistry to muscle physiology, *Annu. Rev. Physiol.*, **42**, 293–309.
Gordon, A. M. (1982) Muscle, in T. Ruch and H. Patton (eds) *Physiology and Biophysics*, vol. **IV**, Saunders, Philadelphia, pp. 170–260.

Hill, A. V. (1970) *First and Last Experiments in Muscle Mechanisms*, Cambridge University Press, London.

Huxley, A. F. (1974) Review lecture. Muscular contraction, *J. Physiol.*, **243**, 1–43.

Partridge, L. D. and Benton, L. A. (1981) Muscle, the motor, in V. B. Brooks (ed.), *Handbook of Physiology*, sect. 1, vol. 2, part 1, Williams and Wilkins, Baltimore, pp. 43–106.

Pollack, G. H. (1983) The cross-bridge theory, *Physiol. Rev.* **63**, 1049.

Roberts, T. D. M. (1978) *Neurophysiology of Postural Mechanisms*, Butterworths, London, Chapters 2 and 10.

Original papers

Bizzi, E., Accornero, N., Chappele, W. *et al.* (1982) Arm trajectory formation in monkeys, *Exp. Brain Res.*, 46, 139–143.

Burke, R. E., Rudomin, P. and Zajac F. E. (1970) Catch property in single mammalian motor units, *Science*, **168**, 122–124.

Day, B. L. and Marsden, C. D. (1982) Accurate repositioning of the human thumb against unpredictable dynamic loads is dependent upon peripheral feedback, *J. Physiol.*, **327** 393–407.

Feldman, A. G. (1974) Change of muscle length due to shift of the equilibrium point of the muscle-load system. *Biofizika*, **19**, 534–538.

Giszter, S., Mussa-Ivaldi, F. A. and Bizzi, E. (1991) The organisation of limb motor space in the spinal cord. In R. Caminiti, P. B. Johnson, Y. Burnod (eds) *Control of Arm Movement in Space*, Springer-Verlag, Berlin, pp. 321–331.

Gordon, A. M., Huxley, A. F. and Julian, F. J. (1966) The variation in isometric tension with sarcomere length in vertebrate muscle fibres, *J. Physiol.*, **184** 170–192.

Hill, A. V. (1938) The heat of shortening and the dynamic constants of muscle, *Proc. Roy. Soc. B.*, **126**, 136–195.

Huxley, A. F. (1957) Muscle structure and theories of contraction, *Prog. Biophy. Chem.*, **7**, 255–318.

Huxley, A. F. and Niedergerke, R. (1954) Structural changes in muscle during contraction, *Nature*, **173**, 971–977.

Huxley, H. E. and Hanson, J. (1954) Changes in the cross-striations of muscle during contraction and stretch and their structural interpretation, *Nature*, **173**, 978–987.

Ismail, H. M. and Ranatunga, K. W. (1978) Isometric tension development in a human skeletal muscle in relation to its working range of movement: the length–tension relationship of biceps brachii muscle, *Exp. Neurol.*, **62**, 595–604.

Joyce, G. C., Rack, P. M. H. and Westbury D. R. (1969) The mechanical properties of cat soleus muscle during controlled lengthening and shortening movements, *J. Physiol.*, **204**, 461–474.

Marsh, E., Sale, D., McComas, A. J. and Quinlan (1981) Influence of joint position on ankle dorsiflexion in humans, *J. Appl. Physiol.*, **51**, 160–167.

Polit, A. and Bizzi E. (1978) Processes controlling arm movements in monkeys, *Science*, **201**, 1235–1237.

Polit, A. and Bizzi, E. (1979) Characteristics of motor programs underlying arm movements in monkey, *J. Neurophysiol.*, **42**, 187–194.

Rack, P. H. M. and Westbury, D. R. (1969) The effect of length and stimulus rate on the tension in the isometric cat soleus muscle, *J. Physiol.*, **204**, 443–460.

Rothwell, J. C., Traub, M. M., Day, B. L. *et al.* (1982) Manual motor performance in a deafferented man, *Brain*, **105**, pp. 515–42.

Wilkie, D. R. (1950) The relation between muscle force and velocity in human muscle, *J. Physiol.*, **110**, 249–280.

The motor unit

3

The previous chapter discussed some of the properties of whole muscles and the importance of these properties in the overall control of movement. This chapter is concerned with the internal organization of single muscles. A muscle does not consist of an homogeneous population of muscle fibres: there are several different types of fibre within muscle, each of which have different mechanical properties. Thus the range of operation of the whole muscle is extended beyond that expected from any one fibre type alone.

3.1 *The concept of the motor unit*

In mammalian skeletal muscle, the primary organization of the muscle is imposed by the CNS. During development, or during re-innervation following nerve injury, each nerve axon forms synaptic contacts with many muscle fibres. At this stage each fibre also receives terminals from many different axons, but as time passes, the terminals from one axon predominate and the others degenerate (but see section on tonic muscle fibres below). In the final state, each muscle fibre is contacted by terminals from a single parent axon. The combination of motoneurone axon, terminal branches and the muscle fibres which they innervate is known as the **motor unit**, a term coined by Leyton and Sherrington in 1925 (Figure 3.1). This is a basic 'quantal' unit of muscular contraction and represents the smallest number of fibres that can be activated by the CNS at any one time.

3.2 *Twitch and tonic muscle fibres*

Striated muscle contains two types of muscle fibre. The great majority of human striated muscle is made up of **twitch** fibres, although **tonic**

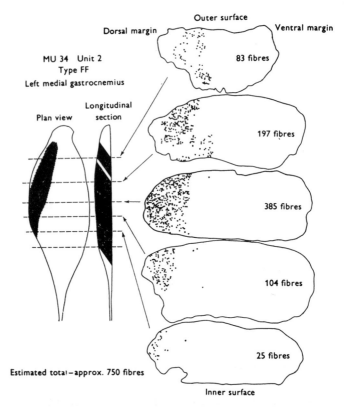

Figure 3.1 Reconstruction of the territory of a single motor unit in the medial gastrocnemius muscle of the cat. A single motor axon, innervating a type FF unit, was stimulated repetitively so as to fatigue the muscle fibres which it innervated. Fatigue depletes fibres of their glycogen reserve (see text). The muscle was then quickly excised and frozen and stained for fibres showing glycogen depletion using the periodic acid–Schiff method. The position of identified fibres is shown in the sections of muscle taken on the right. This was a large unit, with some 750 fibres distributed over a large area of the left part of the muscle (stippled area on plan view). (From Burke and Tsairis, 1973, Figure 1; with permission.)

fibres have been demonstrated in the extraocular, laryngeal, and middle ear muscles. It is also thought that some intrafusal fibres of the muscle spindle may be of the tonic type. The two types are similar in the basic molecular structure, but differ in their nerve innervation and electrical properties. In twitch fibres, synapses with the motor nerve are found only at one end-plate region, where there is a complex infolding of the underlying fibre membrane. This region is organized to produce large end-plate potentials, which can trigger off all-or-none propagated action potentials. These travel to either end of the fibre and the depolarization causes a near synchronous contraction, or twitch, of the whole fibre.

In tonic fibres, there is more than one end-plate region per fibre, but each is simpler than that of the twitch fibre, not having the complex infolded membrane or such a highly branched motor axon. In general, tonic fibres do not conduct action potentials; depolarization spreads by local circuits from each end-plate. Thus, tonic fibre contraction can be graded by the amount of membrane depolarization, unlike that of the twitch fibres in which the action potential always depolarizes the fibre by the same amount. Indeed, tonic fibres rarely produce any tension in response to a single stimulus. Tonic fibres are sometimes known as **slow** fibres, but they should not be confused with the **slow twitch** fibres described below. The remainder of this chapter will be concerned only with twitch fibres (Lewis, 1981).

3.3 *Physiological investigation of the motor unit*

Territory and size of motor units

In animal experiments individual motor units can be stimulated electrically via their motor axon in teased-out strands from ventral root filaments, or by microelectrodes inserted into the motoneurone cell body in the ventral horn of the spinal cord. The distribution of muscle fibres which make up the motor unit can be visualized directly using the glycogen depletion method, which was devised by Edstrom and Kugelberg in the late 1960s.

Single units are stimulated repetitively at high frequency until the fibres within the unit become fatigued and depleted of glycogen. They are then stained selectively by the PAS (periodic acid–Schiff) method. Rather than showing the fibres of a motor unit to be in close proximity, this reveals that they are spread throughout a fairly large territory of muscle (10–30%) and show considerable overlap with fibres from other units (Figure 3.1). Even within a single muscle fascicle, the fibres belong to many different motor units, there being only two to five fibres from each unit. Most fibres do not have direct contact with other fibres of the same unit.

The same method allows counts to be made of the number of fibres per unit. However, a more common way of estimating the **average** number of fibres per unit in different muscles is to count the total number of fibres in the muscle and divide by the number of motor axons in the muscle nerve. In animal experiments, counts of motor axons usually are made one to two weeks after deafferentation by section of the dorsal roots. Obviously, in human material, such pre-treatment is not possible: estimates of the number of motor fibres are made by counting the number of large axons (greater than 7–8 μm diameter) in intact muscle nerves, and then assuming 40–50% of these are sensory axons. In general, motor units are largest in those muscles which act on the largest body masses. There are over 1000 fibres per unit in the medial gastrocnemius, but less than 100 fibres per unit in the extraocular muscles (Edstrom and Kugelberg, 1968; Burke and Tsairis, 1973; Buchthal and Schmalbruch, 1980; Burke, 1981).

Differences in the contraction of motor units

Ever since the time of Ranvier, it has been known that whole muscles may have different contractile properties. In 1873, he showed that red muscles, such as the soleus, were in general slower to contract following an electrical stimulus than pale muscles like the gastrocnemius. This difference between muscles extends to the level of the motor unit. Indeed, an excess of one type of unit gives a muscle its characteristic overall properties.

Speed of contraction and fatiguability (Figure 3.2)
Two main categories of motor unit can be distinguished on the basis of their contraction speed following a single stimulus to the nerve axon. This is known as the **twitch test**. Fast units (F) reach their peak tension and relax more rapidly from it than slow (S) units. These two categories of unit also behave differently in the **sag test**. Brief unfused tetani are given by stimulating the axon at intervals of 1.25 times the twitch contraction time. Fast units show a slow decline in the peak tension over the duration of the tetanus ('sag'), whereas slow units do not.

The group of fast units can be further subdivided on the basis of resistance to fatigue. Repeated tetanic stimuli are given briefly every second and the tension output monitored for each contraction. Slow units produce almost the same output for each contraction. Fast units do not. There is a subgroup with fast twitch contraction times (FF, fast fatiguable) which show a rapid decline in tension output over the first few minutes of the test. Another subgroup (FR, fast fatigue-resistant) has a high resistance to fatigue similar to the S units. Sometimes, another category of fast units is added: F(int), with properties lying between the FF and the FR subgroups. It should be noted that muscle fibres may be fatigued without any failure of electrical transmission at the neuromuscular junction. That is, the fibre action potential can be normal, while twitch force is reduced, indicating that there is fatigue of the force-generating mechanism (see section on human fatigue below).

Twitch tension and specific twitch tension
The major way of classifying motor units relies on the speed of contraction and fatiguability tests above. However, there are many other differences between units, although they are not so useful in distinguishing motor unit **type**. The peak twitch or tetanic tension is larger in fast units than in slow units and, in general, the peak tension declines in

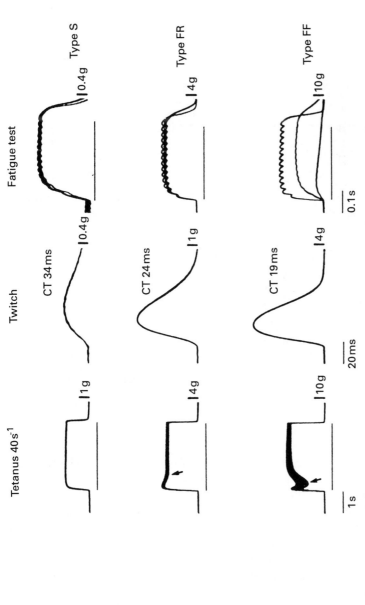

Figure 3.2 Physiological identification of motor unit type. Each row illustrates the tension records produced by activity in a particular type of motor unit. The columns show the different responses of these units to three tests, tetanus, twitch and fatigue. Note the calibration marks for tension are different for each unit. Tetanic tension is smallest for type S units: type FF and FR units show 'sag' (arrow). Twitch size is smallest and twitch duration longest in type S units. Repeated tetani do not produce fatigue in type S and type FR units, but do in type FF units (three superimposed records of the first, 60th and 120th tetanus.) (From Jami *et al*, 1982, Figure 1; with permission.)

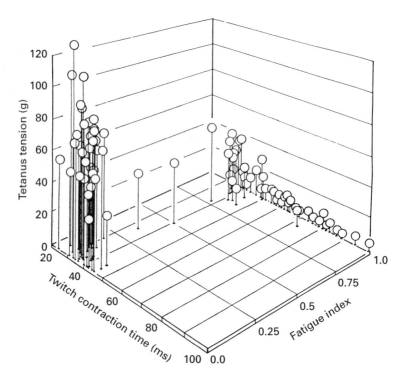

Figure 3.3 Three-dimensional relationship between twitch tension, twitch contraction time and the fatigue index (the ratio between contraction strength in the first and 120th tetanus of a series given regularly at 1 s intervals) in the medial gastrocnemius muscle of the cat. Units showing 'sag' (that is, FR and FF) denoted by open circles, whereas units without 'sag' denoted by stippled circles (that is, S). Note the clear division of the units plotted by open circles into two groups: the type FF with a very low fatigue index, and the type FR with high fatigue index. Two units fall between these groups and may be of the F_{int} category. (From Burke *et al.*, 1973, Figure 5; with permission.)

the order FF > FR > S (see the three-dimensional relationship in Figure 3.3). There is a combination of reasons for this.

First, estimates of the average number of fibres per unit have shown that type S units have a smaller number of muscle fibres than type FF or FR. This is done by counting the total number of muscle fibres with each characteristic histochemical profile (see below) and multiplying by the relative frequencies of each type of unit within the same muscle. In addition, it can be seen microscopically that type S units have fibres with smaller diameter than type FF or FR units.

However, even when these two factors of innervation ratio and fibre size are taken into account, they are still insufficient to explain completely the difference between the force production of fast and slow units. That is, when expressed as the twitch tension per unit cross-sectional area of muscle (which takes into account the differences above), the S units still generate less force than the F units. The type S units are said to have a low *specific twitch tension*. This is thought to be due to differences in the mechanism of force production within the fibrils of the type S units.

Post-tetanic potentiation
Following a tetanus, the peak twitch tension of fast units is increased for a short period. The contraction of slow units in different muscles is either less potentiated or may even be diminished. The reason for this effect and its difference between unit types is unknown. It may be related to the kinetics of Ca^{2+} binding and sequestration following muscle activation.

Size and conduction velocity of motor axons
Although the motor axons supplying extrafusal muscle are all of the alpha category, they have a range of sizes (in the cat, from 10–20 µm) and conduction velocity (cat, 40–100m s^{-1}). It has been known for many years that the motor nerves supplying slow muscles have a lower conduction velocity than those to fast muscles. The same applies at the level of the motor unit. The average conduction velocity of axons supplying the different types of motor unit follows the order FF > FR > S. In addition, it is also known that large diameter axons have large cell bodies within the ventral horn of the spinal cord. Thus, fast units have motoneurones with large cell bodies, while slow units are innervated by smaller motoneurones.

Summary
Muscles are made up of various proportions of these different types of motor unit. Units with a fast contraction time are composed of fibres with a relatively large diameter and are supplied by a fast conducting axon from a large motoneurone. They produce high twitch tensions and are relatively easy to fatigue. Slow units are slow to contract and are made up of rather fewer fibres of small diameter. The motor axons conduct slowly and the units produce only small amounts of force, although they are fatigue-resistant.

The advantage of having the subgroups is that the total range of operation of the muscle is extended beyond that of any single unit type. The relative number of each unit can give distinctive properties to whole muscles which makes them suitable for different kinds of movement. In the cat, prototypical examples of this are the soleus and gastrocnemius muscles. The soleus has a large percentage of type S units. It has a slow contraction time and is not readily fatigued. Gastrocnemius has more fast units and can exert a greater tension than soleus, but only over short periods. Consistent with this difference in contractile properties is the role played by each muscle in normal movement. Soleus is used almost continuously in walking and standing, whereas gastrocnemius is particularly useful in jumping and sprinting.

Even blood flow through the muscle is related to the properties of the motor units. In the cat, the flow to gastrocnemius has a capacity 2.5 times greater than that to soleus. Because the oxygen consumption per twitch of slow units is less than that of fast units, this means that the blood supply to soleus is sufficient to deliver enough oxygen even during a long tetanus at 20 Hz. In contrast, oxygen delivery fails in gastrocnemius at low rates of activation above 4 Hz. At the level of the single unit, capillary density is greater next to the fibres of slow than fast units (Burke *et al.*, 1973; Buchthal and Schmalbruch, 1980; Burke, 1981).

3.4 *Histochemical and biochemical classification of muscle fibres*

Most mammalian muscles contain a mixture of muscle fibre types which can be distinguished by biochemical methods. It is probable that much of the physiological differences between motor unit types can be explained on the basis of differences in the biochemical profile of their constitutent muscle fibres (see below). For many years it had been known that there were different numbers of mitochondria within muscle fibres, suggesting differences in their oxidative capacities. More recently, histochemical methods have been used, which allow visualization of enzyme content, metabolic substrates or structural proteins within individual fibres. Differences in the content of metabolic enzymes suggests preferential use of particular metabolic pathways. Examples of some of the presently used stains are given in the following list:

1. ATPase activity of myofibrillar material. Myosin ATPase has various forms or isoenzymes which break down ATP at different rates. Testing in an alkaline pH of 9.4 separates two main types of muscle fibre. Those which stain darkly are predominant in pale muscle, and those which stain lightly are more common in red muscle. Sometimes acid preincubation at pH 4.6 is used to separate out three different categories.
2. SDH (succinic dehydrogenase) and NADH-D (NADH-dehydrogenase). Both these enzymes are involved in major pathways for oxidative metabolism in the mitochondria.
3. Glycogen and phosphorylase stains. Both stains relate to the capacity for anaerobic metabolism, used when the oxygen supply of muscle is limited. Phosphorylase is an enzyme involved in the breakdown of glycogen.

By applying selective stains for several enzymes to serial sections of muscle, a particular 'histochemical profile' can be obtained for each fibre. Unfortunately, the nomenclature for this classification has been changed repeatedly and a classification evolved for the muscles in one animal may not be appropriate to other species. In this chapter discussion is limited to fibre types in guinea pig, cat and man; rat muscle fibres have a different histochemical profile.

Two main histochemical nomenclatures have been developed: types I, IIA and IIB; and SO (slow, oxidative), FOG (fast, oxidative glycolytic) and FG (fast, glycolytic). The latter nomenclature employs terms relating to the physiological properties of the fibres, and is discussed in the next section. Table 3.1 shows the typical histochemical profile of each of the fibre types. ATPase levels and the capacity for anaerobic respiration separate the fibres clearly into two groups. Type I fibres have a low capacity for anaerobic respiration and a low level of ATPase staining, whereas type II fibres have a high anaerobic capacity and high levels of ATPase staining. Levels of oxidative enzymes can be used to separate two subtypes; type IIA fibres have high levels of both oxidative and glycolytic enzymes; type IIB have high levels of glycolytic enzymes but are poor in oxidative capacity. (Kugelberg and Edstrom, 1968; Burke *et al.*, 1973; Burke, 1981; Lewis, 1981 give further details of the work covered in this section.)

Correlation between histochemical and physiological classifications of motor units

Ever since Ranvier's observation that the colour and histological feature of whole muscle were in many cases correlated with the speed of contraction, it has become generally accepted that the increasingly well-studied histochemical characteristics of individual muscle fibres must be related in some way to the physiological properties of the same fibres. However, this is technically rather difficult to prove, since it involves being able to analyse the histochemistry of individual muscle fibres which comprise a physiologically indentified motor unit.

An important advance was made with the introduction of the glycogen depletion technique. In an elegant series of experiments on the gastrocnemius muscle, Burke *et al.* (1973) studied the physiological properties of individual motor units. The unit was then fatigued by repetitive stimulation of its axon, and serial sections were cut from the muscle. The sections were then stained for glycogen depletion, or tested for the levels of oxidative (SDH), glycolytic (phosphorylase) or ATPase enzymes. Thus the fibres of an individual unit could be identified in one section by their lack of glycogen and characterized in the other sections for their histochemical properties. It should be noted, however, that the

Table 3.1 Relationship between Fibre Type, Motor Unit Type and Histo-chemical Profiles of Muscle Fibres

Fibre type	I SO (Slow, oxidative)	IIA FOG (fast, oxidative glycolytic)	IIB FG (fast glycolytic)
Motor unit type	S	FR	FF
Histochemical profiles			
Myofibrillar ATPase (pH 9.4)	Low	High	High
NADH dehydrogenase	High	Medium-High	Low
SDH	High	Medium-High	Low
Glycogen	Low	High	High
Phosphorylase	Low	High	High
Capillary supply	Rich	Rich	Sparse
Fibre diameter	Small	Medium-small	Large

results from slow twitch units were a little more difficult to interpret because of the problems in fatiguing all the fibres of each unit so that they could be identified by the glycogen depletion method.

There were two important findings: the first was that all the fibres from a single motor unit were of the same histochemical type; the second was that the histochemical characteristics were intimately related to the physiological contractile properties of the unit. Table 3.1 summarizes the important relationships between histochemistry and physiology.

The **speed** of contraction was related to the ATPase and phosphorylase activity of the fibres. Fast twitch fibres had high levels of phosphorylase, and stained darkly for ATPase after alkaline incubation. They were type II fibres. The **fatiguability** of contraction was related to the SDH activity. FF units were made up of fibres which had low levels of SDH (type IIB, or FO), whereas S units had fibres with high levels of SDH (type I, or SO). FR units were intermediate (type IIA, or FOG) (see Table 3.1). As a general rule, the relative susceptibility to fatigue was related to the relative amounts of oxidative and glycolytic enzymes. The intensity of stain for ATPase reflected the twitch contraction time of the unit.

The precise nature of the connection between speed of contraction and ATPase staining is not clear. Although contraction speed is proportional to the intensity of staining in a wide range of muscles, there are instances in which two fibres in a muscle may stain with the same intensity, yet have contraction times which differ by a factor of two. Part of the problem arises because the stain is not specific for myofibrillar ATPase but also stains mitochondrial and sarcotubular ATPases. Thus, differences between fibre types may not necessarily indicate differences in myofibrillar ATPase. Also, it is not clear, as yet, whether contraction speed is a function of myofibrillar ATPase activity. Increased speed of contraction may mean that the sarcomeres slide over each other more quickly, but the ATPase activity may not be the rate-limiting step in the process. It is possible that the rate of Ca^{2+} release from the sarcoplasmic reticulum may limit contraction velocity. In this connection it is interesting to note that the rapid relaxation time of fast units in a single twitch probably is due to an increased rate of reabsorption of Ca^{2+} into the sarcoplasmic reticulum (Burke, 1981, gives a full review of this topic).

3.5 *Some electrophysiological properties of motoneurones*

Motoneurones are the largest cells in the ventral horn of the spinal cord and may have dendritic trees which extend for distances of over 1 mm. In order to understand better the ways in which the nervous system regulates the firing frequency and pattern of these cells, it is important to have some knowledge of their intrinsic electrophysiological properties. The features have been analysed in detail.

Synaptic inputs to motoneurones
The main characteristics of excitatory postsynaptic potential (EPSP) and inhibitory postsynaptic potential (IPSP) generation in motoneurones are well known, since these cells were the subject of the first microelectrode studies in the spinal cord. Synapses are formed on the whole of the soma and dendritic membrane, but action potentials can be generated only on the somatic membrane and at the axon hillock. Of these two sites, the axon hillock has the lowest threshold for spike initiation. The dendrites do not appear to have any of the voltage-sensitive ion channels necessary to produce regenerative action potentials.

When a cell is impaled by a microelectrode, the penetration is invariably made into the soma, since the dendrites and axon are far too small to allow insertion of a needle. Recordings made here of the size and time course of EPSPs give an indication of the dendritic location of the active synapses. Electrical models of motoneurones first suggested, and later it was confirmed, that EPSPs produced by terminals on distal portions of the dendritic tree tend to be smaller and to have slower rise and decay times (Figure 3.4) than synapses nearer the soma. The reason for this is the increased resistance and membrane capacitance between distal dendrites and soma, which attenuates and slows the EPSPs. However, absolute size of an EPSP is not a useful parameter to provide unequivocal evidence of synaptic location, since the absolute size of an EPSP depends on the number and efficiency of active synaptic contacts as well as distance from the soma membrane. Usually, the EPSPs recorded at the soma are analysed only in terms of their rise times to peak amplitude. Longer rise times indicate more distal synaptic inputs.

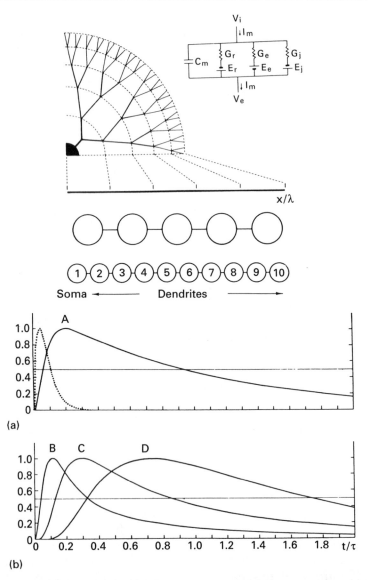

Figure 3.4 (a) Schematic diagram of motoneurone cell body and dendritic tree. Circles below indicate the mathematical transformation of this motoneurone into a series of linked compartments described by circuitry above. (b) Comparison of EPSP shapes recorded at soma. Dotted line shows assumed time course of excitatory conductance change. A is the EPSP expected from equal input into all compartments of the model. For curve B, input was into compartment 1 alone, for C into compartment 4, and for D into compartment 8 alone. Curves are scaled so that their peak amplitudes are equal to 1.0 on the y-axis. The absolute sizes would, of course, be quite different, being, for example, larger in B than in D. The time course is plotted in terms of the membrane time constant τ. (From Rall, 1967, Figures 1 and 2; with permission.)

Input resistance of motoneurones

The input resistance of a whole cell gives a measure of the voltage change produced across the cell membrane by a known current applied at the soma. The usual procedure is to record the cell membrane potential with a microelectrode, at the same time as a current is injected into the soma. Large cells, with large diameter axons (and high conduction velocities), have a lower input resistance than small ones. This means that for a given current, the small motoneurones will be depolarized more than large cells. If the voltage threshold for firing is the same in all cells, then the small cells will reach threshold first. The reason for the different input resistances is caused mainly by the different surface areas of the cells. The specific resistivity (the resistance per unit area) of the cell membrane probably does not vary greatly between motoneurones so that the effect of increasing the cell surface area can be imagined to be analogous to adding more and more resistors **in parallel** between the interior and exterior of the cell. The nett effect is, of course, to **decrease** the total resistance of the pathway. An alternative and more common way of looking at this is to use the **conductivity** of the cell membrane, which is the reciprocal of resistivity. The specific conductivity is expressed in mho per unit area. Conductivities add linearly when the membrane area is increased by a unitary amount.

Firing patterns of motoneurones

In order to understand how fast and how many impulses a motoneurone will fire in response to particular synaptic inputs, cells can be depolarized by a known current applied to the soma through an intracellular microelectrode. The firing pattern can then be analysed in response to carefully controlled input currents. Examples of the response to step changes of current to different levels are shown in Figure 3.5. All the recordings are taken from an intracellular microelectrode. The three traces in row 1 show the response to antidromic invasion of the cell following stimulation of the motor axon. The action potential rapidly repolarizes and is followed by a small delayed depolarization which sits as a small wavelet on the potential before recovering to resting potential.

This contrasts with the spikes recorded following current injection into the soma through the microelectrode. These spikes, seen in row 2, do not have any delayed depolarization. They behave differently to an antidromic spike, and are followed by **an after-hyperpolarization**. If two potentials follow one another, the after-hyperpolarizations can summate. After-hyperpolarization is also seen following action potentials initiated by normal synaptic inputs, and is the usual mode of cell firing. Lack of an after-hyperpolarization with antidromic cell firing represents an unusual mode of activation. The delayed depolarization is thought to

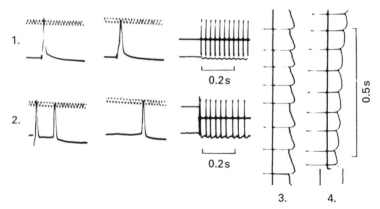

Figure 3.5 Action potentials recorded from a microelectrode inserted into the soma of a motoneurone. In row 1 are three records of potentials produced by antidromic invasion of the motoneurone. The first two records are on a much shorter time-base than the third (note the 50 Hz signal) and show that the spike is followed by a delayed depolarization before returning back to resting potential. The third trace shows a repetitive series of antidromic potentials. Row 2 shows the spikes recorded following depolarizing current injection into the soma. These spikes are followed by an after-hyperpolarization, which can be seen to summate (in the first record) with that following a second spike. The third record of the row illustrates repetitive firing produced by prolonged, steady, current injection into cell (shown as downwards step in the upper trace). At onset of current injection, the cell fires two impulses in very rapid succession and then settles down to a regular firing rate. Summation of the after-hyperpolarization from the first two impulses can be seen. Rows 3 and 4 should be read vertically: 3 shows steady firing of motoneurone caused by current injection into soma, in which the after-hyperpolarizations after each spike are quite clear; 4 shows in detail what happens at the onset of current injection into the cell (downwards step in upper trace). The first two spikes occur rapidly and then firing settles down to a steady rate. After-hyperpolarizations summate as in row 2. (From Granit *et al.*, 1963, Figure 9, parts 4–7; with permission.)

be caused by a wave of depolarization spreading out into the dendrites.

The third record in row 2 illustrates what happens to the firing rate of the cell immediately following a step depolarization of the soma membrane. The cell fires off two impulses in rapid succession and then settles down to a slower steady rate of firing. The same can be seen on a longer time scale in part 4 of Figure 3.5. The relationship between steady firing frequency and applied current (F-I curve) is shown in Figure 3.6 and was first described by Kernell (1965). The curve consists of two approximate straight line segments known as the primary and secondary range of firing. It shall be seen below that the steady discharge of most motoneurones invariably is restricted to the primary range (between 10– 50 Hz in different motoneurones). The secondary range, in which large depolarizing currents must be applied, is rarely

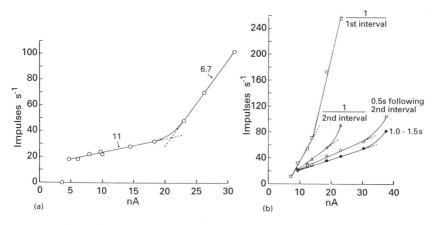

Figure 3.6 Relationship between current injected into soma and firing frequency of a cat motoneurone. (a) The relation for the steady-state firing frequency of the cell; (b) the relations for the first, second and subsequent intervals after onset of stimulation in a different cell. Because neurones fire rapidly at onset of current injection, the initial firing rates are much higher than those recorded later in the train of action potentials. Note the two straight line segments of each curve: these are the primary and secondary ranges of motoneurone firing described in the test. (From Kernell, 1965, Figures 4 and 5; with permission.)

achieved in normal activation. Entry into the secondary range probably is caused by partial inactivation of the spike-generating mechanism due to maintained depolarization of the cell. This reduces the height of the action potentials and also produces a change in the after-hyperpolarization (see below).

The initial rapid adaptation of firing frequency to that seen in the steady state is caused mainly by development of after-hyperpolarization which follows each spike. The after-hyperpolarization builds up over the first two to three impulses and serves to stabilize the motoneurone against excessive depolarization. One consequence of this is that initially it is very easy to push the firing frequency of the motoneurone into the secondary range. However, as the after-hyperpolarization develops, this becomes more and more difficult, and more current is needed. Effectively, by this mechanism, the primary range of the motoneurone is extended over a large range of input currents (Figure 3.6). The duration of the after-hyperpolarization is related to the size of the cell; small cells have longer after-hyperpolarization than large cells. It is also related to the minimal steady firing frequency of the cell. Cells with longer after-hyperpolarizations can discharge steadily at lower rates than those with shorter ones. Thus small cells can maintain slower firing frequencies than large cells, because of their long after-hyperpolarization. Conversely, large cells with a shorter after-hyperpolarization can fire more rapidly within the primary range than small cells. This is of some use since large motoneurones innervate large motor units which must discharge at higher frequencies to produce a fused contraction than their smaller neighbours. Similarly, rapid initial firing rates are used to produce rapid increases in muscle tension (see below), for example at the onset of a contraction (Granit, 1970; Burke, 1981).

3.6 *Control of motor units and their recruitment order*

Most muscles contain a mixture of unit types. Since the contractile properties of each type of unit are different, it should come as no surprise to find that the control of the units is carefully regulated. Earlier this century, Denny-Brown (1929) had shown that slow pale muscles (that is, those with a preponderance of slow units) were activated preferentially in tonic contractions such as the stretch reflex, whereas red, fast muscles were activated only in rapid contractions such as the scratch reflex. More recently, this question of relative threshold, or **recruitment order** has been extended to the individual motor units of a muscle.

Henneman, Somjen and Carpenter (1965) performed most of the original experiments on the recruitment order of motor units. In order to examine the firing pattern of a small number of different units, they recorded the nerve action potentials from ventral root filaments during contraction. The method is superior to recording motor unit potentials from active muscle, since the recording electrodes do not get dislodged by the contraction.

A variety of different types of contraction were examined: reflex contractions produced by stretch or by ear twisting, and contractions produced by electrical stimulation of the brain at various sites. For any ventral root filament that they examined, during any type of contraction, they always found that the axons were recruited in a particular order (Figure 3.7). In general, axons with small action potentials were recruited first in the contraction; only at higher levels of contraction did axons with larger potentials begin to discharge. Since the size of the potentials picked up by the recording electrode is related to the size of the active axon, this meant that small motor axons were recruited before large ones.

The size of a motor axon is related to the number of muscle fibres that it innervates, and also to the size of the cell body within the spinal cord. Thus, at the level of the muscle, this meant that as force built up in the contraction, small motor units were recruited before large ones. At the level of the motoneurone, small neurones began firing before large ones. The rule that Henneman and his colleagues postulated was simple: the

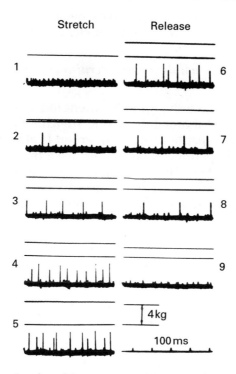

Figure 3.7 Recruitment order of two motoneurones recorded in a ventral root filament during a stretch-reflex-induced contraction of the triceps surae muscle. The amount of tension developed by the muscle is shown by the separation of the upper two beams in each frame. Stretch increases from frames 1 to 5 and then decreases back to zero in frames 6 to 9. A unit with a small action potential is recruited first (frame 2) and is followed by a larger unit (frame 4). Derecruitment proceeds in the opposite order: the large unit drops out first (frame 7), followed by the small unit (frame 9). (From Henneman *et al.*, 1965, Figure 1; with permission.)

recruitment order of a motor unit depended, in all types of contraction, solely on the size of its motoneurone. This had become known as the **size principle**. In addition to investigating the order of activation using facilitatory inputs, they also examined the **derecruitment** of units by inhibitory inputs. This followed the reverse order: large active units always were inhibited before small ones.

Much of this work was started before the physiological classification of motor units into different types. Later studies have confirmed the supposition that in general, motor units are recruited in the order S, FR, FF. Thus, small, slow twitch units are activated first. They are suited to participate in long-lasting but relatively weak tonic contractions. The FR units are activated later, and the FF units last of all. The FF units generate large forces but are easily fatigued, and hence cannot sustain tension over long periods. The whole scheme seems beautifully adapted to allow the CNS to control body muscles in the most logical manner.

Mechanisms responsible for the recruitment order of motoneurones

The mechanisms underlying motoneurone recruitment order have been investigated most thoroughly during stretch reflex contractions produced by monosynaptic input from group Ia afferents. These are the largest fibres in peripheral nerve, and have the lowest threshold for electrical stimulation. They can therefore be excited independently of other smaller fibres in the nerve trunk. A large number of these fibres can be excited simultaneously using this method, and the size of the

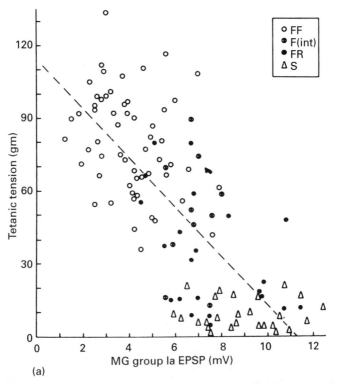

(a)

Figure 3.8 Relationship between motor unit type and size of the composite group Ia EPSP in the cat. The EPSPs were recorded in the ventral horn of the spinal cord from motoneurones innervating medial gastrocnemius. A synchronous Ia input to the motoneurones was achieved by stimulating the peripheral nerve innervating the muscle. (a) The relation between size of group Ia EPSP and the tetanic tension produced in the muscle by each innervated muscle unit. Units which are innervated by motoneurones with large Ia EPSPs produce only small tetanic tensions and vice versa. The different symbols in the figure indicate motor units of different physiological types.

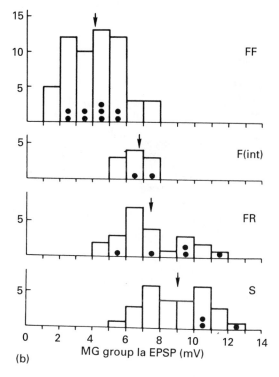

Figure 3.8 Continued. (b) The number of units in each category with Ia EPSPs of a given size. The EPSPs are smaller in the FF units than in the type S units. Black dots indicate data collected from one animal. Arrows indicate mean values. (From Burke *et al.*, 1976, Figures 2 and 3; with permission.)

monosynaptic EPSP produced in different motoneurones may be recorded by a microelectrode. This EPSP is known as a composite or aggregate EPSP since it is produced by the activation of many separate Ia fibres. Eccles *et al.* (1957) initially showed that this EPSP was larger on average in motoneurones innervating grossly slow muscles, such as soleus, than it was in motoneurones innervating grossly fast muscles such as gastrocnemius. Burke and colleagues (1973) have since extended these findings to the individual units within the motoneurone pools of many different muscles. The composite EPSP usually is larger in motoneurones innervating identified type S units than it is in those innervating FF units. Type FR units have EPSPs more nearly related to the size of those in motoneurones of slow units (Figures 3.8 and 3.9).

It is thought that all the motoneurones of the ventral horn of the spinal cord have approximately the same voltage threshold for impulse generation. Thus, the different EPSP sizes can readily explain the preferential recruitment of slow motor units by facilitatory inputs. The

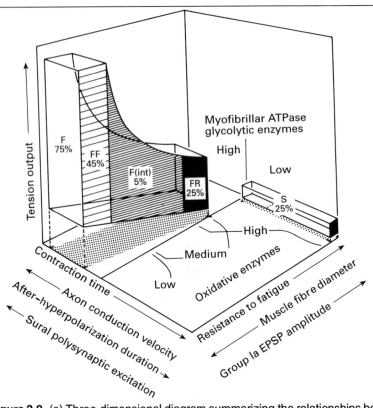

Figure 3.9 (a) Three-dimensional diagram summarizing the relationships between a variety of physiological, morphological and histochemical profiles of the motor units of cat medial gastrocnemius muscle. The unit types FF, F(int), FR and S are shown by the shaded areas, together with percentage figures of their predominance in the muscle. The diagram emphasizes that a great many properties of motoneurones and the motor units that they innervate are correlated one with another.

derecruitment of units by inhibitory inputs also follows the size principle. However, from this statement alone it is not possible to deduce anything about the size of the IPSPs in different motoneurones. Since large motoneurones receive only weak excitation, either a weak or a strong inhibition would be sufficient to reduce their excitation below threshold. In fact, it has been shown that, like the excitatory input, group Ia IPSPs are largest in those motoneurones innervating slow muscle units. Figure 3.9 is a summary diagram.

Although the composite PSP sizes can account for the order of motoneurone recruitment, it is not clear why the PSPs are of different sizes in different motoneurones. The original idea of Henneman and his associates was that the order of recruitment was determined by properties dependent solely on the size of the motoneurone. It was assumed

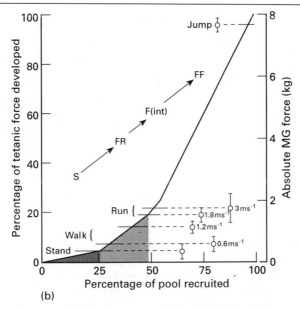

(b)

Figure 3.9. Continued. (b) Diagram summarizing the recruitment of different proportions of the motor unit pool of cat medial gastrocnemius during different types of muscular activity from quiet stance, through walking, to jumping. During standing, only about 25% of the total population of units is active. These are the type S units. Types FR and F(int) are recruited during walking and running. The type FF units (which are 45% of the total population) are only rarely recruited for brief periods of time in very powerful contractions such as needed in jumping. (a) From Burke, 1981 (after Burke, 1975), Figure 35; with permission; (b) from Walmsley *et al.* 1978, Figure 9; with permission.)

that the passive electrical properties of the cell membrane were the same in motoneurones of any size. In particular, the resistance of a unit area of membrane (the specific resistivity) was assumed to be constant. As described above, this means that the input resistance of a small motoneurone will be larger than that of a large motoneurone. Given this assumption, it is possible to explain recruitment order by postulating that a given input to a motoneuronal pool evokes the same synaptic current in all cells. If the current is the same, then small motoneurones with a large input resistance will be depolarized more than large motoneurones. If voltage threshold is the same in all cells, then the small cells will fire impulses before the large cells.

But why should the synaptic current be equal in all motoneurones? It might have been expected that because of their larger surface area, there would be more synaptic boutons on large than on small cells, and that the total synaptic current would be greater the larger the cell. Several possibilities can be put forward to explain the assumption of equal synaptic currents:

1. The synaptic efficiency of boutons on small cells may be greater than those on large cells. This could be because: (a) the input to small motoneurones is located near to the cell body and produces larger soma PSPs than those on larger motoneurones which are located on distal dendrites; (b) the synapses on small cells could produce larger PSPs by being physically larger or produce larger conductance changes than those on large cells; or (c) if large neurones had a disproportionately large soma relative to dendritic area, then dendritic inputs would be less effective than those in small cells.

    There is no strong experimental evidence for any of these possibilities. Recent anatomical studies using horseradish peroxidase (HRP) staining of Ia afferent terminals show little difference in anatomical location on motoneurones of different size. Electrophysiological techniques have confirmed this. Electrical models of the motoneurone have shown that if it is assumed that all the afferent inputs arrive synchronously at the cell, the composite PSP recorded in the soma from inputs located primarily on distal dendrites lasts longer in proportion to its size than the input produced by more proximal synapses (see above). There is no experimental evidence of this type to suggest that the terminals may be located slightly more distal on large than on small motoneurones. Similarly, anatomical studies have shown that there is no difference between the ratio of soma size to dendritic area in large and small motoneurones. At present there is no direct evidence concerning the conductance of single synapses in the membranes of large and small motoneurones. Analysis of the quantal potentials making up the EPSP in motoneurones of different sizes showed that the voltage changes were larger in small than in large cells. However, this was probably not due to any differences in the membrane conductance of single ion channels, but simply to the larger input resistance of small cells.

2. There may be the same **number** of synapses on motoneurones of all sizes. This would mean that the **density** of terminals would be lower on large cells. From an experimental viewpoint this hypothesis is very difficult to verify. It is never possible to stain selectively all terminals of a particular afferent class, or even a known proportion of them, so that they can be counted anatomically. However, some authors have favoured this possibility.

3. There may be an equal density of synapses on all motoneurones (and hence more synapses in total on large cells), but perhaps because the axons branch more often to innervate more terminals, the invasion of the presynaptic nerve impulse into the terminal arborization is less complete for synapses on large than on small cells. This means that proportionately less synapses would be

activated on large motoneurones. Recent studies of the quantal fluctuations in EPSP size in motoneurones of different size suggest that this is a possibility.

In the past, the idea that recruitment order depends solely on size-related properties of motoneurones has come under criticism. For example, direct estimates of total cell area using HRP techniques show that motoneurones innervating different types of motor units have rather small differences in average surface area. This makes it difficult to imagine that the size of a cell is the sole determinant of its recruitment order. Other possibilities put forward to account for the orderly recruitment of motoneurones include the idea that membrane properties may vary between cells. For example, if specific membrane resistance were to vary, this might be the reason why cells of apparently similar surface areas can have different recruitment thresholds. At the present, the matter is unresolved. (Burke, 1981; Henneman and Mendell, 1981; Gustaffson and Pinter, 1984; Zengel *et al.*, 1985, give further information on the work covered in this section.)

Exceptions to the normal order of recruitment

The main reason for using Ia monosynaptic excitation and Ia disynaptic inhibition in the investigation of the mechanisms underlying the order of motoneurone recruitment, is that the methods are technically feasible, rather than being necessarily the most important. There are many other inputs to the motoneurones, but these use polysynaptic pathways which are not as easy to investigate directly. In their original experiments, Henneman and his colleagues (1965) studied the recruitment order of motoneurones elicited by stretch reflex contractions, crossed extensor reflexes, pinna reflexes and by direct stimulation of sites in the motor cortex and brainstem. Except for the stretch reflex, all these inputs use pathways other than the Ia afferent neurone. Despite this, from all the sites, motoneurones were recruited in order of size. One possible interpretation is that all the inputs tested had a synaptic weighting which favoured the small motoneurones just like the Ia connections. However, this is not necessarily true. For example, if there were any *tonic* activity in the Ia pathways, this would *bias* the motoneurone pool by bringing small neurones nearer to threshold than large neurones. Thus, even if other inputs projected with equal synaptic weight to the motoneurones, the tonic Ia firing would assure recruitment of motoneurones according to size.

Despite the generality of the size principle, there are exceptions to the rule which have been a source of debate for some time. These are the inputs from the rubrospinal tract (in the cat) and cutaneous afferents. Both these systems project via polysynaptic pathways so that the composite EPSP produced by stimulation of a large number of axons in a cutaneous nerve, or of a large number of cells in the red nucleus, is more diffuse than that from the monosynaptic pathway. Generally, a train of stimuli needs to be applied, which often produces a combination of excitation and inhibition in the cell. The most rapid effects from both the rubrospinal tract and cutaneous afferents probably traverse a trisynaptic pathway. In both cases, there is a much stronger excitation of fast motor units. Slow units often are inhibited by these inputs, which is the reverse of the situation proposed by the size principle.

The implication of this is that the synaptic weighting to motoneurones of different size can vary with the input source. The arrows on the left in Figure 3.11 show the usual (size principle) weighting of inputs, whereas those on the right show the reverse pattern expected from cutaneous and rubrospinal input. One consequence of this hypothesis would be that unusual patterns of recruitment might be observed when both types of input are combined. Unless the exact synaptic weighting onto each neurone were known specifically, it would be impossible to predict the order of unit activation during combined stimulation. (Articles in Desmedt, 1981, give further details.)

Examples of motoneurone recruitment during normal movement (Figure 3.10)

In many animal experiments, the pattern of activity of whole muscles rather than single units has been investigated. The two favourite muscles for investigation are the predominantly fast medial gastrocnemius and the slow soleus muscle.

In the cat hindlimb, Burke and his colleagues have classified three 'grades' of muscle activity: (1) postural maintenance, as in quiet quadrupedal standing; (2) sustained walking or running; and (3) movements made only in brief bursts, such as jumping, galloping and scratching. The soleus muscle produces almost the same level of force during all these activities, and even during walking, is generating a force near to its maximal output. In contrast, the medial gastrocnemius produces only 8–12% of its maximal output during quiet stance, and only 25% maximal force during walking or running. Only during the third category of movement does it generate maximal force. These are probably the only activities during which the 45% FF units are recruited. Indeed, jumping often is achieved with a lengthening contraction of the muscle which may produce more force than that seen in a maximal tetanic contraction (see Chapter 2).

Such results can be interpreted as a direct consequence of the size principle of motor unit recruitment. If there are common inputs to the medial gastrocnemius and soleus motoneurone pools, then it is not

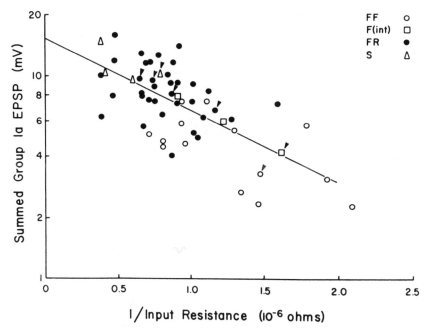

Figure 3.10 Relationship between input resistance and size of group Ia EPSP in motoneurones innervating cat tibialis anterior muscle. The larger the Ia EPSP, the larger the input resistance. Points obtained from experiments on a single cat are marked by arrows. Motoneurones innervating motor units of different physiological types can be distinguished by the symbols. (From Dum and Kennedy, 1980, Figure 2; with permission.)

surprising to find that the soleus, with its 90% type S units, is always recruited before the gastrocnemius and may be fully activated at relatively low force levels. It is obviously a muscle which is well suited to tonic, sustained muscle contraction. However, it is generally not only active during tonic contractions. Even in jumping, the soleus is maximally active together with the gastrocnemius, only in this case the gastrocnemius produces most of the force.

There is one well documented exception to this orderly recruitment of motor units. That is the **preferential** activation of lateral gastrocnemius during rapid paw shakes in the cat. In the example cited above, the lateral head of gastrocnemius is activated together with the medial head, and hence is recruited after the soleus. However, during rapid paw shake, which is a vigorous automatic movement made by the animal to remove an object from its paw, the soleus muscle is not activated at all. The movement is made using only the fast ankle extensor. The practical reason for this may be that the movement is so fast (10–12 Hz) that it cannot be followed by the soleus muscle, so that

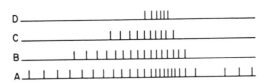

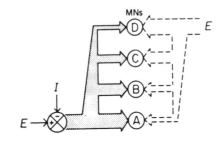

Figure 3.11 Diagram to illustrate recruitment order of motoneurones with different weightings of synaptic input. Four types of motoneurone, labelled A, B, C and D receive synaptic input with an efficacy proportional to the width of the arrows in the figure. The stippled arrows on the left represent the 'conventional' recruitment order of the motoneurones, with both excitatory (E) and inhibitory (I) inputs being most effective on motoneurones type A and least effective on type D. This type of organization produces recruitment and derecruitment sequences like those shown in the upper spike traces. On the right is input from another source (for example, cutaneous afferents) in which the weighting of synaptic input is the reverse of that on the left of the figure. Combining inputs from left and right might lead to dislocations from 'expected' recruitment order as shown in lower set of spike diagrams. (From Kanda *et al.*, 1977, Figure 8; with permission.)

if the soleus were active it would interfere with, rather than assist, the contraction. Interestingly, the movement is triggered preferentially by large, low-threshold cutaneous afferents from the plantar pads, which may, as noted above, be instrumental in inhibiting the slow soleus motor units (Walmsley *et al.*, 1978; Burke, 1981).

3.7 *The study of motor units in human physiology*

The ease with which single motor units can be recorded in normal humans, and the ability of human subjects to perform complex movements with little practice, has made the study of human motor units a particularly rewarding field in the control of movement. At the same time, the neurophysiological investigation of motor units plays a very important role in the diagnosis of neurological disease (Buchthal and Schmalbruch, 1980; Desmedt, 1981; Freund, 1983).

Recording motor units in humans

Concentric needle electrodes were first used by Adrian and Bronk in 1929. Essentially, they consist of a hollow needle with a fine wire inserted down the centre to the tip and insulated from the shaft with resin. The tip of the wire is bared and acts as one electrode, while the barrel of the needle acts as the other. A differential amplifier is used to amplify the potential difference between core and shaft. The diameter of a typical needle usually is of the order of 0.5–1 mm and, as such, will record from only a small portion of the whole muscle when it is inserted.

The recording lead of a conventional concentric needle electrode picks up activity of many individual muscle **fibres**. However, all the fibres of a given motor **unit** will discharge almost simultaneously, so that for each unit, only one aggregate spike is picked up by the electrode. Generally, the spike is of largest amplitude if the needle tip is at the centre of the motor unit territory, indicating a larger density of fibres at this point. Different motor units recorded at the same time from the same electrode usually have spikes of different shapes and sizes because their active fibres have a different spatial distribution about the recording tip. Motor units of human limb muscles have an average territory 5–10 mm diameter, allowing space for the (intermixed) fibres of 15 to 30 motor units.

For even finer localization, other types of electrode with two or more wires in the shaft can be used, and recordings made from the exposed tips with the needle barrel acting as a shielding ground. In such a situation, when the barrel is not used as one electrode, the wires usually will record only those motor unit potentials at a distance less than 1 mm from the recording head.

Histochemistry of human motor units

Human motor units are obviously more difficult to characterize histochemically than those in the cat. The earliest studies inferred the presence of different unit types in humans by examining the properties of whole muscle, or strips of excised muscle which contain a preponderance of one or other type of histochemically identified fibre. When this was done, it was found that muscles which had high levels of

ATPase and lower levels of SDH (that is, like type II fibres) tended to have fast contraction times. Those with slower contraction times had higher levels of SDH and less ATPase.

A small number of experiments have been performed on single motor units in humans which are directly comparable with those performed in animals. For example, Garnett *et al.* (1979) inserted a small needle into the medial gastrocnemius muscle and used it to stimulate selectively a single motor unit. Using this technique it was possible to obtain the average time course of a twitch following a single stimulus. In addition, repeated tetani could be given in order to test the fatiguability of the unit as in cat experiments. With these data it was possible to classify the unit physiologically as type S, FR, or FF. The final procedure was to stimulate the unit tetanically for a continuous period of up to two hours to deplete its stores of glycogen. At this point the muscle was biopsied and the specimens frozen. Serial sections were cut to identify the glycogen-depleted fibres and to characterize them for levels of SDH and myosin ATPase. As expected from cat experiments, physiologically identified FF units were composed of type IIB fibres, whereas the physiologically identified type S units were composed of type I fibres. Only a single FR unit was found, and had type I fibres.

From these studies it appears that motor units in humans may be classified using the same terminology as that used in the cat.

Mechanical properties of human motor units

Electrophysiological recordings of motor units provide a less invasive method of identifying the mechanical properties of single units during a muscle contraction. The technique involves detection of the discharge from a single unit with needle electrodes. This signal is then used to average the force exerted by the unit from the total force produced by the muscle. The idea is that the random fluctuations in muscle force during contraction will cancel out and one will be left with the average twitch force of the unit being recorded (Figure 3.12).

There are three main restrictions. The first is that there should be no fusion of the unit contractions or else the average unit force will not represent that of a single twitch contraction. Thus, the unit should fire at a low frequency (see below). Second, there must be no synchrony between the firing of different motor units in the same muscle. If another unit were synchronized with the unit under study, this would contaminate the average twitch force which was recorded. Third, the unit must be able to be distinguished from all others throughout the time of contraction. This is relatively easy at low force levels. However, with conventional needle electrodes, it is impossible to identify a single unit during contractions of moderate to high force.

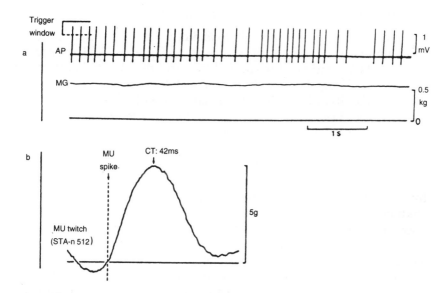

Figure 3.12 Method of measuring the twitch force of a single motor unit in the human first dorsal interosseous muscle. Discharges of a single motor unit are recorded by a needle electrode inserted into the muscle (a), while the subject exerts a fairly constant voluntary contraction of the muscle (second trace, MG). The spike of the selected motor unit then is arranged to trigger an electronic averager of the muscle force record. If all the active units in the muscle are firing asynchronously, the average force twitch time-locked to the unit discharge represents the twitch contraction of that unit (b): note difference in force calibration in this part of the figure compared with that in (a). (From Desmedt, 1981, Figure 1 pp. 97–136; with permission.)

These restrictions can never be met in practice. Motor units never discharge less than 8 Hz continuously, and as such there is always summation of individual twitches. This should therefore lead to an under-estimation of the size of an individual unit twitch. In addition, there is inevitably some synchronization between discharges of different motor units (see below). The fact that more than one unit is firing at the same time should lead to an over-estimation of the single unit twitch. In practice, these two factors seem to cancel each other out. In a study in which the axons supplying single motor units were stimulated selectively through a micro-electrode inserted into a peripheral nerve, the twitch force recorded after a single stimulus was very similar to that observed after averaging as described above (Thomas *et al.*, 1990).

Used alone, the averaging technique has shown that when fatigue is induced during prolonged contractions, units with high initial twitch tensions fatigue earlier than those with low twitch tensions.

Recruitment order of motor units in humans

With this technique it is possible to examine the recruitment order of motor units during voluntary contractions (Figure 3.13). The order in which units are activated is studied first and then the twitch force of each unit is averaged out later. In almost all types of contraction, small slow twitch units are recruited before large units. However, the recruitment threshold of a motor unit, that is, the muscle force at which the unit first begins to fire, is dependent upon the speed of contraction. The faster the contraction, the earlier the unit is recruited, although the **order** of recruitment, with respect to the other units of the muscle, remains constant. The one exception to this occurs in very rapid contractions during which units are recruited so rapidly one after the other that the conduction velocity of the efferent axon becomes an important variable. Slow units, which may have been recruited first at the level of the motoneurone actually may fire after some of their larger colleagues because of the time delay in conduction down their small axons.

One consequence of this change in recruitment threshold is that if very rapid adjustments are made from one level of force to another, there is a large but brief increase in motoneurone drive which is needed to produce the high velocity of contraction. This reduces the threshold force at which new units are recruited. Because of this, discharge rates in all units will increase and large units will be recruited transiently during the change in force. When the steady level of force is reached, the motoneurone drive will fall and the largest units will stop firing.

Another way of looking at recruitment threshold, rather than examining the muscle force at which single units began to **discharge**, is to look at the level of muscle force at which the units actually begin to contribute **tension**. The two thresholds are different because there is a time delay of up to 100 ms between electrical activation of a unit and the peak tension that it produces (see Chapter 2). When this delay is taken into account, then the force level at which the unit's tension is recruited is the same in maintained (but not twitch) contractions at any speed.

Changes in recruitment order

As in the cat, there have been several reports of changes in the 'fixed' recruitment order of human motor units. Electrical stimulation of cutaneous nerves can reverse the usual pattern of motor unit recruitment (Figure 3.14). Studies examining the effect of stimulation on the firing rate of continuously active units have shown that, like the cat, cutaneous input has an overall inhibitory effect on small units and an excitatory effect on large units. In the first dorsal interosseous muscle, stimulation

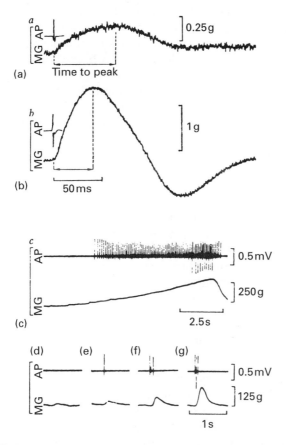

Figure 3.13 Twitch force profile and recruitment order of two motor units from the first dorsal interosseous muscle. Records (a) and (b) show the twitch forces of the two units and their respective action potentials (inserts to left of force record). Unit (a) was a small, slow twitch unit, and unit (b) was a larger, faster unit (note time to peak tension). The spikes of both these units could be recorded and distinguished through a single selective needle electrode. This is shown in (c) in which the subject was asked to produce a slowly increasing voluntary contraction of the muscle. As force rises, the first unit is recruited (it has a predominantly up-going spike potential), and later, at higher force levels, the second unit is brought to discharge (its action potentials can be distinguished by the down-going EMG deflections). In this slowly increasing contraction, unit (a) is recruited at a threshold force of 20 *g*, while unit (b) is recruited at about 280 *g*.

Records (d) to (g) are from rapid ('ballistic') voluntary contractions to different peak forces ranging from 10–110 *g*. Unit (a) is recruited in the small contraction of trace (e). Unit (b) is recruited only in the largest contraction (g). Note that this unit is recruited even though the peak tension of the contraction is only 100 *g*. This is below the threshold force seen in the slow voluntary contraction (c). Thus, the force threshold of this unit depends on the type of contraction which is being made. Its recruitment order, however, remains unchanged. (From Desmedt and Godaux, 1977, Figure 1; copyright 1977 Macmillan Journals Ltd, with permission.)

of the digital (cutaneous) nerves of the forefinger at three times the sensory threshold can reduce the force threshold for recruitment of large units and increase the threshold for recruitment of small units.

Changes in recruitment order also have been seen in certain bifunctional muscles. The abductor pollicis can be used either as prime mover in thumb abduction or as synergist in thumb flexion. Recruitment order of the units depends upon whether the muscle is being used as prime mover or synergist – that is, upon the direction of thumb movement. Such results suggest that the weighting of synaptic input to different motoneurones varies according to the motor command.

One hypothesis to explain this is that the prime mover command might be effected by synapses that are equally distributed to all units in the muscle. This would give rise to a 'normal' recruitment order of the units. However, the command to synergist muscles might use a different set of synapses which are preferentially distributed to certain of the units. Another possibility is that the pattern of afferent input produced by different movements, rather like the known effects of cutaneous inputs, could change the threshold for recruitment of motor units when the muscle is used as a synergist (Datta and Stephens, 1981; Desmedt, 1981).

Firing frequency of human motor units

Unlike the units in many sensory systems, which have firing rates which can vary continuously between 0 and 300 Hz or more, the range of firing rate modulation of human motor units is very small. For the muscles of the forearm, units only discharge continuously at frequencies from 6–8 Hz to 20–35 Hz. This frequency range corresponds to the primary range of motoneurone firing described by Kernell (1965) (see above). It is well matched to the mechanical properties of the motor units, since the contractions of most units are fused at above 40 Hz so that no extra force can be obtained by stimulating at higher frequencies. The only time it is necessary to exceed this limit is to increase the **speed** of a developing contraction. During very rapid contractions, firing frequencies within the secondary range are observed transiently. Such rates of 150 Hz or more do not alter the maximum force of contraction, but they do shorten the time taken to reach peak tension because the individual twitches summate very rapidly.

The lower limit of 6–8 Hz is more difficult to understand. The only way to discharge units at frequencies below 6–8 Hz is for that subject to perform a series of small rapid muscle contractions at the frequency rather than maintain a constant level of contraction.

The factors responsible for the low frequency limit are not clear. Animal experiments in which currents were injected via a microelectrode directly into the motoneurone have shown that even under these conditions, minimal firing rates are of the order of 5 Hz. Thus, the low frequency limitation cannot be a result of the excitatory synaptic input to the motoneurone. Renshaw inhibition has been regarded as a possible candidate (see Chapter 4). The firing probability of Renshaw cells is very high at low motoneurone discharge rates, but gets progressively smaller at higher rates. It could be that the inhibition is so profound as to limit the lowest frequencies of discharge.

Motor unit synchronization

Under most conditions, motor units in a muscle tend to fire asynchronously. Indeed, as described above, if the units were to fire synchronously, force production would be compromised. There are two conditions, though, when motor units fire synchronously: during powerful contractions, and after fatigue. In both circumstances, the synchronization produces a noticeable tremor. The mechanisms responsible for this have not been analysed in any detail. Nevertheless, the effects are considerable. In a maximum voluntary contraction, the synchronized firing of many units at about 40 Hz can produce an audible rumbling sound known as the Piper rhythm, best heard by placing a stethoscope over the belly of the contracting muscle.

A smaller degree of synchronization between units also occurs during normal contraction. This is very difficult to see by simple visual inspection of two single unit spike trains. However, it can be revealed by a technique known as cross-correlation. Two needle electrodes are inserted into a muscle in order to record the firing of two separate motor units. A computer then records the discharge of each unit, and every time unit A discharges, the computer notes the relative time at which the second unit, B, discharges. A cross-correlogram is a graph which plots the probability that unit B will fire at different times before and after the discharge of unit A (Figure 3.15). If the discharge of unit B was completely unrelated to that of unit A, the graph would be flat. The probability of firing would be the same at all times relative to the discharge of unit A. In reality, the graph often shows a small peak around time zero. This means that unit B fires more often at around the same time as unit A than it does at other times (Figure 3.15). In the pair of units illustrated, unit B was six times more likely to fire at the same time as unit A than expected by chance alone. In other words, there was a tendency for the two units to fire synchronously.

The reason the two units fire at the same time is that their parent motoneurones in the spinal cord receive a common synaptic input which brings them to threshold at approximately the same time. The form of the cross-correlogram can be predicted knowing the size and time course of the shared synaptic input. One can easily imagine that if

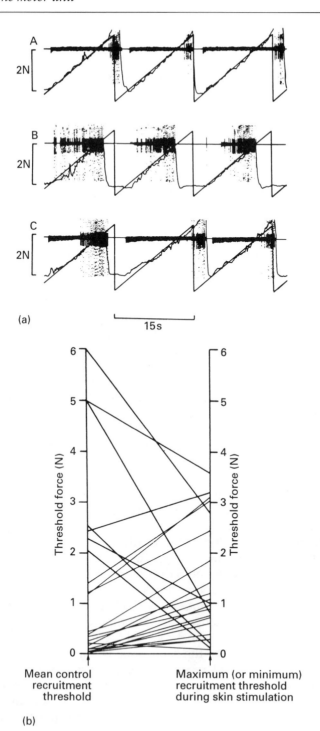

(a)

15s

(b)

Mean control
recruitment
threshold

Maximum (or minimum)
recruitment threshold
during skin stimulation

two motoneurones were totally quiescent and received a very large shared synaptic input, then they would always discharge together at exactly the same time. However, this is not normally the case. The shared synaptic input is usually not large enough on its own to bring a motoneurone to its firing threshold. Under normal conditions, other inputs to motoneurones cause the membrane potential to fluctuate continuously, so that it is sometimes close to and sometimes far from threshold. The result is that two motoneurones will only discharge together if they receive shared synaptic input at a time when chance fluctuations in membrane potential bring them both near to threshold at the same time. Under these conditions, synchronized motor unit firing will be quite rare. The precise time at which the units fire depends on how near each neurone is to its firing threshold when the shared EPSP is received. The spread of possible discharge times is related to the duration of the EPSP, and this is reflected in the width of the cross-correlogram peak.

The size of the cross-correlogram peak is a measure of the amount of shared synaptic input to a pair of motoneurones. Most experiments have been performed on the intrinsic muscles of the hand. Here, peaks are largest for pairs of motor units within the same muscle, but also can be seen between units in antagonist muscles, and even between units in distant muscles. As a general rule, the further the muscles are apart anatomically, the smaller the amount of synchronization between the units, and hence the fewer common inputs that are shared between the

Figure 3.14 Changes in recruitment threshold of motor units in the first dorsal interosseous muscle produced by electrical stimulation of the digital (cutaneous) nerves of the index finger. (a) Nine panels of EMG and force recordings taken as the subject made a gradually increasing voluntary isometric contraction of the muscle. The subject was instructed to follow a target ramp increase in force (the straight, saw-toothed line) with his voluntary contraction (the irregular trace superimposed on the saw-tooth). Two units can be distinguished in the EMG traces: a unit with small action potentials producing the black smudge around the baseline and a unit with much larger potentials, seen as the more distinct large spike. The control state, with no stimulation is seen in row A. During three consecutive ramp contractions the large unit was recruited at a force of about 2.5 N. Row B shows the behaviour of the same units 30 s after switching on a continuous electrical stimulation of the digital nerves of the forefinger. The recruitment threshold of the large unit is now less than 1 N. In contrast, the threshold for the small unit has risen. Row C is a further control series recorded 10 s after the end of stimulation. The recruitment thresholds are similar to those in row A. (b) Summary of data from a number of experiments. Threshold forces at which units were recruited in the control state are plotted on the left; thresholds during skin stimulation are on the right. The trend is for high threshold units to be recruited at lower levels during stimulation. The opposite is true for low threshold units: they are recruited at higher force levels during cutaneous stimulation. (From Garnett and Stephens, 1981, Figures 1 and 4A; with permission.)

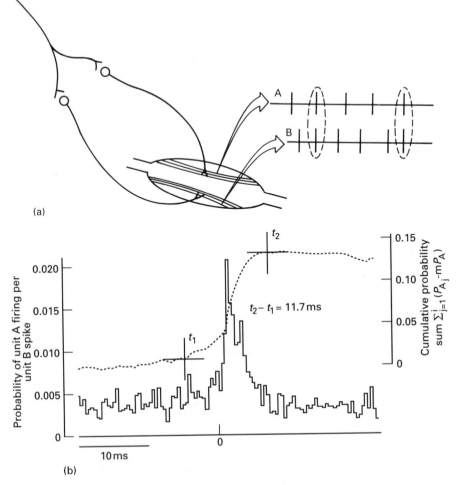

(a)

(b)

Figure 3.15 (a) Schematic diagram showing the recording arrangement for detection of synchronization between motor unit discharge. Two needle electrodes are inserted into a muscle to record the discharge of two different units. A computer then records the spike trains and plots the time of occurrence of spikes in unit B with respect to the spikes in A. The motoneurones in the spinal cord which supply the units are shown as being innervated by a common presynaptic input. (b) An example of the cross-correlogram between two motor units in the first dorsal interosseous muscle. 5763 spikes in unit A were recorded. The y-axis on the left plots the probability that the second unit, B, would fire at different times before or after unit A ($t = 0$ ms). Both units were being discharged at about 10 Hz in a voluntary contraction. The dotted curve which is superimposed is a CUSUM, showing the extra probability of firing over and above baseline levels. It begins to rise from 0 (see y-scale on the right) about 5 ms before $t = 0$ and then levels out 11.7 ms later. This curve defines the width of the cross-correlogram peak (t_1 and t_2). (From Datta and Stephens, 1990, Figure 1; with permission.)

two motoneurone pools. Muscles that are frequently used together as synergists (for example, the finger extensors and the interossei) often have greater common input than other muscle pairs.

Although the technique tells us that motoneurones can share synaptic input, it does not tell us where this input comes from. Motoneurones receive shared inputs from many sources. For example, an Ia fibre from a muscle spindle synapses on almost every single motoneurone in the homonymous motoneurone pool. Similarly, monosynaptic input from the corticospinal tract is also widely distributed amongst the motoneurones. All these common inputs could contribute to synchronization of motor unit discharge. However, at least when subjects make a voluntary contraction, it is thought that the major proportion of shared inputs comes via the corticospinal tract. One piece of evidence that supports this is that cross-correlogram peaks are larger in intrinsic hand muscles than they are in the soleus muscle. This would not have been expected if cross-correlogram peaks were due primarily to shared Ia inputs. Ia inputs to soleus motoneurones are much more powerful (at least as judged by the size of the tendon jerk in that muscle) than those to the intrinsic hand muscles.

A second piece of evidence is that the cross-correlation peaks disappear in stroke patients after lesions of the corticospinal tract. They are replaced by broader peaks which are inconsistent with direct shared synaptic input. Broad peaks are likely to reflect synchronization in the population of neurones which projects onto the motoneurones. In stroke patients, this may mean that there is increased synchronization in a population of spinal interneurones which excites the motoneurones monosynaptically.

A final piece of evidence favouring corticospinal mechanisms in motor unit synchronization comes from the observation that the amount of synchronization varies in different tasks. A good example of this occurs in breathing. Breathing can be thought of as normally being driven by automatic mechanisms in the brainstem. However, voluntary effort can intervene, and this may activate the breathing muscles via a corticospinal pathway. Synchronization between motoneurones in the sternocleidomastoid muscle in the neck during automatic breathing against a resistance (the sternomastoid muscle is an accessory muscle for respiration, and is used only during large efforts) is much less than that seen during voluntary breathing against a resistance. (Datta and Stephens (1990) and Bremner, Baker and Stephens (1991) further discuss the work covered in this section.)

Gradation of force in human muscle contraction

Ever since the first recordings of single unit activity by Adrian and

Bronk (1929), one of the basic questions of muscle control has been whether muscle force is graded by the **number** of active units, or the **rate** at which each unit is stimulated. Either system of control seems feasible. For example, in the extreme case of recruitment (number) control, each unit, when recruited, would fire immediately at high frequency, and produce its maximum force. Under such circumstances, force would be graded by the number of active units. In the extreme case of rate control, every unit could be recruited at the same threshold force. In such a case, every unit would be firing at all levels of contraction and force could be graded only by changing the rate of unit activation. These extremes of recruitment and rate modulation are never seen; each mechanism contributes to the gradation of force. The relative contribution of each of these mechanisms has been analysed in some detail for various human muscles.

One of the problems in these experiments is to be able to follow the firing pattern of a single motor unit during forcible contractions. Only recently have very selective microelectrodes been introduced together with computer decomposition techniques which enable experimenters to distinguish a single unit among many during maximal effort. Some of the first experiments were performed on the small muscles of the hand. In these it was found that during a ramp isometric contraction against a slowly increasing force, small units were recruited first and began to fire at about 9 Hz. As the force increased, these units increased their firing frequency and at the same time new, larger units were recruited which began firing at about the same initial rate of 9 Hz. By the time the force reached 40% maximum voluntary contraction (MVC), all the motor units in the muscle had been recruited and the rest of the force modulation was provided by changes in firing frequency up to a maximum of about 40 Hz.

In the deltoid, a much larger muscle (about 1000 units compared with the 120 of first dorsal interosseous) with a similar proportion of unit types, recruitment proved more important. During slowly increasing contractions new units were recruited up to force levels of 80% MVC. Rate modulation played little part in increasing force since the firing frequency of units only rose to about 25 Hz after initial recruitment at about 13 Hz (De Luca *et al.*, 1982).

This latter observation is of some interest, since at such frequencies, many of the motor unit contractions may still be unfused. If this is so, it means that the MVC was not, in fact, the maximum force that the muscle was capable of exerting. There remains a large potential for increasing the force of contraction still further, which might explain some of the feats performed by people under stress. This mismatch of MVC and the maximal force available in muscle is not seen in the small muscles of the hand. In the first dorsal interosseous and adductor pollicis, MVC is indeed the maximum force of muscle contraction and cannot be exceeded by an electrical tetanus to the nerve.

The reason for the different proportions of rate and recruitment modulation of force in these muscles can probably be related to the size and function of the muscle involved. For example, the first dorsal interosseous is a small muscle with only 120 units. Gradation of force solely by recruitment would perhaps prove far too coarse a mechanism in a muscle used in finely tuned tasks such as writing. In contrast, the deltoid is a large muscle used primarily for generating forceful contractions. Its 1000 motor units may give a sufficiently finely graded contraction to suit its function.

Muscle fatigue

During a maximum contraction of many mixed muscles, it is not uncommon to see a 50% reduction of muscle force over a period of a few seconds. This is known as fatigue. There are three main reasons why this might occur: (1) failure in transmission at the nerve–muscle junction, or in conduction at the fine terminal branches of motor axons; (2) failure of the contractile machinery in muscle; and (3) reduction in the central drive to motoneurones below that necessary to sustain maximal muscle activity. The relative roles of these three phenomena depend on the type of contraction, the muscle group being tested and the motivation of the subject.

Maximum contractions of hand muscles

Failure of neuromuscular transmission was believed to be a major factor in fatigue for many years. Animal experiments using repetitive stimulation of muscle nerves had shown that transmission failure could occur. In humans, such failure can be demonstrated by supramaximal electrical stimulation of a motor nerve at frequencies of 50 Hz or above. A single supramaximal shock produces a mass muscle action potential (M wave) due to the synchronous activation of all the units in the muscle. After 20 s or so of 50 Hz stimulation, the size of the action potential following each shock declines, indicating that action potentials are no longer being generated in all the units of the muscle. There has been a failure of electrical or neuromuscular transmission between the stimulating electrode and the muscle.

As noted above, however, such high rates of unit activity of 50 Hz or more are never seen over prolonged periods in humans. It turns out that the neuromuscular system is so designed that it operates at frequencies just below those at which neuromuscular transmission fails, even during muscle fatigue. This was demonstrated quite elegantly by Merton

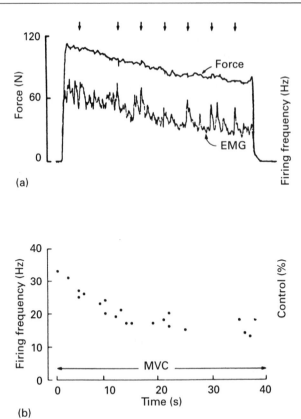

Figure 3.16 Effect on fatigue on (a), force of a sustained maximal voluntary contraction, and (b), discharge rates recorded from single motor units during the contraction. The muscle used was adductor pollicis, a favourite for this type of experiment because it is the only muscle in the thenar eminence supplied totally by the ulnar nerve. In (a) are superimposed traces of muscle force and the surface rectified EMG. Both decline over the 40 s period of contraction. Supramaximal ulnar nerve stimuli given at the times indicated by the arrows fail to produce any further increase in muscle force over and above MVC. Hence, the contraction must have remained maximal throughout this period. In (b), highly selective needle electrodes were used to record short periods of activity from 21 identified single motor units. At the start of contraction, the firing rates are high (>30 Hz) and slowly decline to less than 20 Hz. (From Bigland-Ritchie *et al.*, 1983, Figures 4A and B; with permission.)

(1954), who recorded the size of the maximal motor potential (M-wave) in the adductor pollicis muscle at different times during the course of maximal voluntary contraction. In contrast to results during maximal electrical stimulation, during voluntary muscle fatigue there was no change in the size of the M-wave, indicating that electrical transmission had remained secure.

Fatigue probably is not due to central mechanisms either since throughout the course of an MVC in the small hand muscles, a supra-maximal electrical stimulus of the muscle nerve cannot generate extra force over and above that produced by a well trained and motivated subject (Figure 3.16) ('twitch interpolation test') Thus, fatigue is most probably caused by failure in the production of contractile force, al-though the precise nature of this effect is not known.

During the course of a maximal contraction, the firing rate of the motor units is modulated in a remarkable way. At the start of a rapid contraction they may fire at frequencies up to 150 Hz. Indeed, the instantaneous firing frequency of the first two or three impulses of a train can be as high as 150–200 Hz. Firing rate then declines slowly throughout the rest of the maximal contraction to rates as low as 20 Hz (Figure 3.16). One consequence of this decline is that transmission failure does not occur at the neuromuscular junction. The surprising feature of this decline in motoneurone firing rates is that it is not accompanied by any additional drop in muscle force. Even after 40 s MVC, when units are firing at less than 20 Hz, the muscle force is maximal. The reason for this is that during fatigue there is a change in the mechanical properties of the motor units. The main effect is that the relaxation time of single unit twitches increases. Because of this increase in twitch duration, the stimulation frequency needed to produce a fused tetanus declines. The result is that the same unit can be acti-vated maximally at reduced firing frequencies. Motoneurone firing rates are carefully matched to the changing properties of the motor units.

Figure 3.17 shows an example of the 'muscular wisdom' which matches firing frequencies to changing motor unit properties during fatigue. A and F are maximum voluntary fatiguing contractions of adductor pollicis. B,C,D and E are attempts to mimic the time course of the contraction using electrical stimulation of the muscle nerve. Initial force is high if very rapid trains are used (B,C and D), but declines more rapidly than MVC because of neuromuscular failure. In E, a good fit to MVC is achieved by changing the stimulus frequency during the course of stimulation, in much the same way as motor units have been found to change discharge frequency during a MVC (Merton, 1954; Bigland-Ritchie *et al.*, 1983).

Other types of fatiguing contraction

Although failure of the contractile machinery is likely to be the most important factor in fatigue during maximal voluntary contractions of hand muscles by well-motivated subjects, this is probably not the case for all muscle groups, or during submaximal fatiguing efforts.

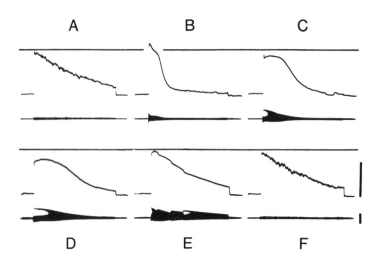

Figure 3.17 Comparison of voluntary and electrically excited contractions of adductor pollicis. In each panel traces are force of contraction (top) and surface (bottom) EMG from the muscle. A and F represent a 95 s maximal voluntary contraction. B,C,D, show contractions produced by electrical stimulation of the muscle nerve at 100, 50 and 35 Hz, respectively. E is an electrical contraction in which stimuli were given at 60 Hz for 8 s, 45 Hz for 17 s, 30 Hz for 15 s, and 20 Hz for 55 s. Only in this latter case does the force output from the muscle resemble that of an MVC. Calibrations are 5 kg and 10 mV. (From Marsden, Meadows and Merton, 1983, Figure 3, pp. 169–211; copyright 1983, Raven Press, New York; with permission.)

1. Even in well-motivated subjects, central fatigue may sometimes be important. For example, it is not difficult to find a group of subjects who are able to activate maximally their tibialis anterior muscle, as judged by the 'twitch interpolation test' (see above); they may fail, however, when they contract the soleus. Whether such differences between muscle groups are related to the strength of their projections from the cerebral cortex is not known.

2. During submaximal contractions, fatigue may result from failure of neuromuscular propagation. If subjects make prolonged efforts at, say, 20% maximum, then the amplitude of the maximal electrically-elicited M-wave may decline by 25% or more. Such a decline is not seen in short maximal contractions and may indicate a time-dependent impairment of neuromuscular propagation. Whether this is due to failure of the neuromuscular synapse, or to a decrease in the amplitude of muscle fibre action potentials is not clear.

(Enoka and Stuart, 1992; Fuglevand *et al.*, 1993 give further details.)

3.8 *Pathophysiology of the motor unit*

By definition, the motor unit has two components: the motoneurone and the muscle fibres which it innervates. Diseases of the motor unit are divided into two categories, depending upon which of these components is affected: (i) **neurogenic diseases** affect the motoneurone, either at its cell body (motor neurone diseases) or along its axon or myelin sheath (peripheral neuropathies); (ii) **myopathic diseases** affect the muscle itself. Electrophysiological study of the motor unit using conventional needle electrodes plays an important part in diagnosis of these disorders (Kimura, 1983).

The normal electromyogram (Figure 3.18)

There is no electrical activity in a normal muscle when subjects are completely at rest. When a concentric needle recording electrode is inserted, there is a brief burst of activity (known as 'insertional activity') due to irritation of the muscle membrane with the needle. This dies away rapidly and the recording becomes silent. Voluntary effort recruits motor units which usually have di- or triphasic action potentials with a duration of 10 ms or so. As voluntary effort increases, more units are recruited and it becomes difficult to distinguish individual muscle action potentials. At maximum effort, single unit discharges are no longer visible and the EMG is said to show a full 'interference pattern'. This is because all the units of the muscle discharge asynchronously and their spikes interfere with each other in the electrical recording.

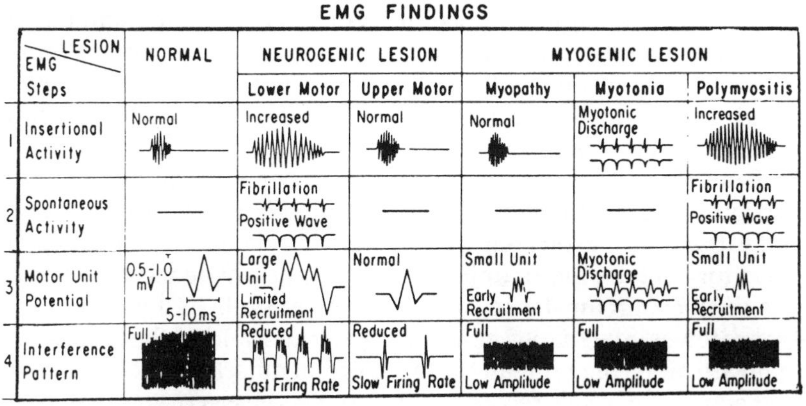

EMG FINDINGS

LESION / EMG Steps	NORMAL	NEUROGENIC LESION		MYOGENIC LESION		
		Lower Motor	Upper Motor	Myopathy	Myotonia	Polymyositis
1 Insertional Activity	Normal	Increased	Normal	Normal	Myotonic Discharge	Increased
2 Spontaneous Activity	—	Fibrillation / Positive Wave	—	—	—	Fibrillation / Positive Wave
3 Motor Unit Potential	0.5–1.0 mV / 5–10 ms	Large Unit / Limited Recruitment	Normal	Small Unit / Early Recruitment	Myotonic Discharge	Small Unit / Early Recruitment
4 Interference Pattern	Full	Reduced / Fast Firing Rate	Reduced / Slow Firing Rate	Full / Low Amplitude	Full / Low Amplitude	Full / Low Amplitude

Figure 3.18 Typical EMG findings recorded with conventional needle electrodes in lower and upper motoneurone disorders, and in myopathic lesions. See text for full explanation. (From Kimura, 1983, Figure 13.4; with permission.)

The electromyogram in neurogenic diseases

Complete lesion of a motor nerve produces remarkable changes in the muscle which it innervates. There is, of course, no voluntary recruitment of motor units. However, several days after the lesion, there is an increase in the level of spontaneous activity in the muscle. In the EMG record this is seen as trains of very small action potentials firing at frequencies of about 15 Hz. These are known as 'fibrillation potentials'. They are small in amplitude because they are due to the activity of single muscle fibres, rather than to the ensemble of muscle fibres which fire together when a motor unit is active.

This spontaneous activity is thought to be due to denervation supersensitivity. When the motor end-plates on a muscle are no longer active, the acetylcholine receptors begin to proliferate on regions of the muscle fibre membrane distant from the end-plate region. This proliferation can produce a 1000-fold increase in sensitivity of the muscle fibre to acetylcholine. The muscle membrane also becomes more sensitive to mechanical deformation such as stretch or pressure. Fibrillation potentials may represent the response of some hypersensitive fibres to small amounts of circulating acetylcholine which normally are ineffective in exciting intact muscle.

If the lesion was not complete, or if regrowth occurs along the damaged segment of nerve, some of the denervated muscle fibres will become reinnervated by remaining or regrowing axons. Intact axons begin to sprout near their terminals on the muscle. These sprouts either form new motor end-plates on denervated fibres, or take over the original end-plates left by dying neurones. When a new neuromuscular connection is formed, the extra acetylcholine receptors which had proliferated over the entire surface of the muscle fibre membrane, are reabsorbed. The new end-plate again becomes the only point on the membrane where acetylcholine receptors are found. Fibre sensitivity to applied acetylcholine declines and spontaneous activity (fibrillation potentials) disappears.

It is not clear what induces the nerves to begin sprouting, nor why the sprouts form terminals only on denervated fibres. However, the nett result is that intact axons remaining after a partial nerve lesion expand the territory of their motor units by reinnervation of denervated muscle fibres. In such circumstances, there are fewer units in the EMG (since there are fewer intact axons innervating the muscle), but the units which remain are much bigger than those of normal muscle. The consequence is that action potentials recorded through needle electrodes may be up to ten times larger than average but the interference pattern during maximal contraction is reduced.

A final feature is that the motor unit action potentials, in addition to being large, also are far longer and more polyphasic than normal. This is because conduction along the axonal sprouts which reinnervate denervated muscle fibres is not secure. There is intermittent conduction block or slowing along the new part of the axon so that some fibres of the unit only fire intermittently or much later than others. The asynchronous discharge of a large number of fibres gives the motor unit action potential its polyphasic shape.

Such an EMG picture of denervation and reinnervation is typical of many neuropathic diseases which affect the lower motoneurone, such as chronic peripheral neuropathies and polio. The picture is quite different for diseases of the upper motoneurone. In these cases, there is usually no denervation of the muscle since the spinal motoneurones remain intact. However, loss of descending excitatory input to the motoneurones makes it more difficult to recruit motor units in a voluntary contraction. With an incomplete lesion of the upper motoneurone, motor units are normal, but cannot be discharged at high frequencies by voluntary effort. In addition, it is impossible to recruit all motor units into contraction, so that the interference pattern is reduced.

In some patients with chronic lesions of the upper motoneurone, there may be a small amount of trans-synaptic degeneration of spinal motoneurones. A proportion of cells deprived of their normal input from supraspinal sources may die. If this occurs, there may be typical signs of denervation (fibrillation potentials) and reinnervation (large, polyphasic units). Some diseases, such as amyotrophic lateral sclerosis (ALS, or motoneurone disease) affect both upper and lower motoneurones.

The electromyogram in myopathic diseases

In most myopathic diseases such as the muscular dystrophies, the change in the EMG is not so marked as in neuropathic lesions. Death of muscle fibres usually is reflected only as a decrease in size of the motor unit action potentials. In extreme cases, only a single fibre of a unit may remain, so that voluntary activity produces a record indistinguishable from a fibrillation potential. Potentials may be more polyphasic than normal.

There are two types of muscle disease which affect the electrical properties of the muscle fibre membranes and produce characteristic changes in the motor unit potentials. These are the myotonias and periodic paralysis.

In **myotonia**, the muscle, once activated, tends to discharge repetitively. In EMG studies this is evident in two ways: (1) insertion of the needle provokes a repetitive 'myotonic' discharge of activated muscle fibres which far outlasts the normal period of insertional activity; and

(2) there is delayed relaxation of muscle contraction and electrical activity following voluntary contraction, percussion or electrical stimulation. This extra activity persists even after nerve block or curarization, implying that the site of the disorder is in the muscle membrane. It is thought that in some forms of myotonia, this could be due to a decrease of Cl⁻ conductance of the membrane.

Following an action potential, the muscle fibre membrane repolarizes and there is a movement of K^+ out of the cell. Because of the complex infolding of the fibre membrane, much of this is into the transverse tubular system, where the ions are effectively trapped. Such an accumulation of K^+ depolarizes the membrane. Under normal circumstances, the amount of depolarization is reduced by movement of Cl⁻ across the membrane. In situations where Cl⁻ movement is decreased, this depolarization can become large enough to initiate a further action potential and so on.

Periodic paralysis is the result of reversible inexcitability of the muscle membrane. Clinically, it is a condition in which patients have episodes of paralysis, lasting an hour or more and during which the muscle membrane cannot be driven to discharge an action potential. Three major subtypes are distinguished according to the level of serum K^+ during an attack: hypo-, hyper- and normo-kalaemic. However, serum K^+ levels play no causal part in producing an attack. A possible clue as to the abnormality comes from the finding that the muscle membrane is substantially depolarized. This may lead to depolarization block and fibre inexcitability.

Changes in firing rate and recruitment order of motor units

The firing rate of a motoneurone depends on the amount of excitatory input which it receives; its recruitment order depends in most instances upon its size relative to the other members of the same motoneurone pool. What happens to these properties when some of the inputs to the motoneurone are lost, as in patients with lesions of the upper motoneurone, or when the motor units are changed by sprouting and reinnervation?

Interruption of descending supraspinal pathways, as in patients with hemiplegia due to motor stroke, decreases the firing rate of spinal motoneurones. Motor unit discharge rates are abnormally low and more variable than usual in spastic muscles (that is, less than 8 Hz). Abnormal patterns of discharge also have been seen in patients with Parkinson's disease (low firing rates and difficulty in recruiting units) and Huntington's chorea (irregular discharges). These changes are probably caused by a decrease in size and increase in variability of the descending motor command onto the spinal motoneurones, rather than

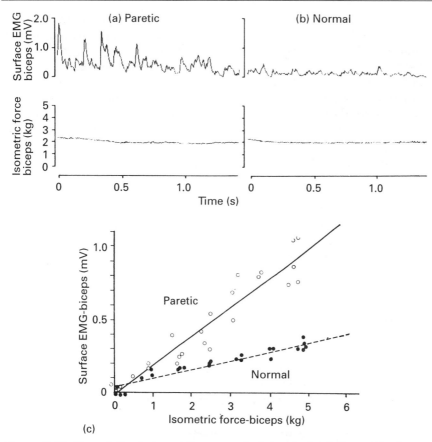

Figure 3.19 Dissociation between the amount of electrical activity and the force exerted in paretic muscle as compared with normal. A patient with hemiparesis due to a capsular stroke was asked to perform isometric contractions of the biceps brachii, while the EMG was recorded from surface electrodes placed over the belly of the muscle. The upper diagram shows some raw data from the paretic (weak) arm (a), and the normal arm (b). In order to exert the same force (bottom traces), the subject produced much larger amounts of EMG activity (rectified signals; top traces) on the paretic side. The lower graph (c) shows the relationship between force exerted and average EMG activity in the same subject for a range of different contractions. Closed circles and dotted line are for the normal side; open circles and solid line are for the paretic side. (From Tang and Rymer, 1981, Figures 2 and 3; with permission.)

a change in the properties of the motoneurones themselves.

In spastic patients, such low firing rates may contribute to muscular weakness. If patients cannot achieve firing frequencies high enough to produce a fused contraction of motor units, then maximum force will not be available from the muscle. Unfused contractions do not produce maximum force. This effect can be seen in the relation between force and EMG in spastic muscles. Figure 3.19 shows an experiment in which a hemiparetic patient was asked to make an isometric flexion at the elbow while the surface EMG was recorded from the biceps muscle. On the paretic (weak) side, far more EMG was needed to produce the contraction than on the normal side. The most probable reason for this discrepancy (taking into account difficulties in electrode placement, muscle size, and so on) is that the contraction of the motor units on the paretic side was not fused. Thus, more units have to be recruited to produce a given level of force, and more EMG is recorded from the surface.

Recruitment order and mechanical properties of motor units usually are unaffected by upper motoneurone lesions. However, changes do appear in diseases affecting the lower motoneurone, which produce signs of denervation and reinnervation in the muscle under study. As noted, reinnervated motor units have potentials which are larger than normal, consistent with the increased number of muscle fibres within each unit. However, averaging the force produced by each enlarged unit in patients with peripheral neuropathy or motoneurone disease gives a rather unexpected result: the twitch tensions are no bigger than in normal muscle, and may in many cases be even smaller, on average, than normal. In other words, the large unit appears to be a less efficient contractile machine than normal. The reason for this change is not known. It could be due to changes in the contractile mechanisms of muscle fibres or to fibrosis affecting the linkage between motor unit contraction and externally monitored force.

The recruitment order is not affected in most diseases of the motor unit. Units with small twitch tensions are recruited earlier than those with large twitch tensions. This is true even when motor unit territory has been expanded through reinnervation. This may be because the number of new fibres reinnervated by existing units is proportional to the size of the original motor unit. Thus, large units may remain large, and small units remain relatively small.

An exception to this has been described in a patient with surgical repair of a completely severed nerve. Regrowth occurred over two years and reinnervation produced a normal spectrum of unit sizes. However, recruitment order was abnormal: there was no consistent size order. It may be that recruitment order was not preserved in this case because large and small motoneurones regrow and reinnervate at the same rate. The situation may then arise that a large axon captures only a small number of fibres, while a small axon captures a large number of fibres. Assuming that no retrograde changes in size of the motoneurone occur, this means that even if the motoneurones retain their recruitment order, the size of recruited motor units in muscle would not be orderly (Milner-Brown, Stein and Lee, 1974; Rosenfalck and Andreassen, 1980; Tang and Rymer, 1981).

Bibliography

Review articles

Buchthal, F. and Schmalbruch, M. (1980) Motor unit of mammalian muscle, *Physiol. Rev.*, **60,** 90–142.

Burke, R. E. (1981) Motor units: anatomy, physiology and functional organisation, in V. B. Brooks (ed.) *Handbook of Physiology*, sect. 1, vol. 2, part 1, Williams and Wilkins, Baltimore, pp. 345–411.

Desmedt, J. E. (ed.) (1981) Motor Unit Types, Recruitment and Plasticity in Health and Disease. *Progress in Clinical Neurophysiology*, vol. 9, Karger, Basel.

Enoka, R. and Stuart, D. G. (1992) Physiology of muscle fatigue. *J. App. Physiol.*, **72,** 1631–1648.

Freund, H.-J. (1983) Motor unit and muscle activity in voluntary motor control, *Physiol. Rev.*, **63,** 387–436.

Granit, R. (1970) *The Basis of Motor Control*, Academic Press, New York.

Henneman, E. and Mendell, L. M. (1981) Functional organisation of motoneuron pool and its inputs, in V. B. Brooks (ed.), *Handbook of Physiology*, sect. 1, vol. 2, part 1, Williams and Wilkins, Baltimore, pp. 423–508.

Kimura, J. (1983) *Electrodiagnosis in Diseases of Nerve and Muscle: Principles and Practice*, F. A. Davis, Philadelphia.

Lewis, D. M. (1981) The physiology of motor units in mammalian skeletal muscle, in A. L. Towe and E. S. Luschei (eds), *Handbook of Behavioural Neurobiology*, vol. 5, Plenum, New York, pp. 1–67.

Original papers

Adrian, E. D. and Bronk, D. W. (1929) The discharge of impulses in motor nerve fibres. Part II. The frequency of discharge in reflex and voluntary contractions, *J. Physiol.*, **67,** 119–151.

Bigland-Ritchie, B., Johansson, R., Lippold, O. C. J. *et al.* (1983) Changes in motoneurone firing rates during sustained maximal voluntary contractions, *J. Physiol.*, **340,** 335–346.

Bremner, F. D., Baker, J. R. and Stephens, J. A. (1991) Correlation between the discharges of motor units recorded from the same and from different finger muscles in man, *J. Physiol.*, **432,** 355–380; 381–399; 401–425.

Burke, R. E. (1975) A comment on the existence of motor unit 'types', in D. B. Tower (ed.) *The Nervous System. The Basic Neurosciences*, vol. 1, Raven Press, New York, pp. 611–619.

Burke, R. E. and Tsairis, P. (1973) Anatomy and innervation ratios in motor units of cat gastrocnemius, *J. Physiol.*, **234,** 749–765.

Burke, R. E., Levine, D. N., Tsairis, P. *et al.* (1973) Physiological types and histochemical profiles in motor units of the cat gastrocnemius, *J. Physiol.*, **234,** 723–748.

Burke, R. E., Rymer, W. Z. and Walsh, J. V. (1976) Relative strength of synaptic input from short-latency pathways to motor units of defined type in cat medial gastrocnemius', *J. Neurophysiol.*, **39**, 447–485.

Datta, A. K. and Stephens, J. A. (1981) The effects of digital nerve stimulation on the firing of motor units in human first dorsal interosseous muscle, *J. Physiol.*, **318**, 501–510.

Datta, A. K. and Stephens, J. A. (1990) Synchronisation of motor unit activity during voluntary contraction in man, *J. Physiol.*, **422**, 397–419.

De Luca, C. J., Lefever, R. S., McCue, M. P., *et al.* (1982) Behaviour of human motor units in different muscles during linearly varying contractions, *J. Physiol.*, **329**, 113–128.

Denny-Brown, D. (1929) On the nature of postural reflexes, *Proc. Roy. Soc. B.*, **104**, 252–301.

Desmedt, E. and Godaux, E. (1977) Fast motor units are not preferentially activated in rapid voluntary contractions in man, *Nature*, **267**, 717–719.

Dum, R. P. and Kennedy, T. T. (1980) Synaptic organisation of defined motor-unit types in cat tibialis anterior, *J. Neurophysiol.*, **43**, 1631–1644.

Eccles, J. C., Eccles, R. M. and Lundberg, A. (1957) The convergence of monosynaptic excitatory afferents onto many different species of alpha-motoneurone, *J. Physiol.*, **137**, 22–50.

Edstrom, L. and Kugelberg, E. (1968) Histochemical composition, distribution of fibres and fatiguability of single motor units, *J. Neurol. Neurosurg. Psychiat.*, **31**, 424–433.

Fugelvand, A. J., Zackowski, K., Huey, K. A., Enoka, R. M. (1993) Impairment of neuromuscular propagation during human fatiguing contractions at submaximal strength. *J. Physiol.*, **460**, 549–572.

Garnett, R. A. F., O'Donovan, M. J., Stephens, J. A. *et al.* (1979) Motor unit organisation of human medial gastrocnemius, *J. Physiol.*, **287**, 33–43.

Garnett, R. and Stephens, J. A. (1981) Changes in the recruitment threshold of motor units produced by cutaneous stimulation in man, *J. Physiol.*, **311**, 463–473.

Granit, R., Kernell, D. and Smith, R. S. (1963) Delayed depolarisation and the repetitive response to intracellular stimulation of mammalian motoneurones', *J. Physiol.*, **168**, 890–910.

Gustaffson, B. and Pinter, M. J. (1984) Relations among passive electrical properties of lumbar α motoneurones of the cat. *J. Physiol.*, **356**, 401–432; 433–442.

Henneman, E., Somjen, G. and Carpenter, D. O. (1965) Functional significance of cell size in spinal motoneurons, *J. Neurophysiol.*, **28**, 560–580.

Jami, L., Murthy, K. S. K., Petit, J. *et al.* (1982) Distribution of physiological types of motor unit in the cat peroneus tertius muscle, *Exp. Brain Res.*, **48**, 177–184.

Kanda, K., Burke, R. E. and Walmsley, B. (1977) Differential control of fast and slow twitch motor units in the decerebrate cat, *Exp. Brain Res.*, **19**, 57–74.

Kernell, D. (1965) High-frequency repetitive firing of cat lumbosacral motoneurones stimulated by long-lasting injected currents, *Acta Physiol. Scand.*, **65**, 74–86.

Kugelberg, E. and Edstrom, L. (1968) Differential histochemical effects of muscle contractions on phosphorylase and glycogen in various types of fibres: relation to fatigue, *J. Neurol. Neurosurg. Psychiat.*, **31**, 415–423.

Marsden, C. D., Meadows, J. C. and Merton, P. A. (1983) 'Muscular Wisdom' that minimises fatigue during prolonged effort in man: peak rates of motoneurone discharge and slowing of discharge during fatigue, in J. E. Desmedt (ed.), *Adv. Neurol.*, vol. 39, Raven Press, New York.

Merton, P. A. (1954) Voluntary strength and fatigue, *J. Physiol.*, **123**, 553–564.

Milner-Brown, H. S., Stein, R. B. and Lee, R. G. (1974) Contractile and electrical properties of human motor units in neuropathies and motor neurone disease, *J. Neurol. Neurosurg. Psychiat.*, **37**, 665–669; 670–676.

Rall, W. (1967) Distinguishing theoretical synaptic potentials computed for different somadendritic distributions of synaptic input, *J. Neurophysiol.*, **30**, 1138–1168.

Rosenfalck, A. and Andreassen, S. (1980) Impaired regulation of force and firing pattern of single motor units in patients with spasticity, *J. Neurol. Neurosurg. Psychiat.*, **43**, 907–916.

Tang, A. and Rymer, W. Z. (1981) Abnormal force-EMG relations in paretic limbs of hemiparetic human subjects, *J. Neurol. Neurosurg. Psychiat.*, **44**, 690–698.

Thomas, C. K., Johansson, R. S., Westling, G. and Bigland-Ritchie, B. (1990) Twitch properties of human thenar motor units measured in response to intraneural motor-axon stimulation, *J. Neurophysiol.*, **64**, 1339–1346; 1331–1338; 1347–1351.

Walmsley, B., Hodgson, J. A. and Burke, R. E. (1978) Forces produced by medial gastrocnemius and soleus muscles during locomotion in freely moving cats, *J. Neurophysiol.*, **41**, 1203–1216.

Zengel, J. E., Reid, S. A., Sypert, G. W. *et al.* (1985) Membrane electrical properties and prediction of motor-unit types of medial gastrocnemius motoneurons in the cat, *J. Neurophysiol.*, **53**, 1323–1344.

Proprioceptors in   **4**
muscles, joints and skin

Proprioceptive organs signal to the CNS information about the relative positions of the body parts. Apart from pressure receptors in the soles of the feet, they do not supply any information as to the orientation of the body with respect to gravity; they only signal the position of one part of the body with respect to another. The receptors involved lie in the muscles (spindles and Golgi tendon organs), the joints and the skin. In this chapter, only the structure and characteristics of the afferent discharge will be summarized for each type of receptor. The possible roles of these receptors in the control of movement will be discussed in subsequent chapters. Despite its age, Matthews's (1972) book on muscle receptors is still one of the best complete reviews on these topics.

4.1 *Muscle receptors: I. The muscle spindle*

Anatomy

The muscle spindle is one of the most well studied and sensitive sensory receptors in the body. It rivals the eye and the ear in its complexity and yet, in comparison, relatively little is known of its role in sensation or even in the control of movement.

Spindles were first identified in muscle about 120 years ago and were seen to consist of both nervous and muscular elements. At first, there was some speculation as to whether the spindles might be 'muscle buds' (the site of formation of new muscle fibres), an inflammatory focus or perhaps a sensory ending. The matter was resolved in 1894 by Sherrington, who cut the ventral root supply to a number of muscles. After allowing some weeks for degeneration of the motor fibres, he found a dense, surviving, afferent supply to the spindles. The spindle was

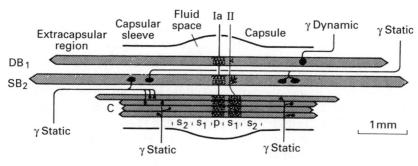

Figure 4.1 Diagram of a typical muscle spindle from a cat hindlimb muscle. The spindle consists of two bag fibres (DB$_1$, SB$_2$: dynamic bag$_1$, and static bag$_2$) and a number of chain fibres (C), encapsulated in a fluid-filled space. Sensory endings are found in the p and s$_1$ regions of the fibres. Primary spiral endings are seen on all fibres of the spindle in the equatorial, p. region. These endings give rise to the large group Ia afferent fibres. Secondary endings consist of spiral terminations on the chain fibres and less extensive spray-like terminations on the primary endings in the s$_1$ region. Secondary endings are rarely seen on the DB$_1$ fibre. They give rise to the group II afferent fibres from the spindle. The motor (γ) supply to the spindle terminates outside the equatorial region of the fibres. The endings sometimes are described as 'plate endings' on the bag fibres and 'trail endings' on the chain fibres. The γ_d efferents innervate the DB$_1$ fibre, while the γ_s efferents innervate the DB$_2$ and chain fibres. The β supply to the spindle is not shown. (From Boyd, 1980, Figure 1; with permission.)

proved to be a 'sensorial end organ'. At the same time, Ruffini (1898) published the first histological diagrams of the nervous innervation of the spindle. He classified, on morphological grounds, three types of nerve ending on the spindle fibres which are still known by the same names today – primary, secondary (both now known to be afferent) and plate (now known to be motor) endings.

The presently accepted structure of the muscle spindle is shown in Figure 4.1. This illustrates the structure of a typical spindle from the tenuissimus muscle of the cat hindlimb, a favourite muscle for such studies. The spindle consists of a bundle of specialized muscle fibres which lie in parallel with the fibres of the main extrafusal muscle. The intrafusal fibres are about 10 mm long, which is much shorter than fibres in the main muscle. At each end they may attach to extrafusal fibres or to tendinous insertions. For about half its length the spindle is contained within a thick connective tissue capsule which expands in its central 2 mm to form a fluid-filled space, giving the structure its characteristic 'fusiform' appearance.

There are two types of muscle fibre in the spindle: generally, there may be two (or three) bag and three to five chain fibres. Like all muscle fibres, they are formed from a syncitium of cells whose cell walls break down during development to form a continuous structure. Up to 100 or so of the nuclei of the bag fibres are collected together at the spindle equator, where they may be three or four deep when seen in cross-section. In contrast, the chain fibres only have a single row of nuclei in the same region. On the basis of other features (see below), the bag fibres have been subdivided into two separate categories: bag_1 and bag_2.

The sensory innervation of the spindle is of two types, as described by Ruffini (1898). The large diameter Ia neurones distribute primary endings to all the muscle fibres of a spindle. These endings occupy the most central region of each fibre. Either side of them may be the secondary endings of the group II afferent neurones. There are few secondary endings on the bag_1 fibres. Primary endings consist of spiral terminations on all types of spindle fibre, whereas secondary endings usually only have spiral terminations on the bag_2 and chain fibres. The secondary innervation of bag_1 fibres consists of less extensive spray-like terminations, which may also be seen at times on the chain fibres.

The motor supply to the spindle fibres consists mainly of the small diameter γ-neurones, which distribute their endings to the poles of the fibres, usually within the connective tissue capsule. The γ-neurones were first subdivided using physiological techniques (see below) into γ_s and γ_d categories. This physiological difference is reflected in differences in their anatomical termination on the intrafusal muscle fibres. The γ_d-neurones innervate only the bag_1 fibres, whereas the γ_s-neurones innervate bag_2 and chain fibres. There is a possibility that there are subgroups of γ_s fibres which innervate just bag_2 or just chain fibres.

Recently, more attention has been directed towards another source of efferent neurones to the intrafusal fibres. In some cases, up to 20% of the motor supply has been shown to arise from branches of motoneurones innervating the extrafusal muscle. Such shared axons are known as β-axons, a confusing term since they have the same axon diameter as the unshared α-axons. Attempts have been made to subdivide them into static β_s and dynamic β_d. The former tend to have a higher conduction velocity than the latter and appear to have endings only on chain fibres. However, the relative roles of the γ and β systems in spindle control have yet to be understood. It is worth mentioning, though, that β-innervation is the only type of innervation seen in amphibian spindles. The additional system found in mammals is therefore thought to be an evolutionary 'improvement' in design. It allows for the possibility of independent control of extrafusal and intrafusal motor systems.

Figure 4.1 is the 'typical' spindle structure characteristic of many limb muscles in both cat and man. Other types of spindle are known, some of the most complex being found in the small paravertebral muscles of the cervical cord. These muscles, many of which are so small that they have never been given specific names, have some of the highest densities of spindles in the body. In many cases, the spindles are 'atypical'; they may lack the bag₁ fibre or may be found in large arrays of up to 10 spindles, sharing intrafusal fibres and capsules (Boyd, 1962, 1976, 1980, 1985; Boyd and Ward, 1975; Boyd *et al.*, 1981; Matthews, 1981; Hulliger, 1984).

Physiology of spindle afferent responses

The first systematic study of muscle spindle physiology was made by B.H.C. Matthews in the early 1930s. He subdivided the nerve to a variety of muscles in the cat until only one afferent fibre remained intact. Recording from such filaments, he classified three types of response (A,B,C) which he presumed came from three different receptors. The C endings were rare and fired impulses both during stretch and release of the muscle, but not during sustained contraction produced by electrical stimulation of the muscle nerve. The responses disappeared when the fascia was removed completely from the muscle and it is possible that they arose from free nerve endings (see below).

The other two classes of response were studied in much greater detail (Figure 4.2). Both endings increased their firing rate in response to muscle stretch, but the A endings had a lower threshold than the B endings and frequently discharged spontaneously at a low rate. The crucial difference between them was seen in their responses to an electrical muscle twitch. The A endings silenced at the onset of the twitch, whereas the B endings increased their rate of firing. This is

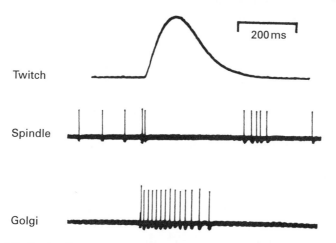

Figure 4.2 Contrasting responses of a spindle ending and a tendon organ during a twitch contraction of the muscle. The spindle ending pauses during the twitch, while the tendon organ discharges. (From Matthews, 1972, Figure 3.2; with permission.)

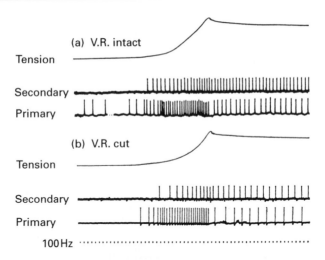

Figure 4.3 Contrasting responses of a spindle primary and secondary ending to a rapidly applied stretch shown in the presence (a) and absence (b) of fusimotor activity. The endings were in the soleus muscle of a decerebrate cat, and the afferent fibres had a conduction velocity of 85 m s⁻¹ (primary) and 44 m s⁻¹ (secondary). Note the dynamic sensitivity of the primary ending during the course of stretch. (From Matthews, 1972, Figure 4.2; with permission.)

exactly the behaviour predicted for muscle spindle receptors and Golgi tendon organs. Spindles, lying in parallel with the main muscle, would be unloaded by the extrafusal contraction and reduce their discharge, while tendon organs, in series with the extrafusal muscle, would be excited by the twitch. This 'twitch test' is still used today in both human and animal work to differentiate spindle and tendon organ afferent fibres. Matthews also found that the A endings (spindle) were sensitive to both the rate of stretch of the muscle as well as its extent. The receptors were said to have both dynamic and static sensitivity. Later studies, some 30 years later, improved on this description and distinguished between the responses of the primary and secondary ending.

If recordings are taken from spindle afferent fibres during a maintained muscle contraction or stretch, it is not possible to distinguish between the discharges from primary or secondary endings since they both fire at approximately the same rate. Under these circumstances they may be differentiated only on the basis of the conduction velocity of their afferent fibres. The Ia fibres in the cat conduct at about 80–120 m s⁻¹ and the group II fibres from 20–70 m s⁻¹ (values are slightly less in man; maximum sensory nerve conduction velocity is never greater than 80 m s⁻¹). During muscle stretch, their responses are considerably different (Figure 4.3). The primary ending is very sensitive during the dynamic phase of stretch. At the onset of a ramp stretch the discharge

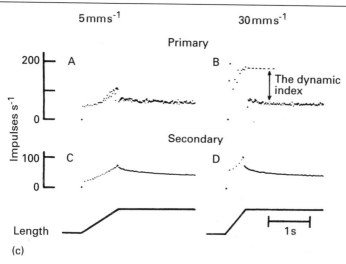

(c)

Figure 4.3 Continued. (c) Comparison of the responses of de-efferented primary and secondary endings to ramp stretches of different velocity. The receptors were located in the tibialis posterior muscle of a decerebrate cat. The firing rate is plotted as dots in an instantaneous frequencygram. Each spot represents the occurrence of an action potential, and its height above the baseline is proportional to the time interval since the immediately preceding spike. Measurement of the 'dynamic index' is shown in record B. It is the difference in frequency between the discharge near the end of the dynamic phase of stretch and that occurring when the muscle has been maintained at its final length for 0.5 s. The time marker only applies while the muscle is at a constant length; during the phase of dynamic stretching, it has been somewhat expanded. (From Brown, Crowe and Matthews, 1965, Figure 1; with permission.)

frequency suddenly jumps to a new value and then continues to increase throughout the remainder of the stretch (see Figure 4.3 a). The end of stretch produces a decline in discharge over a period of 0.5 s or so to a new steady, or static level of activity. The decline in discharge during the first 0.5 s after the end of the dynamic phase is termed the dynamic index, and is a convenient way of separating primary and secondary responses (Figure 4.4, bottom). In contrast, the final discharge level, measured some 0.5 s after the end of stretch (the static or position sensitivity of the ending) is approximately the same for both types of ending.

The behaviour of the primary ending is thought to be due to the mechanical response of the bag$_1$ fibre (which is not shared with the secondary ending) to stretch. The poles of the bag$_1$ fibre are stiffer than the equatorial, sensory region. Cine films of single intrafusal fibres show that when a stretch is imposed on a bag$_1$ fibre, a much larger proportion of the extension is taken up by the sensory region than the poles, and at this stage the afferent discharge can reach very high peak frequencies.

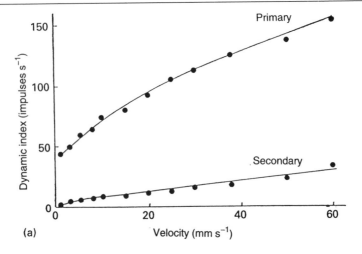

(a)

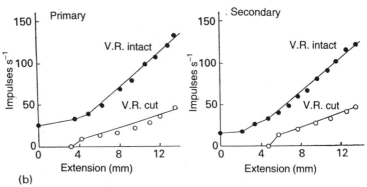

(b)

Figure 4.4 (a) The relationship between dynamic index and velocity of stretching for a primary and secondary spindle ending in the soleus muscle of a decerebrate cat. De-efferented preparation. (From Matthews, 1972, Figure 4.4; with permission.) (b) The relationship between frequency of static firing and the degree of muscle extension with (VR intact) and without (VR cut) spontaneous fusimotor activity. Measurements were made from receptors in the soleus muscle of a decerebrate cat, 0.5 s after completion of a ramp stretch to the appropriate muscle length. Note the increased sensitivity of both types of ending in the presence of spontaneous fusimotor drive. (From Jansen and Matthews, 1962, Figure 4; with permission.)

When the fibre is held in the stretched position, the poles 'creep' back towards the sensory spiral. This movement reduces the degree of extension of the equatorial region and results in the decay of primary afferent discharge (Figure 4.5).

It is sometimes said that the primary endings are **velocity**- sensitive. Indeed, this is true, but one should be quite clear what part of the discharge profile in Figure 4.3 represents velocity sensitivity. Ramp

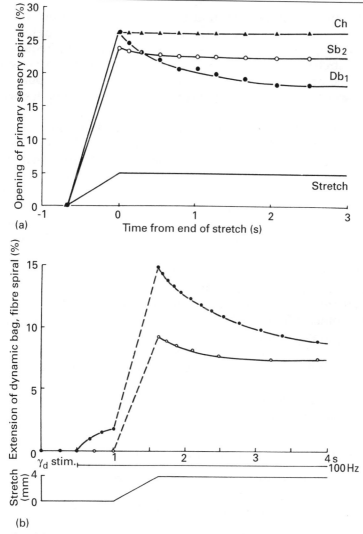

Figure 4.5 (a) Intrafusal 'creep' in the primary sensory spiral of a dynamic bag$_1$ fibre in an inactive spindle following a ramp and hold stretch. During stretch, the dynamic bag$_1$ (Db$_1$) spiral extends a great deal; at the end of stretch the spiral creeps back towards the spindle equator. There is no creep in the static bag$_2$ (Sb$_2$) or the chain (Ch) fibres. (From Boyd, 1976, Figure 4; with Permission.) (b) Comparison of intrafusal 'creep' in the sensory spiral of a dynamic bag$_1$ fibre following a ramp and hold stretch with (filled circles) and without (open circles) γ_d stimulation at 100 Hz. Note how the spiral is extended more, and shows greater creep during γ_d stimulation. From Boyd, Gladden and Ward, 1981, Figure 1; with permission.)

stretches were applied to the spindle, which is the same as giving a **step** change in velocity: the velocity of stretch suddenly increased from zero to, say, 30 mm s^{-1}. If the primary ending were only sensitive to velocity, then a ramp stretch should produce a step increase in firing, from one value at rest to another steady value for the duration of the stretch. What actually happens is that there is a step change in firing frequency at onset of stretch, but superimposed on this is a slower ramp increase in firing. The primary ending is not only velocity-sensitive, but also is length-sensitive, and this ramp increase represents the length sensitivity of the ending during stretch.

Sometimes, an 'initial burst' of firing is seen in the primary response at onset of stretch (Figure 4.3c, part B): the initial discharge of the first two or three impulses overshoots the rate seen during the remainder of stretch. This initial burst may be referred to as an 'acceleration' response. It varies considerably under different conditions. It is most prominent in spindles which are stretched from a resting position; changes in acceleration during an ongoing stretch do not produce an 'acceleration' response. If stretch is applied some seconds after a period of intense fusimotor activity, the initial burst is enhanced. Further stretches applied without intervening fusimotor activity then fail to evoke the response. These features are believed to be a consequence of the nonlinear characteristics of the ending discussed in detail below.

Some other differences between the responses of primary and secondary spindle endings to different types of muscle stretch are shown in Figure 4.6. In general, primary endings are much more sensitive to rapidly changing stimuli; they are particularly responsive to tendon taps and to sinusoidal stretching at relatively high frequencies. Secondary endings show little response to either type of input (Matthews, 1933; Boyd, 1980; articles in Barnes and Gladden 1985).

The effect of γ activity on spindle afferent responses

Following the description of the two types of spindle afferent response, two types of γ-efferent action soon were described. The technique used was to record the response to stretch of single afferent fibres in dissected

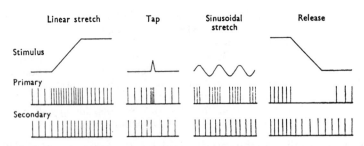

Figure 4.6 Diagrammatic comparison of the typical responses of primary and secondary spindle endings to large stretches applied in the absence of fusimotor activity. (From Matthews, 1964, Figure 2; with permission.)

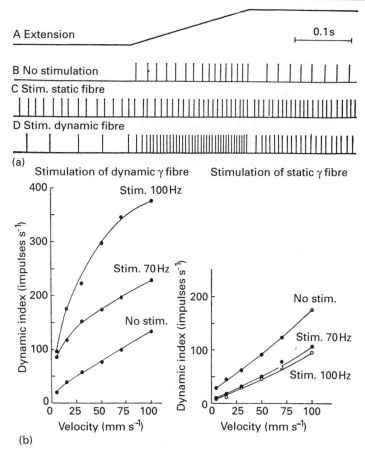

(a)

(b)

Figure 4.7 (a) Different effects of stimulation of a single static or dynamic fusimotor fibre on the response of a primary ending to ramp and hold stretching. Soleus muscle of a decerebrate cat. In C and D, stimulation started before the onset of the traces and was given at 70 Hz in each case. (b) Different effects of static and dynamic fusimotor stimulation on the dynamic index of a primary spindle ending: γ_d stimulation increases the dynamic sensitivity, while γ_s stimulation decreases it. (From Crowe and Matthews, 1964, Figures 1 and 3; with permission.)

dorsal root filaments while stimulating single γ-efferent fibres within ventral root filaments. Primary endings were found to respond in two different ways to stimulation of γ-neurones (Figure 4.7). Stimulation of some filaments (γ_s) during ramp and hold stretches was found to increase the resting level of discharge and the static sensitivity of the ending, while decreasing the dynamic sensitivity of the ending. Stimulation of the other category of γ fibres (γ_d) also increased the resting discharge, but not so much as the γ_s stimulation. Its main effect was to produce a much greater dynamic response of the primary ending. One way of indicating this difference is to plot the dynamic index (see above) of the ending during stimulation of γ_s and γ_d fibres at various frequencies (Figure 4.7 a): γ_d stimulation increases the dynamic index whereas γ_s stimulation decreases it.

The response of secondary endings usually is unaffected by γ_d stimulation, supporting the idea that γ_d fibres terminate only on the bag$_1$ fibres. Stimulation of γ_s fibres has a similar effect to that seen on the primary endings – an increase in resting discharge and a slightly greater sensitivity to static length changes.

The mechanism of action of the γ-efferents on the discharge from the sensory endings is thought to be a result of the contractile properties of each of the types of intrafusal muscle fibre. Contraction of all three types of fibre is limited to the capsular sleeve or extracapsular region. Activation of any fibre will therefore result in stretch of the central region and a consequent increase in the resting discharge rate.

Neither of the bag fibres can conduct action potentials like the extrafusal muscle. When they are activated by γ-axons, non-propagated depolarizations are produced at the end-plates. In the absence of the amplifying capacity of an action potential such stimuli produce relatively small mechanical contractions of the fibre. This is particularly true of the bag$_1$ fibre, which contracts very little even when stimulated tetanically (70 Hz) by a γ_d-axon. There may be only a 5% extension of the equatorial, sensory region. This explains why γ_d stimulation has relatively little effect on the resting discharge of primary endings. Although γ_d stimulation produces a rather small contraction of the bag$_1$ fibre, during muscle stretch it increases the stiffness of the poles of the fibre quite considerably. When an active bag$_1$ fibre is stretched, a much greater proportion of the extension is taken up by the primary sensory spiral than if the fibre is inactive. At the end of stretch, the 'creep' back to static length also takes much longer (Figure 4.5). The result is that both primary afferent discharge during stretch, and the dynamic index at the end of stretch are increased by γ_d stimulation.

In contrast to the behaviour of the bag$_1$ fibre, the bag$_2$ fibre, which is innervated by γ_s-axons, may contract strongly and stretch the sensory region by 20% or so. This can increase resting discharge by 20–30 Hz in a primary ending. However, the effect of γ_s stimulation on the bag$_2$ fibre seems limited simply to increasing the basal firing level of its sensory endings (both primary, and to a lesser extent, secondary), with little effect on either dynamic or static responses to stretch.

Chain fibres are capable of conducting action potentials, yet even when activated by a γ_s-axon, the depolarization seems confined to the outer regions since the sensory, equatorial region is stretched by up to 20%. Single stimuli produce twitch contractions, which may give small,

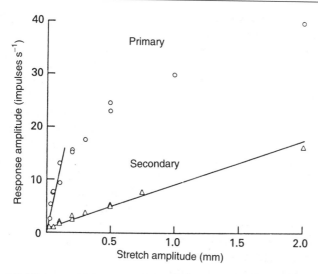

Figure 4.8 High sensitivity of primary endings to very small changes in length and their non-linear behaviour as larger stretches are given. Secondary endings are less sensitive and more linear over a wide range of stretch amplitudes. This graph was constructed by giving continuous sinusoidal stretches at 1 Hz to the soleus muscle of a decerebrate cat with intact ventral roots and spontaneous fusimotor activity. The muscle receptors had a steady background level of discharge which was modulated by the sinusoidal stretch input. The change in firing frequency is plotted on the y-axis, and the amplitude of the sinusoidal stretch (that is, half the peak-peak extent) on the x-axis. (From Matthews and Stein, 1969, Figure 4; with permission.)

transient extensions of the sensory spiral. Such stimuli can 'drive' the primary ending (with one impulse per twitch) to fire at up to 75 Hz. Secondary endings do not show this effect. Sustained stimulation increases the basal rate of sensory discharge in both primary and secondary endings and also (by an unknown mechanism) increases the static length sensitivity of the chain fibre endings (Crowe and Matthews, 1964; Boyd and Ward, 1975; Banks *et al.*, 1978; Boyd, 1980, 1985; Matthews, 1981; Hulliger, 1984).

Response of spindle endings to very small displacements

In the description of spindle responses given above, both primary and secondary endings emerge as relatively insensitive endings, giving an increase in firing frequency of only about 5 Hz for every millimetre change in length. However, if much smaller stretches are given, the response of the primary ending becomes relatively much greater and may reach up to 500 imp s^{-1} mm^{-1}. In general terms, the primary ending is exquisitely sensitive to very small displacements and much less sensitive to large ones. It is, in engineering terms, a non-linear receptor.

In contrast, the secondary ending is fairly linear throughout the physiological range (Figure 4.8).

The analysis of the response to small amplitude stretches was first performed by Matthews and Stein (1969) who used sinusoidal stretch analysis. This is an engineering technique in which the sensitivity of a receptor is plotted in terms of its response to different frequences of sinusoidal inputs. The reason for using this approach is that it provides a complete description of spindle response to all types of input. Plots of dynamic and static sensitivity are useful, but do not describe, for example, how the spindle will discharge to continuously changing inputs. The theory of sinusoidal analysis is that all inputs to a spindle can be broken down by Fourier analysis into sinusoidal components of different frequencies and amplitudes. Knowing the response to such sinusoidal inputs when given alone then allows one to predict the response to complex inputs by adding up the contributions from the various component frequencies.

There is, however, a rather serious limitation to this form of analysis: that the behaviour of the receptor is linear. In sinusoidal analysis, a system is tested with an input of constant amplitude but of varying frequencies. It is assumed that if the amplitude were halved, then the response would also be halved. Unfortunately, this is not true for large stretches applied to a spindle. However it is true when very small stretches are applied.

Figure 4.8 shows the region of linearity for primary and secondary endings. Secondary endings are linear over most of the range tested, but primary endings are linear only over a very small range. Within this range, the primaries are very sensitive indeed, equivalent to $150 \, \text{imp s}^{-1} \, \text{mm}^{-1}$ at 1 Hz.

Within their linear range, the sensitivity of the primary ending increases as the frequency of the input is increased. This is shown in Figure 4.9, in which sensitivity is plotted against frequency of stretching for a sinsusoidal input of constant amplitude. The horizontal part of the curve is equivalent to the static sensitivity of the ending (that is, there is no dependence on frequency of stretch). The sloping part indicates a dependence on stretch velocity (that is, frequency), beginning at about 1.5 Hz. When analysed in this way over their linear range, primary and secondary endings seem remarkably similar. Two points are quite unexpected: (1) secondary endings are sensitive to velocity: the reason this was not appreciated before is that (at least in the linear range) it is some ten times smaller than the sensitivity of the primary ending (see shift of the curves in (Figure 4.9); and (2) over their linear range, the primary endings are remarkably sensitive to static stretch. Like the velocity component, there is a tenfold difference between primary and secondary response.

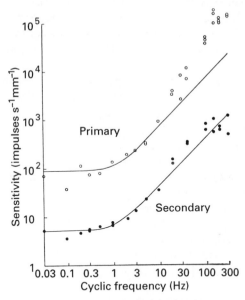

Figure 4.9 Comparison of sensitivity to sinusoidal stretching within the linear range of a primary and secondary ending in the soleus muscle of a decerebrate cat with intact ventral roots. The sensitivity (y-axis) is the amplitude of response divided by the amplitude of stretching (that is, the slope of the lines in Figure 4.8). Continuous lines are theoretical vector sums of a response to velocity (sloping region) and length (horizontal region) components of stretch. Deviation of points from the primary ending above this line at high frequencies of stretch may be taken as indicating an 'acceleration sensitivity' of the primary ending. (From Matthews and Stein, 1969, Figure 5; with permission.)

Such studies emphasize how differently endings can behave, particularly the primary ending, for stretches of small amplitude. The initial high sensitivity of the primary ending is believed to be due to the presence of cross-bridges between actin and myosin in the poles of the intrafusal fibre. These resist any extension with a large short-range stiffness, until some critical deformation when they break and reform. Up to this critical point, most of the stretch will therefore be taken up by the central, poorly striated, portion of the fibre, resulting in a larger proportional deformation of the primary ending than during larger displacements.

The functional significance of this behaviour is that the CNS may be informed of very small changes in muscle length. In the decerebrate cat, a change in length of the soleus muscle of only 50 μm (that is, 0.1% of the muscle length) may produce a change in discharge of some 20 imp s^{-1} from an ending which gave only a 50 imp s^{-1} discharge when the muscle was fully extended. On being given a large stretch, the sensitivity of the ending falls, but when the new ('operating') length is reached, the sensitivity is reset rapidly within a few seconds to a high value once again. This prevents saturation of the response, and effectively increases the working range of a highly sensitive device.

Short-range stiffness of the muscle spindle also has been invoked to explain the existence of the initial burst or 'acceleration' response of spindle primary endings. If a spindle is stretched to a constant length, cross-bridges detach and reattach at the new length over a period of several seconds. If fusimotor stimulation is given, this process is speeded up by increasing the rate at which cross-bridges recycle. When the cross-bridges have formed at the new length, their short-range stiffness changes the mechanical properties of the spindle. The initial portion of a stretch is taken up predominantly by the equatorial (sensory) region of the spindle and produces the initial burst of afferent activity. As stretch continues, the short-range bonds break, and the initial burst declines (Matthews and Stein, 1969; Emonet-Dénand and Laporte, 1981; Matthews, 1981; Hulliger, 1984).

After-effects

The output of a muscle spindle to a given stretch input is not always the same. The response depends upon the movement- and contraction-history of the parent muscle. For example, if a muscle is subjected to a series of rapid stretches, then the response to a subsequent slow test stretch depends on the length at which the muscle was held immediately after conditioning. If the muscle was held for several seconds at a long length and then returned to the initial length before the test stretch is given, the response is much smaller, and has a longer latency, than if the muscle is immediately returned to the initial length after conditioning. The same effect can be seen if a burst of fusimotor stimulation is given instead of a series of rapid conditioning stretches.

The after-effects persist for many minutes, and, like the initial burst phenomenon discussed above, have been attributed to the formation of stable cross-bridges between actin and myosin filaments in passive intrafusal fibres. Repetitive stretches or fusimotor stimulation result in a detachment of stable cross-bridges which then reform when the stretch or stimulation is stopped. If the muscle is held stretched while the bridges are reforming, and is then shortened several seconds later, the intrafusal fibres are unable to shorten passively themselves because of the stiffness of the stable cross-bridges. Because of this the spindles fall slack, and their response to a slow test stretch is small and delayed in time until the test stretch has lengthened the spindle sufficiently to remove the slack. The opposite happens if the conditioning is applied to an already short muscle. No slack develops and the spindle sensitivity remains high (Morgan, Prochazka and Proske, 1984; Proske, Morgan and Gregory, 1992).

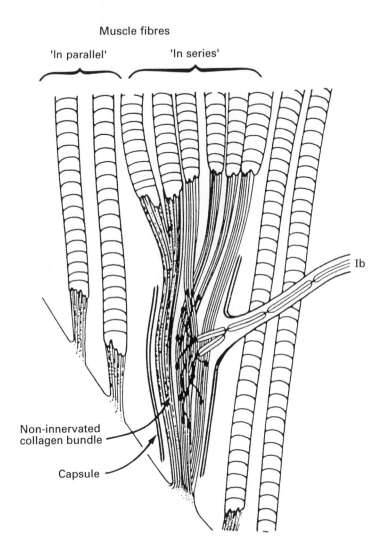

Figure 4.10 Schematic representation of a tendon organ. The capsule is shown in longitudinal section. Sensory terminals are represented by knobs at the end of the unmyelinated branches of the Ib axons. Note the unequal distribution of terminals to collagen bundles within the capsule, leaving some of the bundles completely uninnervated. (From Jami, 1992, Figure 1; with permission.)

4.2 *Muscle receptors: II. Golgi tendon organs*

Anatomy

Tendon organs were first described in detail by Golgi in 1880, although attention had been focused upon them some five years previously by the discovery of tendon reflexes. Despite their name, they rarely lie in the tendon itself, but are usually found at the musculotendinous junction. The mammalian tendon organ is composed of a spindle-shaped connective tissue capsule which encloses a number of collagen strands and the terminals of a large diameter afferent nerve fibre (Figure 4.10). The collagen strands are derived from the main tendon at one end, and at the other attached to about 10–20 individual muscle fibres which may come from several different motor units. Within the spindle sheath, the collagen strands are less densely packed than those outside and are divided into separate fascicles by transverse septae. The afferent nerve fibre is of large diameter (group Ib; 8–12 μm; 70–120 m s^{-1} in the cat) and retains its myelin sheath within the capsule before breaking up into a series of fine, non-medullated branches which are closely applied to the collagen fascicle. The Ib fibres do not innervate every fascicle within a tendon organ. Thus, contraction of muscle fibres which insert onto an innervated fascicle may produce a good response from the ending. Contraction of others which insert onto non-innervated fascicles may either produce no response, or may even unload the force exerted on innervated fascicles and decrease receptor discharge (see below). In addition to the muscle fibres attached in series with the tendon organ, a large number of fibres run in parallel with the receptor, and insert around its tendinous end. Like the non-innervated fascicles within the tendon organ bundle, contraction of these fibres on the outside can reduce the force acting on nearby tendon organs and produce unloading effects.

The effective stimulus for tendon organ endings is thought to be mechanical deformation. Tendon organs occur in a variety of different sizes. The stiffer the tendon organ, the less easy it is to deform the endings within and the higher its threshold. Mechanical deformation of the ending might occur in two ways. First, when force is applied to the organ, the collagen fibres within may be straightened out and crowded together, and compress the nerve terminals between them. Squeezing the nerve endings may therefore be one form of stimulus for the afferent nerve. A second possibility is that if the sensory terminals are very closely applied to the collagen strands, then stretch of the fascicles themselves may deform the endings in a similar manner as is thought to happen in the muscle spindles.

There are usually more spindles in a muscle than tendon organs (often the ratio is 2:1), but in some cases, like the cat peroneus longus or brevis muscle, this ratio may be about equal.

Physiology of Golgi tendon organ responses

As noted above, B.H.C. Matthews was the first to distinguish the responses of muscle spindles and tendon organs in recordings from single afferent fibres. In his hands, the tendon organs of the soleus muscle had a high threshold both to passively applied stretch or active muscle contraction. For the next 30 years, Golgi tendon organs were believed to be high threshold receptors with a function limited to signalling dangerously high muscle tensions.

Since then, that view has changed entirely, and tendon organs have become accepted as very sensitive muscle receptors with a more widespread role in movement control. Part of the reason for this change has been the finding that, particularly in certain muscles such as the cat soleus, much of the passively applied force on a muscle is not transmitted through the tendon organs but via the surrounding connective tissue. When this is dissected away, tendon organ sensitivity can be increased considerably. In fact, tendon organs are, despite Matthews' original assertion, very sensitive to actively generated muscle force. This is because in a contracting muscle, all the force produced by the muscle fibres is exerted through the tendon organs at the musculotendinous junctions.

As an example of the sensitivity of tendon organs to active force, Houk and Henneman (1967) recorded the responses of single tendon organs to tetanic stimulation of single α-motor fibres. They found that each individual organ could be excited by the contraction of just four to 15 motor units (Figure 4.11). Since a maximum of only 25 muscle fibres insert into each tendon, then each motor unit must have been capable of exciting the tendon organ with the contraction of only one or two of its single muscle fibres. Houk and Henneman (1967) estimated that if the contraction of a single fibre can excite a tendon organ, the threshold of sensitivity would be about 0.1 g.

The tendon organ exhibits two important non-linearities in its response to forces applied via motor unit contraction. First, if a tendon organ is given a resting discharge by passive stretch of the muscle, or by contraction of an in-series muscle fibre, then additional contraction of a small number of other units sometimes is found to reduce the firing or even silence the endings. Such an effect arises from the unloading of force on the tendon organ by parallel contraction of neighbouring fibres which do not insert into the receptor. Such unloading effects may explain why the force threshold for producing tendon organ discharge is higher for some units (even when contracted alone) than others. A

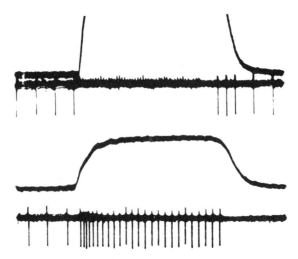

Figure 4.11 Sensitivity of tendon organ discharge to the contraction of particular groups of motor units within a muscle. The two sets of record are from a tendon organ in the soleus muscle of an anaesthetized cat. In each record the upper trace represents the force exerted by the muscle through its tendon and the lower trace the discharge of the tendon organ recorded in a dorsal root filament. The ventral roots were divided into filaments each containing motor axons to several motor units in the soleus. Stimulation of one filament (bottom record) elicits a vigorous tendon organ discharge even though the contraction was very small (18 *g*). Stimulation of another filament produces a larger force (off scale), but the receptor, which initially was discharging due to positive stretch, drops below threshold throughout the contraction. This is due to unloading of the receptor by parallel contraction of neighbouring motor units. Sweep duration is 1 s. (From Houk and Henneman, 1967, Figure 1; with permission.)

unit will have a high threshold if some of the muscle fibres of which it is composed tend to unload the tendon organ whilst others are inserted directly into it.

A second peculiarity of tendon organ responses is that the firing does not always summate linearly to equal increments of force produced by different motor units. For example, if an individual tendon organ were caused to fire at 5 Hz by the separate tetanic contraction of two different motor units, then it would fire at less than 10 Hz if both units were activated simultaneously. It is thought that this behaviour arises from the microphysiology of the receptor. Each tendon organ has several terminals, and deformation at each of these sites produces a local receptor potential. It is thought that the receptor potentials from each terminal summate at some common impulse generating site on the distal part of the afferent axon. Summation of these potentials should be linear, but this may not be true of the relationship between receptor potential and afferent firing rate. Thus, the two units may each elicit a potential of 1 mV and this may cause the axon to fire at 5 Hz. However, a combined receptor potential of 2 mV may only produce firing at 8 Hz. Another possible explanation is that there may be a non-linear relationship between the force applied to a tendon organ terminal and the receptor potential which is produced. Thus, if the two motor units were to be linked to the same receptor (and such cross-linkages have been observed anatomically), then a combined force of 10 g from activity of both units at the same time would produce a smaller receptor potential than expected from the summation of two forces of 5 g each.

Such non-linear effects mean that the discharge of any one Golgi tendon organ is not closely related to the force of the contraction which it experiences. However, the territory of a single motor unit can be quite large, and it is common for a single unit to have fibres which innervate more than one tendon organ. In such a situation, the question arises as to whether the force of a motor unit contraction would be signalled better by the average discharge of all tendon organs than by the discharge of a single organ. The answer seems to be no. There are only about ten Golgi tendon organs in the peroneus tertius muscle, and in some experiments it has been possible to record from all of them. The net result in such tests has been to show that the average discharge of all the tendon organs does not signal the actual strength of muscle contraction.

Thus, in Figure 4.12, one motor unit activiated four of the ten tendon organs. When the force exerted by this unit was increased by increasing the rate of stimulation of its motor axon, the average response of all the tendon organs reflected the contraction level quite well. However, when other units were examined, the authors found that the relationship between force and firing rate was different for different motor units. Another unit is illustrated in Figure 4.12 which innervated six of the ten tendon organs. This unit produced a similar tetanic force at a stimulation frequency of 40 Hz as the previous unit, but the average discharge from the tendon organs was much greater. Finally, when both units were stimulated together (see section of record at 40 Hz), the force of contraction increased considerably, yet the firing rate of the tendon organ remained approximately the same as when the second unit was activated alone. The conclusion is that even the ensemble discharge of all Golgi tendon organs in the muscle does not accurately reflect the steady force of contraction which is being exerted.

It is now thought that the discharge of Golgi tendon organs may be more important in signalling variations in contractile force than the steady level of muscle contraction. The dynamic sensitivity of their discharge is evident in Figure 4.12, in which the individual force variations are followed quite closely by the average discharge of the receptor population. Indeed, during contractions of single motor units, individual tendon organs often fire in a 1:1 relationship to the discharge of the

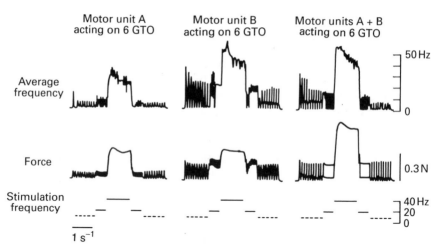

Figure 4.12 Instantaneous average frequencies of discharge of all the Golgi tendon organs in the peroneus tertius muscle of a cat during contraction of two different motor units which were stimulated either separately (left and middle) or together (right). The receptors were stimulated at frequencies of 10, 20 or 40 Hz. The force of the contraction exerted through the tendon is shown in the middle traces. The six receptors activated by unit A included the four receptors activated by unit B. When unit A was stimulated alone, the average firing frequency of the tendon organs reflected the force profile quite well. However, when unit B was stimulated alone, the force exerted at 40 Hz was the same as that exerted by unit A, yet the discharge of the tendon organs was much greater. When both units were activated together, the force exerted (especially at 40 Hz) was much greater than when each unit was activated alone yet the tendon organ discharge was not greatly changed. Note that during stimulation at 10 and 20 Hz, the average firing frequency of the tendon organs was lower when both units were stimulated together than when unit B was stimulated alone. This must have been caused by some unloading effects from unit A. (From Jami, 1992, Figure 6; with permission.)

motor unit, signalling the ripples in contractile force much better than the steady level. It may be that the nervous system obtains an estimate of the static force of muscle contraction by combined processing of discharge from other types of receptors which occurs at the same time as that from the tendon organs (reviewed by Jami, 1992).

4.3 *Muscle receptors: III. Other types of ending*

Two other types of nerve ending found in muscles deserve mention. These are Paciniform corpuscles and free nerve endings. Little is known about the physiological responses or the central effects of either type of ending, although they form an appreciable proportion of the total sensory innervation of the muscle. There are twice as many free nerve endings in muscle as in any other type of sensory receptor, and the number of Paciniform corpuscles may be anything up to 30% of the number of muscle spindles.

Paciniform corpuscles are similar in structure, although rather smaller than the Pacinian corpuscles found in the skin. They are most frequently observed at the musculotendinous junction in the vicinity of tendon organs and are supplied by a large diameter (group II) medullated fibre which may innervate several separate corpuscles. Very little work seems to have been performed specifically on the Paciniform corpuscles in muscle. It is generally accepted that they are rapidly adapting end organs with a sensitivity to high frequency vibration like true Pacinian corpuscles in skin. Their central connections have not been studied.

Free nerve endings are found in close association with almost every structure in muscle (spindles, tendon organs, muscle fibres, fascia, fat and blood vessels, excluding capillaries). It is not known whether their function depends on their location, but the fact that a single fibre may innervate receptors in several different regions argues against this possibility. All non-medullated fibres and the majority of small (less than 5 μm) group III medullated axons terminate as free nerve endings. Recording from slowly conducting muscle afferents (less than 1 m s^{-1} for non-medullated, less than 24 m s^{-1} for group III medullated) has shown that these endings are rarely excited by classical proprioceptive stimuli such as muscle stretch or contraction or muscle vibration. Instead, they are principally activated by high threshold mechanical stimulation of the muscle, such as pinching or pricking, and have been termed 'pressure-pain' receptors. In addition, they also respond to other nociceptive stimuli like ischaemia and injection of hypertonic saline. It is possible, however, that some of these fine endings may be more specifically responsive to humoral or metabolic stimuli and hence play a role in mediating local cardiovascular or respiratory reflexes.

4.4 *Joint receptors*

Anatomy

Three main types of ending are associated with the synovial joints of the body:

(1) free nerve endings, which are the most numerous type of joint receptor, are found throughout the connective tissue;

(2) Golgi endings similar to tendon organs are found in the joint ligaments;

(3) Ruffini endings (see below) are found in the joint capsule.

Paciniform-like corpuscles also have been described in the joint capsule: there are no nerve endings on the cartilaginous surfaces of the joint or in the synovial membranes. Golgi endings are innervated by large diameter (10–17 μm) group I medullated fibres; Ruffini endings by slightly smaller (5–10 μm) group II fibres and free endings by group III or non-medullated nerves.

The Ruffini ending is similar to the same ending found in the skin. It consists of a small capsule which encloses a number of spray arborizations from a single afferent fibre (Figure 4.13).

Physiology

Initial experiments on the responses of joint receptors to passive movement of the joint were performed in the 1950s. The results were quite promising. When single group I–II afferent fibres were recorded from the articular nerves of the cat knee joint, it appeared that individual fibres had a distinct 'turning curve'. That is, the response of a fibre was limited to a small range of joint angles, with maximum firing occurring at a particular point within that range (Figure 4.14). The discharge rate was very slow to adapt and hence this population of receptors was

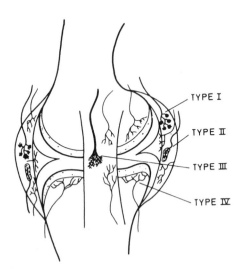

Figure 4.13 Diagram of location and type of receptors commonly found in joints. Type I endings are Ruffini endings; type II, Paciniform endings; type III, Golgi endings; type IV, free nerve endings. (From Brodal, 1981, Figure 2.2; with permission.)

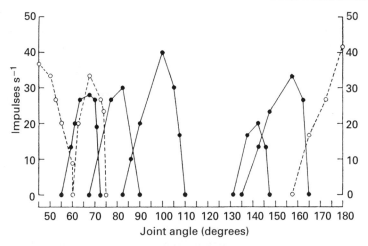

Figure 4.14 Adapted firing frequencies of a number of knee joint afferents at different positions of the knee. Points joined by lines are from a single receptor (solid lines from one cat, dotted lines from another cat). Other experiments in the same series showed a predominance of fibres which fired at the two extremes of joint position rather than in the mid-range. (From Skoglund, 1956, Figure 19; with permission.)

thought to be particularly suited to provide information to the CNS about static joint position.

This view has been challenged, however. In a reinvestigation of knee joint receptor sensitivity, Clark and Burgess (1975) showed that only a small proportion of the joint afferents actually had responses like those described above. The majority had no response to joint angles in the mid range, but only fired at the extremes of joint rotation. Moreover, many of these receptors did not even distinguish between extremes of flexion and extension, so that it became difficult to see what role, if any, they could play in signalling static joint position. Later work has extended these findings to other joints with the same result. The generally accepted conclusion at present is that joint receptors with group I–II afferent fibres (that is, Golgi and Ruffini endings) are responsive only to deformation of the joint capsule or ligaments. At joint positions where no stress is placed on the capsule, the receptors remain virtually silent. The temptation therefore, has been to conclude that joint receptors serve only to indicate extremes of joint rotation, presumably with some protective function. However, in the living animal muscle, activation may place different stresses on the joint capsule and ligaments and change the firing pattern of joint afferents from that seen in the experimental preparation. Thus, the precise role of these receptors is still subject to much discussion.

The firing pattern of non-medullated and group III joint afferents has not been investigated in such detail. In the cat knee joint, many of the fibres are activated by movements within the normal limits, although a large proportion only fire in response to noxious mechanical stimuli. The firing of such afferents may be a 'warning' signal that the joint is about to leave its normal working range. (Clark and Burgess, 1975; Schiable and Schmidt, 1983, give further details on joint receptors.)

4.5 *Cutaneous mechanoreceptors*

Three types of receptor are found in the skin: thermoreceptors, nociceptors and mechanoreceptors. Much of the interest in the study of these receptors obviously lies in the field of sensory physiology and, particularly with reference to man, in the psychophysics of sensation and sensory discrimination. Nevertheless, it is quite clear that mechanoreceptors also play a very important role in control of movement. This is particularly true of the densely innervated regions of the hand and foot. For example, signals from the pressure sensors in the sole of the foot are used in balance. Peripheral neuropathies which cause loss of sensation in the extremities may be accompanied by Romberg's sign: patients are unable to stand unassisted with their feet together when they close their eyes. Although the neuropathy may involve muscle receptor input from the intrinsic muscles of the foot, it is believed that the major contribution to pressure sensation comes from cutaneous receptors. Without information on the differential distribution of pressure on the soles of the feet, the reflexes from the intact vestibular system of these patients are unable to maintain balance alone. In the same way, any complex manipulation of objects in the hand is grossly impaired by cutaneous anaesthesia: writing with a pen, or picking up coins from a flat surface or doing up the buttons of one's shirt, all become difficult, if not impossible, even when vision of the hand and object is allowed. Visual input cannot substitute for the loss of cutaneous mechanoreceptor signals.

Despite the importance of cutaneous input, it is still not clear precisely how these signals are used in the control of movement. In this section some of the anatomical features and physiological properties of the four main types of mechanoreceptor will be detailed. Much of the work has been done on the receptors of the monkey hand: recent work on the physiological responses of human cutaneous mechanoreceptors will be summarized in the next section.

Of the four types of receptor, two are found close to the epidermal – dermal junction within a few hundred micrometres from the surface of the body. These are the Merkel discs and the Meissner corpuscles. In the

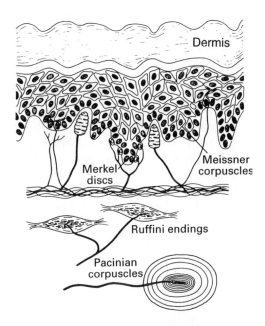

Figure 4.15 Location of the four main types of cutaneous mechanoreceptors in the glabrous skin of the hand. Ruffini endings are in the dermis; Pacinian corpuscles are located deeper, in the subcutaneous tissues. Merkel discs and Meissner corpuscles are found nearer the surface at the dermal–epidermal junction. (From Brodal, 1981, Figure 2.1A; with permission.)

glabrous skin of the human palm, the Merkel discs are found at the base of the epidermal infoldings in the dermis, whereas the Meissner corpuscles are found at the tip of the dermal protrusions. In the monkey, Merkel discs are located just underneath dome-like projections in the stretched skin known as dome corpuscles. The two other types of receptor, Ruffini endings and Pacinian corpuscles, are found in the deeper layers of the skin (Figure 4.15).

Merkel discs are concave, flattened, disc-like formations of cells within the stratum granulosum of the epidermis. They do not contain keratin and hence are differentiated from the rest of the epidermal cells. A single nerve axon innervates the whole group of cells. It emerges from the subdermal plexus and loses its myelin sheath at the dermal–epidermal junction, branching many times to each Merkel disc. In the monkey, records from single axons show that at rest there is little or no discharge. The receptor is sensitive only to localized vertical pressure on the surface of the 'touch dome', and does not respond to any lateral stretch of the skin. The response to a maintained indentation has a dynamic and static phase. In the dynamic phase, very high rates of discharge may be reached, but these slow over about half a second or so to reach their final steady rate, which may be constant for up to 10 minutes. It is not known to what extent the initial adaptation is due to tissue movement under the stimulus probe, or whether this represents a fundamental receptor property.

Meissner corpuscles are ovoid and are found with their long axis oriented perpendicular to the surface of the skin within a dermal papillum. From two to six separate axons innervate each Meissner corpuscle and form a complex network within the structure. Each axon may innervate more than one corpuscle. Krause end bulbs, found in the lips, tongue, conjunctiva and so on, are similar in structure and are thought to be analogous to the Meissner corpuscles of glabrous skin. The receptive field of single axons is small, and the discharge to maintained pressure adapts extremely rapidly. The discharge lasts only a few seconds.

Ruffini endings are found within the dermal layer, and consist of nerve terminals from a single axon intimately associated with collagen fibrils which merge with dermal collagen. The whole structure is surrounded by a fluid-filled space which is enclosed by a thinly lamellated capsule. These receptors are responsive to stretch of skin over a wide area (up to 25 cm^2), and have a slowly adapting response to continuous stimulation. Stretch along the long axis of the structure stimulates, and stretch at right angles inhibits, the discharge. This gives the receptor some directional sensitivity, which is usually only ascribed to joint receptors. However, Ruffini endings in the skin also seem likely candidates to signal some aspects of limb position to the nervous system via their response to skin stretch.

Pacinian corpuscles are found within the subdermal fasciae. They are the largest receptors in the skin, and may be 1–4 mm in size. A single axon terminates within a large number of concentric cytoplasmic lamellae which are separated by fluid-filled spaces. The bare nerve terminal is sensitive to the mechanical deformation which it receives through the 'onion skin' capsule. This filters any slow frequency components of the signal, leaving the Pacinian corpuscle sensitive only to very rapidly changing stimuli. Such rapidly changing pressure stimuli frequently travel in a wave of vibration through the tissues of the skin and the bones, and are picked up by the Pacinian corpuscle from over a wide area.

4.6 *Recordings from human afferent nerve fibres*

One of the most important advances made in technique over the past 20 years has been the ability to record single unit activity from nerve fibres in awake, cooperative human subjects. The technique, which was pioneered by Vallbo and Hagbarth (1968) in Sweden involves the production of fine tungsten microelectrodes with a tip diameter of 1–15 μm which can be inserted through the skin and into underlying nerve trunks. The electrodes are insulated, apart from a length of 10–15 μm at the tip. Usually, the procedure is not painful, although paraesthesias are elicited when manipulating the electrode. Permanent nerve damage is extremely rare.

Activity has been recorded from all classes of myelinated fibres, both afferent and efferent, although there is some debate as to whether reliable records have been made from γ fibres to muscle spindles. The largest potentials are seen with activity in the largest afferents: the Aα class from skin receptors and group Ia spindle afferents. Vallbo has suggested that the shape and polarity of the potentials from these large fibres indicates that the impulse is recorded through a low-impedance pathway between the electrode tip and the inside of the fibre, and hence that the fibre has been impaled by the electrode. The potentials are believed to be recorded extracellularly with smaller fibres.

The major problem with the technique is the stability of recording. Single units may only be held for 15 minutes or so before the electrode moves within the nerve and the unit is lost. Set against this disadvantage is the ability of the human subject to cooperate with the experimenter, either by reporting sensations or by producing voluntary muscle movements. In the former case, for example, it has been possible to show that at the limit of detection a human subject can distinguish the firing of a single impulse in an afferent fibre, caused by minute mechanical stimuli at the finger tips. This section will outline only some of the experiments relating to the discharges of proprioceptive sensory endings in muscle and skin.

Muscle spindles

Much of the work performed on human muscle spindles has been devoted to analysing their discharge during movement and in pathological disease states in man. This will be summarized in the next chapter. However, there are several features of their passive behaviour which will be mentioned here.

Muscle spindles are more abundant in the intrinsic muscles of the hand than they are in proximal muscles. For example, in the cat, it is estimated that they are the equivalent of 90 spindles per gram of tissue in the interosseous muscles compared with only 5 spindles per gram in the medial gastrocnemius. It is not entirely clear why this should be the case. However, the size of motor units in intrinsic hand muscles is much smaller than in large postural muscles. Thus, when expressed as the number of spindles per motor unit, the figure becomes comparable for all muscles.

Because the activity in large nerve fibres tends to be picked up more easily by microelectrodes, most recordings have been made from group Ia axons. The responses of what have been presumed to be group II secondary endings are not so well documented. In general, human spindle endings have been shown to behave very similarly to those in the cat (see, for example, Edin and Vallbo, 1990). In a subject at rest, the primary endings have an impressive sensitivity to small dynamic changes in muscle length. They may discharge even in response to the arterial pulse or to respiratory movements. However, their static position sensitivity is, in contrast, extremely low. In the finger–wrist flexor muscles, only 10% of spindles show any discharge at all at a comfortable rest position of the hand. During imposed static wrist movements, they only change their firing rate at the order of 1 imp s^{-1} mm^{-1} increase in muscle length. Such behaviour is very similar to the behaviour of deefferented spindle endings in the cat. Without any γ-bias, any resting discharge disappears and length sensitivity, in the hip and knee extensor muscles, becomes about 4 imp s^{-1} mm^{-1} extension (as opposed to about 10 imp s^{-1} mm^{-1} in the decerebrate cat).

At first sight, 4 imp s^{-1} mm^{-1} in the cat still seems higher than 1 imp s^{-1} mm^{-1} in man. However, if the human figure is scaled up by a factor of four, to allow for the length difference between the cat extensor muscles and human forearm muscles, then the sensitivities are about the same. The rationale behind this is that the same external stretch applied to muscles of different sizes causes smaller internal length changes per unit length in larger muscles. Thus, much of the stretch applied to a large muscle will be taken up by the muscle fibres in series with a spindle rather than being transmitted directly to the spindle as in a small muscle. The main conclusion of such observations is that, in humans at rest, there is no background fusimotor activity to muscle spindles.

A difference between the behaviour of primary and secondary spindle endings in cat and in humans has been found. In the decerebrate cat, vibration at about 200 Hz is known to be a selective stimulus for the primary spindle endings. Secondary endings are insensitive to such a rapidly changing stimulus. However, in the small number of secondary endings which have been recorded so far in man, such a dramatic difference does not seem to occur. Primary endings can follow vibration, giving one impulse per cycle at frequencies up to 200 Hz (they may discharge at frequencies of up to 800 Hz during muscle stretch). Secondaries only follow vibration up to 100 Hz, but thereafter are still excited potently by vibration. The reason for this difference is not known. It may be due in part to the methods used to vibrate cat spindles *in vitro* and human spindles *in vivo*. In cat experiments, vibration is applied longitudinally to the muscle or isolated spindle, whereas in humans, vibration is usually applied transversely across the tendon or muscle belly. In this case, vibration may produce a less potent mechanical displacement in humans than in the cat.

The advantage of performing experiments on humans is that subjects easily can be instructed either to relax completely and have movement imposed upon them by the experimenter, or they can make the same movements with a voluntary contraction. During passive movement, spindle discharge increases monotonically with applied stretch. During active movements, as in the cat, their behaviour is more complex. Two opposing factors influence spindle discharge: fusimotor drive and changes in muscle length. The general rule seems to be that the slower the contraction of a muscle, and the heavier the load being lifted, the more fusimotor drive will predominate over muscle shortening. In slow voluntary movements, the net outcome can be a virtually constant Ia spindle firing rate which is practically independent of muscle length. In faster contractions, the spindles are unloaded by shortening of the extrafusal muscle. In these circumstances, only the spindles in antagonist muscles can provide signals related to muscle length, (for example Hulliger, Noth and Vallbo, 1982).

Golgi tendon organs

As with the muscle spindle, human observations are in accord with previous findings in the cat. Tendon organs are distinguished from spindle endings by their difference in response to active muscle contraction. In response to a tendon tap, they fire during the active phase of contraction, whereas spindles fire in response to the initial stretch. During electrically induced muscle twitches, they respond during the twitch, in the period when spindles are usually silent. Their threshold to active contractions is very low and they appear to monitor the active tension in muscle. As in the cat, there is evidence to suggest that they may sample preferentially the contraction of particular motor units in series with them. During a steadily rising isometric contraction there may sometimes be discontinuities in the relationship between instantaneous impulse frequency versus muscle force attributable to the recruitment of additional motor units into the contraction.

Cutaneous mechanoreceptors

Corresponding to the four anatomical types of receptor, four physiological types of afferent response can be distinguished in recordings from cutaneous nerves innervating the glabrous skin of the hand. There still is some uncertainty as to the exact correspondence between physiology and anatomy in human studies. In general, the assumptions have been based on parallel detailed studies of identified end organs in animal experiments.

The four types of response are subdivided on the basis of their rate of adaption to novel, sustained stimuli and the size of the receptive field (Figure 4.16). FAI (fast adapting; type I) units and SAI (slowly adapting; type I) units have small receptive fields of about 13 mm^2 area. There are several points of maximal sensitivity within the field, suggesting that the axon terminates with a number of end organs spread within a small area of skin. These two types of receptors are very sensitive to mechanical stimuli (especially the edges of objects) which cause indentation of the skin: FAI units have a threshold of about 10 μm, whereas SAI units have a threshold of about 50 μm. They are thought to correspond to Meissner corpuscles (FAI units) and Merkel discs (SAI units). Each FAI unit fires impulses only in response to rapidly changing or moving stimuli and can follow sinusoidal displacements up to about 100 Hz. Such stimuli give rise to a localized sensation of 'flutter' in man. SAI

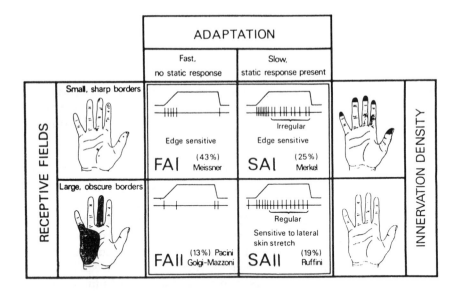

Figure 4.16 Types of mechanoreceptive afferent units in the glabrous skin of the human hand and their distinguishing properties. The type I units (FAI and SAI) have small, well-defined receptive fields (typically 10 mm^2). Each dot in the drawing of the hands represents a single sensory unit. In contrast, the type II units are less densely located and respond to more remote stimuli. FAII units are responsive to transient mechanical stimuli whereas SAII respond to lateral stretch of the skin. (From Westling, 1986; with permission.)

units have a maintained discharge to constant pressure. Both types of unit are extremely dense on the tips of the fingers, declining substantially towards the proximal part of the palm.

FAII (fast adapting; type II) and SAII units (slowly adapting; type II) respond to stimuli applied over a much wider area, which in some cases can be almost as large as the whole palm. However, within this field there is only a single point of maximal sensitivity, suggesting a single terminal end organ. The FAII units are probably Pacinian corpuscles. The SAII units are believed to correspond to Ruffini endings. They respond particularly well to stretch of the skin and have a directional sensitivity, increasing their discharge to stretch in one direction and decreasing (or not responding) to stretch in the opposite direction. The end organ is less sensitive to direct mechanical indentation of skin (threshold of about 250 µm). Its response to stretch may be an important factor in joint position sense in view of the deformation produced in skin by movement at joints and by contraction of underlying muscles. The Pacinian corpuscles are exquisitely sensitive to rapidly changing stimuli (threshold skin indentation of about 10 µm), particularly high frequency vibration at 300–400 Hz. A given unit may be driven to respond, for example, by a single tap with a pencil applied anywhere on the skin of the hand and fingers. Both FAII and SAII units are less common in the glabrous skin of the palm than the FAI and SAI units. They are distributed with a fairly even density over the whole hand. With the hand in a rest position there is little background discharge in any of the cutaneous receptors except the SAII mechanoreceptors. Approximately one-third of the SAII fibres have a background discharge (Vallbo *et al.*, 1979).

Joint receptors

Very few recordings have been made from joint receptors in humans. One of the reasons for this is that it is difficult to identify them with certainty. Since both cutaneous and muscle afferent fibres respond to changes in joint position, the criterion for an afferent to be classified as a joint receptor is that it must be located within the joint. In practical terms, this means that the afferent must discharge in response to focal pressure on the joint itself in addition to movement of the digit. Using these criteria, Gandevia & Burke (1992) found that joint receptor behaviour was similar to that described in the cat. Very few afferents were active throughout the whole range of joint movement. Many only discharged at the extremes of joint movement, often in more than one axis of joint rotation. Thus, joint receptors as a population would provide only a poor indication of joint position.

Table 4.1 Perceptual responses attributed to selective microstimulation of single afferent fibres innervating the human hand

Receptor class		Occasions when a sensation elicited (%)	Single stimulus perception	Character of sensation
Cutaneous	FAI (Meissner)	90	+	Localized tap, flutter or vibration depending on stimulus frequency and duration
	FAII (Pacinian)	85	?+	Diffuse vibration
	SAI (Merkel)	80	+	Local pressure or indentation
	SAII (Ruffini)	10	-	(Rarely) joint movement when receptive field over distal interphalangeal joint
Muscle	Spindle	<10	-	None for single afferent. When a population of fibres stimulated, a sense of movement
	Tendon organ	1 afferent of only 3 sampled	-	1 afferent of 3 produced a sensation of muscle lengthening
Joint		70	-	Focal deep pressure, joint movement or stress

Adapted from Gandevia and Burke, 1992

Sensations produced by stimulation of single afferent fibres in humans

The same microelectrode as used to record impulses from single afferent fibres in man can also be used as a stimulating electrode. With the microelectrode positioned so as to record optimally from a single afferent fibre, stimuli of 0.2–0.3 V for 0.25 ms are capable of activating that fibre without discharging any other fibres in the vicinity. Table 4.1 shows the sensations obtained using this method after stimulation of each type of afferent.

For cutaneous afferents, even a single stimulus can evoke a distinct sensation localized to the receptive field of the afferent fibre. The sensation has a quality which depends upon the type of sensory ending which is innervated by the fibre. The only exception to this is the SAII class of afferent which seem rarely to evoke any conscious sensation when stimulated in isolation. The results in the table are for cutaneous afferents innervating the distal portions of the digits. The probability of evoking sensations from a single afferent innervating this area is very high, but decreases substantially for fibres with receptive fields on more proximal portions of the skin. Presumably the ability to discriminate activity in single afferent fibres is important in producing the high spatial resolution that we have of objects manipulated between the fingertips.

The situation for muscle afferent fibres is quite different. Microstimulation of muscle afferents at the same intensity as would have produced sensation from cutaneous afferents, produces no sensation at all. This is true whether single stimuli or trains of up to 100 Hz are given. One possible reason for this difference between the conscious appreciation of cutaneous and spindle afferent input is that under normal circumstances, the discharge from any spindle depends very much on the context in which it has been recorded. Thus, as we have seen, the discharge may be quite different for active as compared with passive movements. It can be argued that the input from a single spindle in isolation is effectively meaningless, and hence has no direct access to conscious perception unless it is combined with other inputs. Interestingly, it has been suggested that stimulation of a population of spindle afferent fibres may produce a sensation of illusory movements.

Stimulation of joint receptors produces a sensation of deep pressure projected to the receptive field of the afferent, or of joint motion or stress in one or more planes. In most cases, the sensation corresponds to the procedure which was most effective in changing the natural firing pattern of the afferent fibre in question (Vallbo *et al.*, 1984; Gandevia and Burke, 1992).

Sensations of limb position

The sense of limb position is not provided uniquely by any one class of afferent input. Cutaneous, joint and muscle inputs all contribute.

1. Muscle afferents

In the past there was some debate whether muscle spindle input could access conscious sensation. Indeed, as noted above, stimulation of single spindle afferent fibres failed to produce any sensation in normal man. However, there are now several lines of evidence that muscle spindle input probably does reach consciousness. One important observation was made during psychophysical experiments in humans using muscle vibration. Blind-folded subjects were seated at a table with their elbows resting and forearms vertical. One forearm was moved passively by the experimenter and the subjects were asked to indicate the position of that arm by matching it with their other, free arm (Figure 4.17). This is performed very accurately under normal circumstances. However, if vibration was applied over the belly or the tendon of biceps while the arm was being moved by the experimenter subjects would consistently overestimate the angle of extension at the elbow. That is, they perceived the vibrated arm to be further extended than it actually was. Since vibration is known to be a powerful stimulus for spindle receptors, it was concluded that the increased spindle input from biceps during vibration was interpreted by CNS as being caused by increased stretch of the biceps muscle. This led subjects to believe that the elbow was extended.

A second set of observations which confirms this conclusion has been made during routine operations at the wrist or ankle under local anaesthetic. Several investigators have pulled on the long tendons of the digits in such a way as to cause stretch of the muscle without any distal movements occurring. After initial negative reports it is now accepted that in most cases some sensation of movement is perceived and can be referred to the appropriate digit. Indeed, one brave Australian investigator (Dr D. I. McCloskey) even underwent local surgery at the ankle to be able to report for himself, under controlled laboratory conditions, the sensations of movement which accompany pulls on the tendon of the extensor hallucis longus.

The conclusions from tendon pulling experiments were confirmed by complementary experiments in intact individuals. At both the knee and the interphalangeal joint of the fingers, anaesthesia of the joint capsule (to prevent discharge of joint receptors) and the overlying skin (to prevent discharge from cutaneous receptors) does not abolish the ability to detect imposed passive movement. The movement threshold is higher than normal, particularly if the muscles are relaxed during the

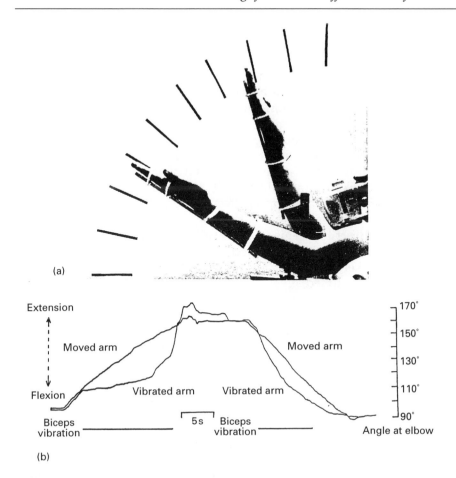

Figure 4.17 Effect of muscle vibration on the sense of joint position at the elbow. (a) Posed picture to illustrate the magnitude of the difference in position of the vibrated and tracking arm that can occur while the subject believes that he is managing to keep them aligned. The scale is marked in 10° divisions. The right arm was vibrated to produce a tonic vibration reflex in biceps and flexion at the elbow. With eyes closed, the subject was required to match the position of arm by moving his left arm. He perceives the right elbow to be more extended than it is. (b) Experiment in which one arm was moved passively by the experimenter while the subject tracked its position with the other arm. Shortly after tracking began, vibration was applied to the tracking arm. This had the effect that the subject underestimated the degree of extension of the moved arm. That is, he perceived the tracking arm to be more extended than it actually was. On turning off the vibrator, the subject matched his elbow angles accurately once again. The process was repeated while the experimenter moved the passive arm in the opposite direction. (From Goodwin, McCloskey and Matthews, 1972, Figures 2 and 8; with permission.)

test and the displacements which are applied are slow. This may explain why initial reports suggested that sensation was abolished with cutaneous and joint anaesthesia. However, when the muscles are active, sensitivity is increased, particularly to fast displacements. It is presumed that contraction removes any slack in the muscle tendon which might reduce transmission of length changes to the muscle belly. In addition, it will increase the background spindle discharge, which may be important if muscle shortening is to be detected.

The evidence is now quite strongly in favour of a contribution of spindle input to conscious appreciation of joint position of movement. However, there is one caveat to the conclusion. In all the experiments described above, some cutaneous as well as muscle receptor input is activated (by skin vibration in the first case, and by movement of muscle bellies underneath the skin, at a distance from the joint, in the second). It is therefore possible that cutaneous input may contribute to the sensations described. This is a difficult argument to refute with certainty. However, one indirect piece of evidence suggests that cutaneous input is unlikely to account for all the observations above.

It is often assumed, given the fine control that we have of the hand, that the sense of position or movement of the digits must be superior to that in other parts of the body. This is not the case. When measured in terms of the angular rotation about the joint, the threshold for detection of movement at the interphalangeal joints of the fingers, is greater than the threshold for detection at the shoulder joint. In one sense, this is not surprising. If we are to control our hand accurately on the end of our arm, an essential prerequisite is that the proximal joint be stabilized so that we can accurately move more distal joints. The question is how is this apparently superior discrimination at proximal joints achieved? The answer is that it is simply a matter of how we express the threshold of detection. If it is expressed, not as the angular change about the joint, but as the relative change in muscle length, then detection threshold is approximately equal for all muscles of the upper limb. The implication is that signals related to how fast or how much a muscle is stretched are important for detection of passive movement. Such signals would obviously be best provided by muscle spindles and not by cutaneous receptors.

2. Joint receptors

Recordings from joint receptors suggest that, at least in the mid-range of joint movement, they provide only a poor signal of joint position. Nevertheless, there is evidence that some of our conscious appreciation of joint position can be provided by joint receptor inputs. If the fingers of the hand are held extended and the middle finger is then flexed at the **proximal** interphalangeal joint, the long flexor and extensor muscles are effectively disengaged from the **distal** interphalangeal joint of the middle finger, and hence movement of the joint can produce no muscular afferent input. If the skin is anaesthetized locally, imposed movements can still be detected (although much less accurately than usual), indicating that receptors in the joint must be capable of signalling joint movement.

3. Cutaneous afferents

Using the anatomical peculiarity of the distal interphalangeal joint described above the muscles can be disengaged and the joint capsule anaesthetized by infiltration of local anaesthetic. Imposed movements can still be detected, although with a higher threshold than usual. It is presumed that Ruffini endings (SAII) in the skin may be particularly sensitive to joint displacement (Edin and Abbs, 1991, for example).

The relative contribution of cutaneous, joint and muscle receptors to the sense of limb position depends upon the muscle examined, and on whether we test the sense of **position** or the sense of **movement**. When joint position sense is examined clinically, the digits are moved relatively fast, and the subject detects the imposed movement of the joint. However, if the joint is moved very slowly (at less than 1 degree per minute) then no **movement** is detected; subjects only report a change of **position** of the joint. At the knee and hip, the sense of joint **position** is dependent on muscle receptor input, and is substantially decreased by muscle paralysis. The situation is different for the distal interphalangeal joint of the middle finger. When the muscles are disengaged, the sensation of joint **position** is the same as it is in the normal position when the muscles are engaged. At this joint, muscular signals are not necessary for the detection of joint position: they only become important when the sense of **movement** is examined. At the interphalangeal joint, movement threshold is reduced when muscles are disengaged, perhaps indicating a particular role of spindle afferents in monitoring the velocity of joint displacement. (Reviewed by Gandevia and Burke, 1992; Gandevia, McCloskey and Burke, 1992.)

Bibliography

Review articles and books

Barnes, W. J. P. and Gladden, M. H. (eds) (1985) *Feedback and Motor Control in Invertebrates and Vertebrates*, Croom Helm, London.
Boyd, I. A. (1980) The isolated mammalian muscle spindle, *Trends Neurosci.* **3**, 258–265.
Boyd, I. A. (ed.) (1985) *The Muscle Spindle*, Macmillan, London.
Brodal, A. (1981) *Neurological Anatomy in Relation to Clinical Medicine*, Oxford University Press, Oxford.

Gandevia, S. C., McCloskey D. I. and Burke, D. (1992) Kinaesthetic signals and muscle contraction. *Trends Neurosci.* **15,** 62–65.

Hulliger, M. (1984) The mammalian muscle spindle and its central control, *Rev. Physiol. Biochem, Pharmacol.,* **101,** 1–110.

Jami, L. (1992) Golgi tendon organs in mammalian skeletal muscle: functional properties and central actions. *Physiol. Rev.,* **72,** 623–666.

Matthews, P. B. C. (1972) *Mammalian Muscle Receptors and Their Central Actions,* Arnold, London.

Matthews, P. B. C. (1977) Muscle afferents and kinaesthesia, *Br. Med. Bull.,* **33,** 137–142.

Matthews, P. B. C. (1981) Review lecture: Evolving views on the internal operation and functional role of the muscle spindle, *J. Physiol.,* **320,** 1–30.

Vallbo, A. B., Hagbarth, K. -E., Torebjork, H. E. *et al.* (1979) Somatosensory, proprioceptive and sympathetic activity in human peripheral nerves, *Physiol. Rev.,* **59,** 919–957.

Original papers

Banks, R. W., Barker, D., Bessou, P. *et al.* (1978) Histological analysis of muscle spindles following direct observation of effects of stimulating dynamic and static motor axons, *J. Physiol.,* **283,** 605–619.

Boyd, I. A. (1962) The structure and innervation of the nuclear chain muscle fibre system in mammalian muscle spindles, *Phil. Trans. Roy. Soc. B.,* **245,** 81–136.

Boyd, I. A. (1976) The mechanical properties of dynamic nuclear bag fibres, static nuclear bag fibres and nuclear chain fibres in isolated cat muscle spindles, *Prog. Brain Res.,* **44,** 33–50.

Boyd, I. A. and Ward, J. (1975) Motor control of nuclear bag and nuclear chain intrafusal fibres in isolated living muscle spindles from the cat, *J. Physiol.,* **224,** 83–112.

Boyd, I. A. Gladden, M. H. and Ward, J. (1981) The contribution of mechanical events in the dynamic bag$_1$ intrafusal fibre in isolated cat muscle spindles to the form of the Ia afferent axon discharge, *J. Physiol.,* **317,** 80–81.

Brown, M. C., Crowe, A. and Matthews, P. B. C. (1965) Observations on the fusimotor fibres of the tibialis posterior muscle of the cat, *J. Physiol.,* **177,** 140–159.

Clark, F. J. and Burgess, P. R. (1975) Slowly adapting receptors in cat knee joint: can they signal joint angle? *J. Neurophysiol.,* **38,** 1448–1463.

Crago, P. E., Houk, J. C. and Rymer, W. Z. (1982) Sampling of total muscle force by tendon organs, *J. Neurophysiol.,* **47,** 1069–1083.

Crowe, A. and Matthews, P. B. C. (1964) The effects of simulation of static and dynamic fusimotor fibres on the response to stretching of the primary endings of muscle spindles, *J. Physiol.,* **174,** 109–131; **175,** 132–151.

Edin, B. B. and Abbs, J. H. (1991) Finger movement responses of cutaneous mechanoreceptors in the dorsal skin of the human hand, *J. Neurophysiol.*, **65**, 657–670.

Edin, B. B. and Vallbo, A. B. (1990) Dynamic response of human muscle spindle afferents to stretch, *J. Neurophysiol.*, **63**, 1297–1306; 1307–1313; 1314–1322.

Emonet-Denand, F. R. and LaPorte, Y. (1981) Muscle stretch as a way of detecting brief activation of bag fibres by dynamic axons, in A. Taylor and A. Prochazka (eds) *Muscle Receptors and Movement*, Macmillan, London.

Gandevia, S. C. and Burke, D. (1992) Does the nervous system depend on kinesthetic information to control natural limb movements? *Behav. Brain Sci.*, **15**, 615–633.

Goodwin, G. M., McCloskey, D. J. and Matthews, P. B. C. (1972) The contribution of muscle afferents to kinaesthesia shown by vibration-induced illusions of movement and by the effects of paralysing joint afferents, *Brain*, **95**, 705–748.

Hulliger, M., Noth, E. and Vallbo, A. B. (1982) The absence of position response in spindle afferent units from human finger muscles during accurate position holding. *J. Physiol.*, **322**, 167–179.

Jansen, J. K. S. and Matthews, P. B. C. (1962) The effects of fusimotor activity on the static responsiveness of primary and secondary endings of muscle spindles in the decerebrate cat, *Acta Physiol. Scand.*, **55**, 376–386.

Houk, J. C. and Henneman, E. (1967) Responses of golgi tendon organs to active contractions of the soleus muscle of the cat, *J. Neurophysiol.*, **330**, 466–481.

Matthews, B. H. C. (1933) Nerve endings in mammalian muscle, *J. Physiol.*, **78**, 1–53.

Matthews, P. B. C. (1964) Muscle spindles and their motor control, *Physiol. Rev.*, **44**, 219–288.

Matthews, P. B. C. and Simmons (1974) Sensations of finger movement elicited by pulling upon flexor tendons in man, *J. Physiol.*, **239**, 27–28.

Matthews, P. B. C. and Stein, R. B. (1969) The sensitivity of muscle spindle afferents to small sinusoidal changes of length, *J. Physiol.*, **200**, 723–743.

Morgan, D. L., Prochazka, A. and Proske, U. (1984) The after-effects of stretch and fusimotor stimulation on the responses of primary endings of cat muscle spindles, *J. Physiol.*, **356**, 465–477.

Proske, U., Morgan, D. L. and Gregory, J. E. (1992) Muscle history dependence of responses to stretch of primary and secondary endings of cat soleus muscle spindles. *J. Physiol.*, **445**, 81–95.

Ruffini, A. (1898) On the minute anatomy of the neuromuscular spindles of the cat, and on their physiological significance, *J. Physiol.*, **23**, 190–208.

Schaible, H.-G. and Schmidt, R. E. (1983) Responses of fine medial articular nerve afferents to passive movements of knee joint, *J. Neurophysiol.*, **49**, 1118–1126.

Skoglund, S. (1956) Anatomical and physiological studies of knee joint innervation in the cat, *Acta Physiol. Scand.*, **36** (suppl. 124), 1–101.

Vallbo, A. B. and Hagbarth, K.-E. (1968) Activity from skin mechano-receptors recorded percutaneously in awake human subjects, *Exp. Neurol.*, **21**, 270–289.

Vallbo, A. B., Olsson, K. A., Westberg, K.-G. *et al.* (1984) Microstimulation of single tactile afferents from the human hand, *Brain*, **107**, 727–749.

Westling, G. (1986) Sensorimotor mechanisms during precision grip in man. *Umea University Medical Dissertations, New Series*, **171**.

Reflex pathways in the spinal cord 5

5.1 Classification of nerve fibres

Reflexes are most easily observed and analysed when the spinal cord receives a synchronous volley of afferent input. Because of this, the afferent volley usually has been provoked by electrical stimulation of nerves, rather than by natural stimulation of peripheral receptors. The result has been that most reflex stimuli, especially in animal experiments, are described in terms of the intensity of electrical stimulation of the nerve, rather than in terms of which sensory receptors have been activated. Fortunately, in muscle nerves there is a fairly close relationship between the electrical stimulation threshold of a fibre and the sensory receptor which it innervates.

In 1943, Lloyd proposed a system of classification of muscle afferent fibres on the basis of fibre diameter (which is inversely related to electrical threshold). The diameters fall into four main groups, which Lloyd labelled I–IV. The largest are group I (that in the cat have a fibre diameter of 12–21 μm); these also have the lowest threshold to electrical stimulation. They are followed by group II (6–12 μm), group III (1–6 μm) and the unmyelinated group IV. The largest group I fibres (Ia) arise from primary spindle endings. Golgi tendon organs have a slightly smaller average diameter (Ib), although there is considerable overlap with the Ia spectrum. Spindle secondary endings have fibres of the group II class.

Unfortunately, the afferents in cutaneous nerves are not classified by the same system. They share a classification with somatic and autonomic motor fibres which was originally based, not on histological measurement of fibre diameter, but on the various peaks which could be identified in the compound nerve action potential. Three main peaks (A, B and C) can be distinguished corresponding to fibres conducting at different velocities. Histologically, myelinated somatic axons correspond to the fastest conducting A fibres; myelinated autonomic axons to the slower conducting B fibres; and unmyelinated axons to the slowest conducting C fibres. The axons which comprise the A group have a wide spectrum of fibre diameter (1–22 μm) and were originally subdivided into Aα, Aβ, Aγ and Aδ in descending order of conduction velocity. However, the subgroups Aβ and Aδ were probably artefacts of the recording technique and are not now used. In the cat, the Aα category have fibre diameters in the range of 6–17 μm and the Aδ from 1–6 μm. Thus Aα cutaneous fibres correspond approximately to muscle fibres in groups I and II; Aδ to group III.

In the cat, group I muscle afferents have larger diameters than the corresponding Aα fibres. As a result, group I muscle afferents have the lowest threshold and fastest conduction of all peripheral nerve fibres. In man, there is little difference between the two: the largest muscle afferents conduct at the same velocity (and have the same threshold) as the largest cutaneous afferents and motor efferents. In man, the maximum conduction velocities range between 50 and 70 m s^{-1}, whereas in the cat, group I fibres have velocities up to 120 m s^{-1}.

5.2 *Anatomy of group I and II projections in the spinal cord*

Using the horseradish peroxidase (HRP) method it has become possible to visualize directly the localization of Ia afferent terminals within the spinal cord. Single Ia afferent fibres are identified electrophysiologically in the dorsal root of anaesthetized cats by their rapid conduction velocity (> 80 m s^{-1}) and their responses to muscle stretch and muscle contraction. The fibres are then injected with a small amount of HRP, and the animal maintained for the next 12 hours or so.

Over this period, the HRP is transported both anterogradely (towards the axon terminals) and retrogradely (towards the cell body in the dorsal root ganglion). The cat is then killed and the HRP is located by means of its reaction product with diaminobenzidene. This appears as a dark granule in the light microscope, and allows identification of all the synaptic terminals of that axon. HRP stain is visible at the electron microscope level, allowing even finer localization of the sites of synaptic termination.

Figure 5.1, A and B show the pattern of termination of single Ia fibres from the triceps surae muscle of the cat, reconstructed from serial sections of spinal cord. Soon after entry into the spinal cord, the axon bifurcates into an ascending and descending branch which travel in the dorsal columns. Every millimetre or so, over one or two segments either side of the point of entry, the axon gives off collaterals which descend

Afferent Fibres

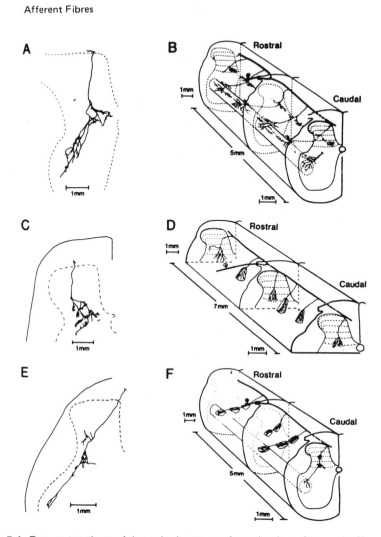

Figure 5.1 Reconstructions of the spinal pattern of termination of group Ia (A and B), group Ib (C and D) and group II (E and F) muscle spindle afferent fibres. A, C and E show transverse sections of the cord with course and branching pattern of a single HRP-stained axon. B, D and F are three-dimensional representations showing the overall pattern of projection. A single fibre is shown entering the dorsal root and bifurcating into an ascending and descending axon with terminals in specific laminae of the spinal grey matter. (From Fyffe, 1984, Figure 7; with permission.)

into the dorsal horn. There are three main areas of termination: Rexed's lamina VI, the intermediate region (lamina VII), and lamina IX. Lamina VII is just dorsal and medial to the motor nuclei, whereas lamina IX comprises the motor nuclei themselves. In the latter, the synaptic contacts

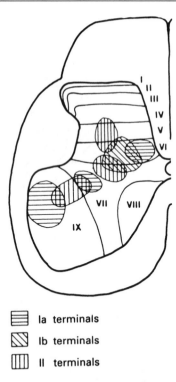

▤	Ia terminals
▨	Ib terminals
▥	II terminals

Figure 5.2 Summary diagram of main areas of termination of muscle afferent fibres from cat lateral gastrocnemius near their point of entry into the spinal cord. (From Brown, 1981, Figure 13.7; with permission.)

appear to be made on to the soma and proximal dendrites of identified α-motoneurones. Ib fibres have a more restricted pattern of termination. Like Ia fibres, their axons bifurcate on entering the cord, giving off a rostral and caudal branch running in the dorsal columns, but the main area of termination is in the intermediate laminae V–VII, as fan-shaped arborizations (Figure 5.1, C and D)

The anatomy of group II spindle afferents has been studied in detail. Other categories of group II fibre have not been investigated so well. The spindle secondary afferents have a more variable morphology of termination than those of the Ia and Ib fibres. In general, they bifurcate into a rostral and caudal branch on entry to the cord. Collaterals then descend to terminate in three main regions: laminae IV–dorsal VI, lamina VI and dorsal VII and, in the ventral horn, lamina VII and IX (Figure 5.1, E and F). Figure 5.2 compares the regions of termination of group Ia, Ib and II fibres (Brown, 1981).

5.3 *Reflex pathways from Ia muscle spindle afferents*

Monosynaptic excitation

The monosynaptic reflex from Ia afferents to motoneurones is probably the best known of all spinal reflexes (Figure 5.3). Proof that only a single synapse was involved came from latency measurements in the 1930s and 1940s. Electrical stimulation of the dorsal root at low intensities can produce an efferent volley in the ventral root with a delay of about 1.5 ms. Increasing the stimulus intensity increases the size of the response and recruits later (probably polysynaptic) events, but does not change the latency of the earliest volley. This figure of 1.5 ms includes both the synaptic delay and the conduction time of the fibres within the cord.

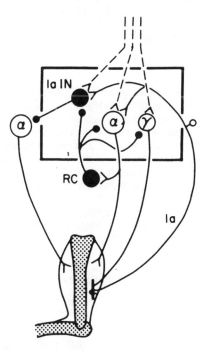

Figure 5.3 Schematic diagram showing principal connections of Ia afferents and Renshaw cells (RC). Inhibitory interneurones are in black. Ia afferents project monosynaptically to the homonymous α-motoneurones and disynaptically, via the Ia inhibitory interneurone (IaIN) to the α-motoneurones of the antagonistic muscle. Renshaw cells are excited by axon collaterals of α-motoneurones and project back to the same motoneurones, the γ-motoneurones and the associated Ia inhibitory interneurones. The box around the Ia inhibitory interneurones, and α and γ-motoneurones emphasizes that these neurones receive very many of the same input connections. Dotted lines represent input from higher centres. (From Hultborn, Lindstrom and Wigstrom, 1979, Figure 1; with permission.)

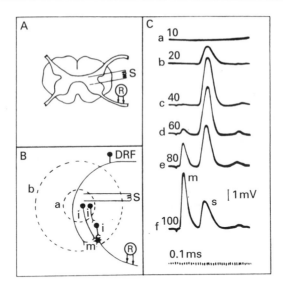

Figure 5.4 Renshaw's method of estimating synaptic delay in the spinal cord. A shows arrangement of stimulating (S) and recording (R) electrodes in the grey matter and ventral roots. B shows spread of stimulating current at low and high intensities (dotted circles a and b, respectively). At low intensities, the axons of dorsal root fibres (DRF) or local spinal interneurones (i) might be activated in the vicinity of the stimulating electrodes. Motoneurones (m) could only be excited trans-synaptically. At high intensities, the motoneurones might be excited directly. C shows the records obtained from the ventral root at different stimulus intensities from 10–100 mV. At high intensities, two responses can be distinguished labelled as m (direct motoneurone stimulation) and s (trans-synaptic motoneurone activity). A very small third wave is present which might have been produced by disynaptic excitation of motoneurones. (From Patton, 1982, Figure 8.4; with permission; after Renshaw, 1940.)

In order to remove the latter factor from estimation of the synaptic delay, Renshaw (1940) performed an experiment in which he estimated the minimal synaptic delay needed to activate a spinal motoneurone. He inserted an electrode into the grey matter of the cord just dorsal to the motor nuclei of the ventral horn. Stimulating here, he found that it was possible to activate the motoneurones at two different latencies (Figure 5.4). At low intensities, there was a delay of about 1 ms before impulses emerged through the ventral root, whereas the delay was reduced to 0.2 ms with much higher intensities. Renshaw interpreted this experiment as follows. At low intensities, only those axons near the stimulating electrode were excited. These fibres then activated motoneurones synaptically, and produced the ventral root volley with a delay of 1 ms. At high intensities, the motoneurones probably were stimulated directly by current spread and produced the ventral root volley with a delay of 0.2 ms. Thus, the difference in latencies (that is, 0.8 ms) between high and low intensity stimulation represents the minimum synaptic delay time. Such a short delay is consistent with monosynaptic transmission. Given that the minimum interval between dorsal root stimulation and ventral root output is 1.5 ms, then only one synaptic relay can be involved.

During the course of such experiments, the conduction velocity of the responsible afferent fibres was measured and found to be of the order (in the cat) of 120 m s^{-1}, which is compatible with the largest muscle afferent (Ia) fibres. (In humans, the Ia fibres conduct more slowly at up to 80 m s^{-1}.) However, the final proof that it was muscle spindle receptors which had monosynaptic access to the motoneurones, rather than any other type of receptors with large afferent fibres, only came some 15 years later, after the advent of intracellular recording. Natural muscle stretch, small enough to activate only primary spindle endings, finally was shown to produce monosynaptic excitatory postsynaptic potentials (EPSPs) in spinal motoneurones. The shape of these EPSPs, in terms of rapid rise time and short decay is consistent with the anatomical location of the Ia terminals on the soma and proximal dendrites of the motoneurones, as shown by the HRP method.

Intracellular recording has been used to investigate the distribution of Ia monosynaptic effects within the spinal cord motor nuclei. The distribution can be quite wide, consistent with the wide anatomical distribution of terminals within the cord. Ia fibres from one muscle provide **homonymous** excitation to motor neurones innervating the parent muscle, and **heteronymous** excitation to those supplying other muscles. At one time it was thought that heteronymous projections were limited mainly to muscles which acted as mechanical synergists at the same joint as the parent muscle. This group of muscles was referred to as the Ia synergists. However, the situation is more complex. For example, heteronymous monosynaptic input can span more than one joint: there are strong connections in both man and the cat between the quadriceps muscles in the thigh and the soleus muscles in the calf. Conversely, some muscles acting at the same joint are not connected. In the cat, soleus receives input from both medial and lateral gastrocnemii, but medial gastrocnemius receives very little input from soleus. In man the situation is reversed: there are large numbers of connections from soleus to the medial gastrocnemius and very few in the reverse direction. Although the homonymous Ia excitation of a motoneurone is usually larger than that from other sources, the efficacy of heteronymous Ia inputs can vary considerably. In the muscles of the baboon's forearm, the heteronymous Ia EPSPs evoked in flexor digitorum communis are considerably larger than those produced by homonymous stimulation.

Polysynaptic excitation

Besides the well-studied monosynaptic excitation, there are, of course, many other routes by which the Ia afferent fibres can influence the activity of spinal motoneurones. Those limited to spinal cord circuitry are discussed in this chapter; supraspinal pathways are covered in Chapter 8.

The more complex pathway which such effects traverse, the more susceptible they are to changes in the state of the animal or to changes in the levels of other afferent inputs. They probably have a different function from that of the monosynaptic pathway since there is more scope for interaction with other descending and spinal input.

Disynaptic inhibition

The other major reflex action of Ia afferents is the disynaptic inhibition of antagonist motoneurones. This was investigated extensively by Lloyd in the 1940s using the technique of monosynaptic testing. Single, low-intensity electrical stimuli were given to the muscle nerve supplying gastrocnemius in order to produce monosynaptic excitation of the gastrocnemius motoneurones in the spinal cord. The size of the reflex was monitored by recording the size of the ventral root discharge at various times before and after a conditioning stimulus was applied to the peroneal nerve which supplies the antagonist anterior tibial muscles. Peroneal nerve stimulation produced inhibition of gastrocnemius monosynaptic reflex, with maximum inhibition occurring if the peroneal nerve was stimulated some 0.5 ms before the gastrocnemius nerve. Thus, for inhibition to occur, the antagonist nerve volley had to arrive at the spinal cord before the agonist test volley.

Araki, Eccles and Ito (1960) later used intracellular microelectrodes to show that the latency of the Ia inhibitory postsynaptic potential (IPSP) produced by stimulation of antagonist nerves was some 0.8 ms longer than the onset of an EPSP evoked by stimulation of homonymous afferents in nerve filaments from the same dorsal root. This delay was attributed to the presence of an interposed interneurone in the inhibitory pathway, and the effect became known as **disynaptic** inhibition (see Figure 5.2). The extra time delay caused by an extra synapse is due to: (1) slowing of nerve impulses in the fine terminals of the presynaptic afferents; (2) the time taken for the postsynaptic neurone to reach firing threshold; and (3) the actual time taken for transmitter to diffuse across the synaptic cleft (which is of the order of 0.2 ms).

Intracellular recording has revealed an extensive distribution of Ia inhibitory effects. A given motor nucleus draws its Ia inhibition mainly from its mechanical antagonists and also from other muscles connected to them in Ia synergism. This is known as **reciprocal Ia inhibition** because, in many instances, the inhibition to antagonist muscles is reciprocated by a similar inhibition from the antagonist Ia afferent fibres. Interestingly, there is no reciprocal inhibition between adductors and abductors.

The spinal interneurone which mediates disynaptic Ia inhibition has been termed the **Ia inhibitory interneurone**. It is one of the few interneurones whose input and output connections have been analysed in detail. Many other types of afferent fibres, besides the Ia afferent fibres converge onto it and will be discussed below. In electrophysiological experiments in which cells are impaled by microelectrodes in the spinal cord, the Ia inhibitory interneurone is identified by the fact that it is the only interneurone which (1) produces monosynaptic inhibition of motoneurones, and (2) receives monosynaptic Renshaw inhibition (see Section 5.6 on the Renshaw cell).

The anatomical location of these interneurones was demonstrated in an elegant series of experiments by Jankowska and Lindstrom (1971). After considerable effort, they succeeded in introducing a microelectrode into neurones in Rexed's lamina VII and identified some cells which received monosynaptic excitation from Ia fibres, and disynaptic inhibition from motoneurones (via Renshaw cells). These cells, in turn, could be stimulated and shown to evoke a monosynaptic inhibition of spinal motoneurones. Hence they are prime anatomical candidates for the physiologically identified Ia inhibitory interneurones. They lie in the second main region of Ia afferent fibre termination, dorsal and medial to the motor nuclei.

Convergence in the Ia inhibitory pathway

Because the Ia inhibitory interneurones lie in well defined disynaptic pathways within the spinal cord, and are activated by the largest diameter afferent fibres, they have proved relatively easy to study. They are one of the best characterized interneurones of the cord. They receive input from an enormous number of systems, including the corticospinal tract, flexor reflex afferents, cutaneous afferents, Ia afferents and Ia inhibitory interneurones of antagonist muscles.

Most of these inputs have a common pattern of action as illustrated in Figure 5.5. Thus, they produce excitation of agonist α- and γ-motoneurones whilst at the same time projecting via the Ia inhibitory interneurone to inhibit the antagonist α-motoneurone.

Because of this common pattern of connections, α-motoneurones and linked Ia inhibitory interneurones (and γ-motoneurones) are said to form a functional **unit** in the spinal cord. Inputs to this unit produce α–γ coactivation and reciprocal inhibition of antagonist muscles. This is

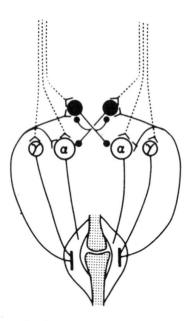

Figure 5.5 Schematic drawing of parallel inputs (dotted lines) to α- and γ-motoneurones and Ia inhibitory interneurones (larged filled circles) of one muscle. Note mutual inhibition of 'opposite' Ia inhibitory interneurones. (From Hultborn, Illert and Santini, 1976, Figure 3; with permission.)

termed **α–γ coactivation in reciprocal inhibition.** The result is that inputs to motoneurones, such as homonymous Ia afferents, or corticospinal tract projections, not only activate the α-motoneurone: they also excite the Ia inhibitory interneurones, which secondarily produce disynaptic inhibition of antagonist motoneurones and the 'opposite' Ia inhibitory interneurones. In addition, γ-motoneurones are excited by some inputs (for instance, descending corticospinal tract inputs). With this combination of excitation and inhibition (which is relatively 'hard-wired' in the spinal cord), isotonic contraction of the agonist muscle will be accompanied by fusimotor activity to maintain Ia input from shortening spindles. Also, although the antagonist muscle will be stretched, Ia firing will not evoke unwanted reciprocal inhibition of the agonist, since the antagonist Ia inhibitory interneurones are turned off. In addition, Ia input from the stretched antagonist is unlikely to evoke any monosynaptic excitation of antagonist motoneurones because they are actively inhibited by the agonist Ia inhibitory inter-neurones (Hultborn, 1976; Hultborn, Illert and Santini, 1976; Hultborn and Pierrot-Deseilligny, 1979; review by Baldissera, Hultborn and Illert, 1981).

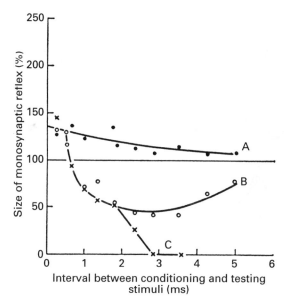

Figure 5.6 Laporte and Lloyd's demonstration of Ib inhibition from synergist muscles. Monosynaptic reflex volleys were recorded in the ventral root after low intensity stimulation of the nerve to the plantaris muscle of a cat. The ventral roots were cut distal to the recording site to prevent antidromic invasion of the spinal motoneurones from activity in motor fibres. The test stimulus was preceded by a conditioning stimulus to the synergist muscle, flexor digitorum longus. Curve A shows the effect of a very weak conditioning stimulus. This was assumed to be mainly Ia in origin, and gave a pure monosynaptic facilitation. Curve B was obtained with slightly higher stimulation intensities. It was presumed to activate Ib fibres as well as Ia and inhibited the monosynaptic reflex with a latency some 0.5 ms longer than that needed to produce monosynaptic excitation. The curve represents the effect of Ib disynaptic inhibition combined with Ia monosynaptic facilitation. Curve C was obtained with yet stronger stimuli which probably activated group II fibres (from Laporte and Lloyd, 1952, Figure 3; with permission.)

5.4 *Reflex pathways from Ib tendon organ afferents*

Laporte and Lloyd (1952) made the first systematic study of Ib effects by using graded electrical stimulation of single muscle nerves. Using monosynaptic testing, they evoked reflexes in motoneurones of agonist and synergist muscles. They then measured the effect on the size of the evoked ventral root volley of single graded stimuli applied to a synergist muscle nerve at different times before and after the test stimulus. Weak stimuli to a synergist muscle nerve produced only facilitatory effects on the monosynaptic reflex. These were interpreted as being due to heteronymous, monosynaptic, Ia excitatory projections. However,

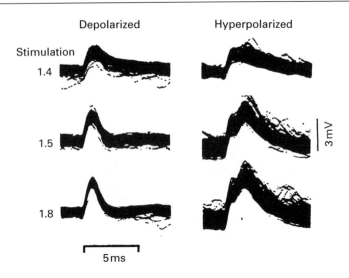

Figure 5.7 Homonymous Ib inhibitory effects shown by intracellular recordings from a motoneurone supplying the gastrocnemius muscle of a cat. The ventral roots were cut in this preparation to prevent antidromic invasion of the motoneurone. Stimulation at low intensities (1.4 times group I threshold) produces a fairly pure EPSP in the records on the left. At higher intensities (1.8 times threshold activates almost all the group I fibres), the effect of the Ib IPSP manifests itself as a more rapid return of the initial EPSP towards baseline. In the records on the right, the cell was held hyperpolarized by passing a steady current through the microelectrode. The amount of hyperpolarization was greater than the reversal potential of the IPSP. In these records, the Ib IPSP can be seen as a second depolarizing wavelet superimposed on the EPSP. (From Eccles, Eccles and Lundberg, 1957, Figure 1; with permission.)

when the intensity of the conditioning stimulus was increased slightly, an inhibition was revealed superimposed on the time course of facilitation. This inhibition began some 0.5–1 ms later than the facilitation, and was attributed to a disynaptic inhibition produced by Ib figures (figure 5.6).

Stimulation of antagonist muscle nerves at these intensities was found to produce a di- or trisynaptic facilitation of the agonist. They termed such effects the **inverse myotatic reflex**, since the Ib actions seemed to be the exact opposite of those of the Ia fibres. Laporte and Lloyd also investigated the distribution of these effects, and found them to be particularly powerful from the Ib fibres of extensor muscles. Their actions may be summarized as: (1) disynaptic inhibition of motoneurones projecting to synergists, and (2) di- or trisynaptic excitation of motoneurones to antagonists.

Microelectrode studies confirmed that Ib fibres could produce inhibitory effects on homonymous motoneurones (Figure 5.7). On the basis of latency measurements, these effects were confirmed to be mediated by di-(or sometimes tri-) synaptic actions. However, such studies also revealed that the pattern of Ib projections was more widespread than originally envisaged by Laporte and Lloyd. The most recent work has shown that the Ib projections may be much more diverse than those of the Ia fibres. They may span more than one joint and often can be found to have actions opposite to those described as the inverse myotatic response. Such physiological findings of widespread action pose considerable problems in the interpretation of the role of Ib afferents in movement control (see Chapter 6).

The source of some of the interneurones within the Ib pathways has been localized to within Rexed's lamina V and VII, the area of the main Ib termination in the spinal cord. Many of these cells are monosynaptically excited by Ib fibres and project into the motor nuclei of the ventral horn. They are a different set of neurones to those mediating the Ia reciprocal inhibition and have been termed **Ib interneurones**. (Brink *et al.*, 1983 give details of further recent work, and Baldissera, Hultborn and Illert 1981, give a complete electrophysiological review.)

Convergence onto Ib interneurones

Like the Ia inhibitory interneurones, the Ib interneurones receive input from a wide range of sources. Of particular interest is the finding that Ib interneurones are facilitated at short latency by low threshold cutaneous and joint afferents. One possible consequence of this would be that if a limb movement were suddenly interrupted by a physical obstruction, the cutaneous input might facilitate autogenic inhibition of the agonist muscles and interrupt the contraction. This is not necessarily in contradiction to the servocontrol of movement via the stretch reflex. It is simply part of the general hypothesis of this work on spinal cord mechanisms that under certain conditions, reflexes in different pathways might prevail. A relevant example might be that in reaching out to touch a delicate Meissen statuette, it might not be advisable for the CNS to employ the servo assistance mechanism to overcome the obstacle. Facilitation of the Ib pathway might be judged much more advisable in this instance.

Joint afferent input may also be of some functional importance. From the study of joint receptors, it is known that joint afferent discharge is particularly prominent at the extremes of joint movement. It may be that their action in facilitating the Ib interneurones could decrease agonist muscle force once the limit of a movement is approached. At the same time, the excitatory effects of Ib interneurones onto antagonist muscles might also contribute to this 'braking' process.

Many descending supraspinal systems converge onto the Ib

interneurones, allowing the possibility of some kind of control of these pathways under different conditions. Recently, it has also been discovered that Ia afferents project onto the Ib interneurones, although the functional significance is not clearly understood.

5.5   *Reflex pathways from group II muscle afferents and 'flexor reflex afferents'*

Group II muscle afferent pathways

The earliest work on the group II projection was performed, like that of other afferent systems, using graded electrical stimulation of muscle nerves in combination with monosynaptic testing of the motoneurones of other muscles. In contrast with the group I effects, the group II effects from flexor and extensor muscles of the cat hindlimb were found to be the same. Stimulation of group II fibres from either set of muscles produced excitation of flexors and inhibition of extensors. Intracellular recording showed that the excitatory effects were produced by a disynaptic pathway and the inhibitory effects by a trisynaptic pathway. Some of the excitatory interneurones appeared to be located in the ventral part of Rexed's lamina VII, in a region of dense termination of group II afferents.

The experiments were performed on the hindlimb of spinalized cats. Quite different results are seen in cats lesioned at other levels of the central nervous system. In decerebrate animals, the interneurones mediating the group II muscle spindle reflexes are almost completely inhibited and no group II reflexes can be obtained. In contrast, after lower pontine lesions, group II inputs evoke effects opposite to those seen in spinal animals: flexor muscles are inhibited rather than excited. The conclusion is that there is more than one reflex pathway from group II fibres to motoneurones, and that they can be selected differentially by supraspinal inputs.

Group II afferent fibres, even in muscle nerve, do not come exclusively from one type of receptor. Muscle spindle secondaries, joint receptors, Pacinian corpuscles and other receptors all have afferent fibres in the group II range. This complicates the interpretation of any experiment in which group II fibres have been investigated purely by electrical stimulation. Matthews (1969) has argued that group II afferents from muscle spindle secondary endings have different actions from those predicted from electrical stimulation of whole muscle nerve. Thus, electrical stimulation techniques suggest that only primary spindle endings contribute to the stretch reflex in extensor muscles. The secondaries, like all group II afferents, should only produce inhibition

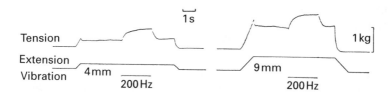

Figure 5.8 Demonstration of a possible role for group II spindle secondary endings in production of the stretch reflex in the soleus muscle of a decerebrate cat. On the left, the muscle was stretched by 4 mm to produce a background stretch reflex and then vibration was applied to the tendon in order to evoke further excitation of the primary group Ia afferents and a larger reflex tension. On the right, a larger stretch of 9 mm was given so as to produce a reflex tension equivalent to that seen when the vibrator was applied on the left. If all spindle primary endings had been 'driven' by vibration, then it should be impossible to exceed this tension by applying vibration in the experiment on the right. However, this was not found to be the case. There was no occlusion between stretch and vibration. The response to vibration was similar despite the different levels of stretch reflex contraction. (From Matthews, 1969, Figure 3; with permission.)

of the extensor muscles. In order to test this hypothesis using more natural stimulation, Matthews examined the effect of two different types of muscle spindle stimulation on the size of the stretch reflex recorded in the cat soleus muscle.

In one experiment, the muscle was held slightly stretched, producing a weak stretch reflex contraction. Vibration at 200 Hz was then applied to the tendon. Since this is an extremely effective stimulus for primary endings (especially in the cat), Matthews argued that it would make all the primaries in the muscle fire at the same frequency. The effect was a marked increase in the reflex muscle tension. Next, without vibration, the muscle was stretched to a point where the reflex tension equalled that seen with vibration in the previous experiment. On the basis that only group Ia fibres contribute to extensor muscle excitation, it was argued that with this new degree of muscle stretch, they should all be discharging at 200 Hz. However, when a vibrator was applied, the tension rose even further above this level (Figure 5.8). Matthews concluded from this that the spindle secondary endings must have be contributing to the stretch-imposed reflex contraction. This idea of group II mediated autogenetic excitation of muscle received further support from later anatomical and physiological data showing that some secondary spindle endings can have monosynaptic connections onto spinal α-motoneurones. The pattern of group II afferent terminations in the spinal cord (see Figure 5.2) is consistent with this.

However, the findings of Matthews' experiments have been questioned by other workers. For example, Jack and Roberts found that vibration was not always capable of driving all Ia endings maximally.

Some of them seem to be unloaded during the very contraction which they evoke. Only by applying a large stretch to the muscle can driving of primaries be guaranteed. Thus they argue that Matthew' results are explicable on the basis of incomplete vibratory driving of primary endings in the first part of his experiment.

In great contrast to the detailed information available about the operation of the spindle secondary ending, it is still unclear whether any of the actions of group II spindle afferents predominate. In the intact animal, it seems likely that the actions will vary according to the control of spinal interneuronal machinery.

Flexor reflex afferent (FRA) pathways

The group II muscle afferent pathways above are sometimes regarded as part of a much larger system of reflex pathways termed the flexor reflex afferent (FRA) system. The arguments for this classification depend upon results obtained using graded electrical stimulation of afferent fibres rather than natural stimulation of one or other type of peripheral receptor. In the spinal cat, electrical stimulation of a large class of groups II and III muscle, cutaneous and joint afferents all produce the same action: excitation of ipsilateral flexors and inhibition of ipsilateral extensors. Weaker, opposite effects can be observed in muscles on the contralateral side of the body. That is, excitation of extensors and inhibition of flexors. This is sometimes described as ipsilateral flexion accompanied by crossed extension. In the intact, animal, these reflexes would withdraw the limb from the stimulus, while at the same time increasing the support provided by the contra-lateral limb to bear the weight of the animal. Because these different classes of afferents converged onto the same interneurones, they were given the collective name of FRA.

The problem with this terminology is that it is easy to assume that the group II and III muscle and cutaneous/joint afferents **only** produce the flexion reflex. This is quite untrue. These afferents can have other actions. Precisely which action they have depends upon which particular reflex pathways are facilitated in the preparation being studied. As described for the group II afferents above, in the mid-collicular decerebrate cat, the flexion reflex is virtually suppressed, whereas this is facilitated in the spinal preparation. There are also many reflexes mediated by cutaneous afferents which do not fit into the general category of FRA action. For example, gentle pressure on the cushion of the cat hindlimb will activate muscles which plantar flex the toes, whereas a pin-prick applied to the same area would produce withdrawal of the foot by dorsiflexion. Thus the quality of the stimulus (and, in other cases, the point of application) is an important feature in determining the type of reflex response.

Despite these arguments, there are two points about the FRA system that deserve mention. First, the FRA interneurones provide a strong input into several ascending sensory systems, including the spinocerebellar pathways in addition to their actions on motoneurones. The purpose of this, which is probably also repeated in other systems, is to inform higher centres of the state of the spinal interneuronal machinery. Thus, not only do supraspinal structures receive direct sensory input about the external environment, they also receive information about what is happening within the spinal cord itself.

The second feature of the FRA system, which also appears to be important in spinal man, is that activity of FRAs can produce flexion reflexes through two different pathways, one acting with short latency, and the other at long latency. The short latency pathway has a central delay in the cat of only a few milliseconds. In humans, the total reflex latency in human leg muscles is about 60 ms. The long latency pathway has a central delay in the cat of 30–50 ms. In humans it produces reflexes in the leg with minimum latencies of 100 ms or more. There is a peculiar interaction between the short and long latency reflexes such that the short pathway inhibits the long pathway. So, if a long train of stimuli is given to the FRAs, the latency of the early response remains the same as if only a short train were given. However, the onset of the late response is delayed, by an amount related to the length of the train.

In spinal cats, the late response is usually difficult to obtain, but becomes much more conspicuous after treatment with monoaminergic precursors such as L-dopa. At the same time, transmission in the short latency pathway is suppressed (Figure 5.9). When the late responses are prominent it is possible to demonstrate one of the most interesting features of their organization. First, we must remember that stimulation of FRAs in one leg produces flexion of the ipsilateral leg coupled with extension of the contralateral leg. If a second stimulus is given later to the contralateral leg, we find that the flexion response in that leg has been suppressed by the preceding crossed extensor response after the first stimulus. In other words, late flexion reflexes on one side inhibit development of flexion reflexes on the other. As time passes, the inhibition wanes, and flexion reflexes can be elicited by the second stimulus.

The point about this is that here we have a system with powerful interaction between the two legs which many people have compared with a primitive 'stepping generator.' Imagine giving a natural FRA stimulus to one leg. As the leg flexes and is lifted from the ground the FRA input declines and the reflex wanes. Conversely, input from FRAs will increase in the opposite leg during the period of extension. At first, this FRA input will be unable to produce its own flexion response because of inhibition remaining from the initial stimulus. Later, though, flexion will begin to occur and any remaining flexor activity in the first leg will be inhibited and be replaced by extension. Thus the system can produce a type of walking, with flexion of one leg being followed at an interval by flexion of the other etc. (Baldissera, Hultborn and Illert, 1981; Lundberg, Malmgren and Schomberg, 1987; give a review of this topic.)

Before L-dopa

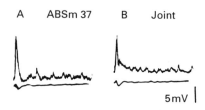

After L-dopa 100 mg^{-1}

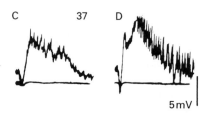

Recovery

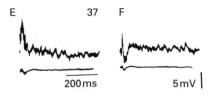

5.6 *The Renshaw cell*

In the 1940s, Renshaw demonstrated that antidromic impulses in motor axons could reduce the excitability of α-motoneurones projecting to the same or synergistic muscles. This phenomenon was called **recurrent inhibition** and has been shown to be due to activation of a group of interneurones by motor axon collaterals. These interneurones, named Renshaw cells after their discoverer by J. C. Eccles, have been identified electrophysiologically and lie in the ventral horn of the spinal cord medial to the motor nuclei. They receive monosynaptic excitation from α-motoneurones and have a monosynaptic inhibitory connection back onto the homonymous and synergistic motoneurones (see Figure 5.2). Each Renshaw cell may receive collaterals from many α-motoneurones. However, because such collaterals only distribute within a distance of less than 1 mm from their parent cell bodies, the motoneurone input is relatively local.

Renshaw inhibition is distributed not only to motoneurones but to homonymous and synergistic γ-motoneurones and Ia inhibitory inter-neurones, as well as to other Renshaw cells. The projection to the Ia inhibitory interneurone is unique since it is the only spinal interneurone interposed in a direct pathway to α-motoneurones which also receives monosynaptic Renshaw inhibition. The convergence of Renshaw and Ia afferents onto Ia inhibitory interneurones is used as a means of identi-fying these cells in electrophysiological experiments.

The duration of the Renshaw-mediated antidromic IPSP motoneurones is relatively long, 40–50 ms or more as against about 15 ms for disynaptic Ia inhibition (see above). The reason for this is that a single antidromic volley in a motor nerve produces a burst of high frequency discharges in Renshaw cells. The duration of the bursts depends upon the intensity of stimulation. A comparison of the intra-cellularly recorded IPSP in a motoneurone and the firing of a Renshaw cell is shown in Figure 5.10.

Figure 5.9 Actions of flexor reflex afferents (FRA) on α-motoneurones before and after activation of the reticulo-spinal noradrenergic pathways by intravenous injection of L-dopa. Upper traces of each pair are intracellular records from three different motoneurones innervating the posterior biceps-semitendinosus muscle (AB;CD;EF). Lower traces are records from the dorsal root entry zone of the L7 segment. Records were taken after short trains of stimuli were given either to the anterior biceps-semimembranosus (ABSm) nerve at 37 times the threshold for the largest afferent fibres, or after stimulation of the posterior knee-joint nerve. A and B show predominant polysynaptic short latency EPSPs from FRAs in the absence of L-dopa. C and D show the depression of short latency effects and appearance of long latency excitation (compare A and C or B and D). Lower records show the appearance of short latency components about two hours after the administration of L-dopa. Note the relatively long time scale of these records. (From Jankowska *et al.*, 1967.)

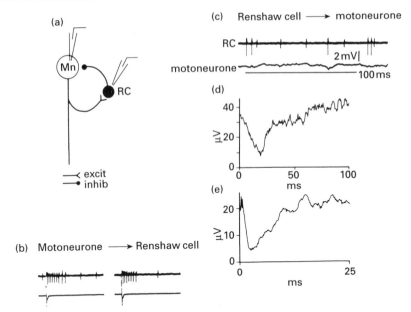

Figure 5.10 Relations between individual motoneurones and individual Renshaw cell. (a) Illustration of experimental arrangement and neuronal circuit. The excitatory input of single motoneurones (Mn) onto individual Renshaw cells (RC), and the inhibitory effect of the Renshaw cell onto the motoneurone was investigated with two microelectrodes: one used for intracellular stimulation and recording from the motoneurone, and the other used for extracellular recording from a Renshaw cell. (b) Two samples of burst firing in a Renshaw cell (upper traces, large spikes) following a single action potential in the motoneurone (see spike at start of lower trace). (c) Slow repetitive firing of a Renshaw cell (upper trace) and its effect on the intracellular membrane potential of a motoneurone (lower trace). Time scale in (c) is the same as for (b); (d) Average change of membrane potential in the motoneurone following each discharge of the Renshaw cell. This is the same motoneurone/Renshaw cell coupling as shown in (b) and (c). Note the remarkably long time course of this unitary IPSP. (e) A more typical time course of a presumed unitary IPSP from a Renshaw cell in a different motoneurone. (From Baldissera, Hultborn and Illert, 1981, Figure 2; with permission.)

The function of the Renshaw cell is still a matter for debate. In the most general terms, its inhibitory feedback would tend to reduce the average resting potential of motoneurones. This would (1) reduce the frequency of α-motoneurone discharge below that expected in a system without recurrent inhibition, and (2) reduce the sensitivity of motoneurones to excitatory inputs. Some authors have suggested that with these actions, Renshaw cells might function to enhance the 'contrast' within a motoneurone pool. It would be a function analogous to the lateral inhibitory mechanisms of sensory systems.

To understand this idea, imagine what happens when a Ia discharge

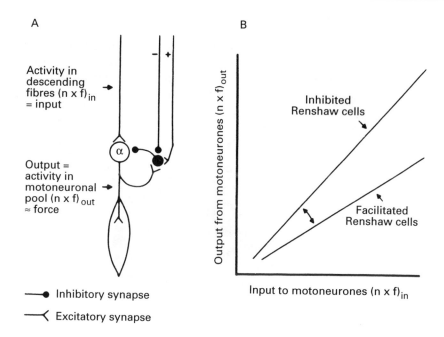

Figure 5.11 Diagram illustrating the hypothesis that Renshaw cells might serve to regulate the 'gain' of motoneurones. A, input and output connections of motoneurone with supraspinal (−, +) control of Renshaw cells. B, graph showing how facilitation of Renshaw cells will decrease the slope of the input-output relationship of motoneurones, making them less sensitive to changes in size of the input signal. (From Hultborn, Illert and Wigstrom, 1979, Figure 1; with permission.)

is produced by stretch of a single muscle. The Ia terminals are widely distributed to homonymous and heteronymous motoneurones, and the excitatory field is fairly large. The homonymous motoneurones would be activated most strongly and would exert strong recurrent inhibition on themselves and on the neighbouring heteronymous motoneurones. If the heteronymous motoneurones had been only weakly excited by the Ia input, then strong Renshaw inhibition might suppress their firing. In doing so, their own recurrent inhibition on the homonymous pool would be removed. The strongly activated motoneurones will therefore receive only their own recurrent inhibition, but because they are highly activated, their output (though diminished) will still be expressed. The nett effect would be that Renshaw cell activity would suppress activity in heteronymous motoneurones and restrict the spatial limits of excitatory inputs. Renshaw cells receive other inputs from supraspinal centres, and it is possible to imagine that the effectiveness of the 'motor contrast' mechanism could be controlled to suit different circumstances.

Another hypothesis of Renshaw cell action relies heavily on the possibility of supraspinal control of Renshaw excitability via the known anatomical projections from various brain centres (Figure 5.11). This views the Renshaw cell as a mechanism for controlling the sensitivity of the α-motoneurone (and Ia inhibitory interneurone) to other excitatory inputs. If the Renshaw cells were facilitated by descending inputs from the brain, the maximum output of the motoneurones would be

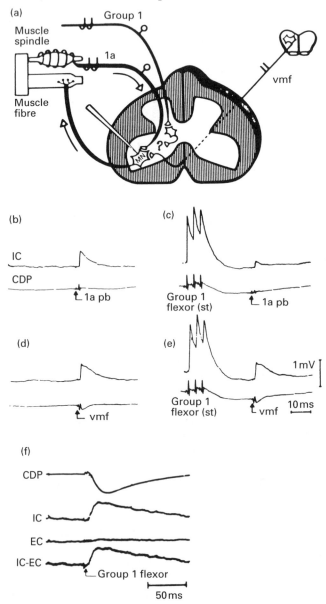

reduced. Strong excitatory inputs to the motoneurone would produce strong Renshaw inhibition and prevent maximum excitation of the cell. To produce a given level of motoneurone output, there would have to be a much greater amount of excitatory input to the motoneurone than there would have been if the Renshaw cells were not facilitated. The slope of the relation between input and output could therefore be controlled by the effectiveness of Renshaw inhibition. With the Renshaw cells facilitated by descending commands, inputs to the motoneurone could play over a large range but cause only small changes in motoneurone output. At the level of the muscle this would mean that resolution of force changes would be high. If the Renshaw cells were inhibited, the motoneurone firing range would be large, and the resolution of muscle force changes by a given input would be reduced (Baldissera *et al.*, 1981).

5.7 *Presynaptic inhibition*

Presynaptic inhibition is a means of changing reflex transmission which can affect even monosynaptic pathways. The effects on the terminals of Ia, Ib and cutaneous afferent fibres have been investigated. All have their synaptic potency changed by presynaptic actions.

Presynaptic inhibition is caused by activity in axo-axonic synapses

---

Figure 5.12 Presynaptic inhibition in Ia afferent fibres. (a) shows the experimental arrangement for demonstrating presumed presynaptic inhibition of Ia afferent fibres. An intracellular electrode records the potential from inside a single motoneurone. This motoneurone can be activated either by stimulating Ia fibres from a muscle spindle in the periphery or by stimulation of the ventromedial fasciculus in the brainstem in order to produce activation of the motoneurone by a descending pathway. The upper records in (b), (c), (d) and (e) are intracellular records from the motoneurone, the lower records are the cord dorsum potential (CDP) recorded from the surface of the spinal cord (not shown on the diagram in (a)). In (b), a single stimulus was given to the posterior biceps nerve at low threshold in order to evoke a monosynaptic EPSP in the motoneurone (the input to the spinal cord can be seen in the cord dorsum potential (Ia pb). In (c) this stimulus was preceded by a train of three stimuli given to group I fibres in the semitendinosus nerve (st). This produced a negative shift of the cord dorsum potential, and a decrease in the size of the monosynaptic Ia EPSP. The same arrangement is shown in (d) and (e), except that the EPSP in the motoneurone was produced by stimulation of the ventromedial fasciculus. In this case, the intracellular record shows that the response is unchanged by a preceding group I flexor volley. In (f), recordings are shown from an intracellular electrode (IC) inserted into the afferent terminal of a Ia fibre. Stimulation of group I flexor nerves produces a depolarization which is not visible in the extracellular records (EC). However, the depolarization is visible in the cord dorsum potential. Note the long time course of this effect corresponding with the long duration of presynaptic inhibition. (From Rudomin, 1990, Figures 1 and 2; with permission.)

near the terminals of the afferent nerve fibres. Release of GABA at these synapses can produce long-lasting depolarization of the afferent terminal, which leads to a decrease in the number of quanta of transmitter released per nervous impulse. It can be detected by placing electrodes over the central end of a cut dorsal root and recording the depolarizing potential which spreads electrotonically from the site of generation in the nerve terminals. This potential is known as the dorsal root potential. An alternative method is to stimulate the terminals within the cord with an electrical impulse. This sets up an antidromic volley which can be recorded in the dorsal roots. Paradoxically, because presynaptic inhibition involves depolarization of nerve terminals, a given stimulus will produce a larger antidromic volley in the presence of presynaptic inhibition.

Both $GABA_A$ and $GABA_B$ receptors on the afferent terminal seem to be involved: presynaptic inhibition is antagonized by bicuculline and picrotoxin (both $GABA_A$ antagonists) and is augmented by baclofen (a $GABA_B$ agonist).

The circuitry involved in producing presynaptic inhibition of group Ia and Ib fibres has been analysed in some detail (Schmidt, 1971; review by Rudomin, 1990). Figure 5.12 shows the situation for group Ia afferents. In cat hindlimb, Ia fibres from either flexor or extensor muscles are presynaptically inhibited by activity in group I fibres from flexor muscles. The records at the bottom were made from a fine electrode inserted into the tip of the Ia axon. They show that presynaptic inhibition is accompanied by depolarization of the afferent terminal.

Presynaptic inhibition in cells can be modulated by other inputs. For example, cutaneous input from the same limb has no direct presynaptic effect on group Ia fibres. However, it can reduce the amount of presynaptic inhibition evoked by group I fibres from flexor muscles. Descending, supraspinal pathways are arranged to have similar effects on presynaptic inhibition of group Ia fibres. Stimulation of the vestibulospinal tract produces clear presynaptic inhibition on its own, which is reduced by stimulation of reticulospinal, rubrospinal or corticospinal pathways. Like the cutaneous effect, none of these three pathways produces any direct presynaptic effects of their own. They presumably modulate activity in the interneurones mediating presynaptic inhibition.

Group Ib fibres have a slightly different organization of presynaptic inhibition than group Ia. They are inhibited by other Ib fibres and also by all four descending systems noted above. Cutaneous input suppresses the presynaptic inhibition of some Ib fibres and facilitates inhibition of others. Presynaptic actions on group II fibres are less clear. They seem to receive inhibition from group I and group II fibres as well as from the corticospinal tract.

Bibliography

Review articles and books

Baldissera, F., Hultborn, H. and Illert, M. (1981) Integration in spinal neuronal systems, in V. B. Brooks (ed.), *Handbook of Physiology*, sect. 1, vol. 2, part 1, Williams and Wilkins, Baltimore, pp. 509–597.

Brown, A. G. (1981) *Organisation in the Spinal Cord,* Springer-Verlag, Berlin.

Fyffe R. E. W. (1984) Afferent fibres, in R. A. Davidoff (ed.), *Handbook of the Spinal Cord*, vols 2 and 3, Marcel Dekker, New York.

Patton, H. (1982) Spinal reflexes and synaptic transmission, in T. Ruch and H. T. Patton (eds.), *Physiology and Biophysics,* Vol. 4, W. B. Saunders, Philadelphia, pp. 261–302.

Rudomin, P. (1990) Presynaptic inhibition of muscle spindle and tendon organ afferents in the mammalian spinal cord. *Trends Neurosci,* **13**, 499–505.

Original papers

Araki, T., Eccles, J. C. and Ito, M. (1960) Correlation of the inhibitory post-synaptic potential of motoneurones with the latency and time course of inhibition of monosynaptic reflexes, *J. Physiol.,* **154**, 354–377.

Brink, E., Jankowska, E., McCrea, D. A. *et al.* (1983) Inhibitory interactions between interneurones in reflex pathways from group Ia and group Ib afferents in the cat, *J. Physiol.,* **343**, 361–373.

Eccles J. C, Eccles, R. M, Lundberg, A. (1957) Synaptic actions on motoneurones caused by impulses by golgi-tendon organ afferents, *J. Physiol.,* **138**, 227–252.

Hultborn, H. (1976) Transmission in the pathway of reciprocal Ia inhibition to motoneurones and its control during the tonic stretch reflex, in S. Homma (ed.), *Progress in Brain Research*, vol. 44, Elsevier, Amsterdam, pp. 235–255.

Hultborn, H. and Pierrot-Deseilligny, E. (1979) Changes in recurrent inhibition during voluntary soleus contractions in man studied by an H-reflex technique, *J. Physiol.,* **297**, 229– 251.

Hultborn, H., Illert, M. and Santini, M. (1976) Convergence on interneurones mediating the reciprocal Ia inhibition of motoneurones, *Acta Physiol. Scand.,* **96**, 193– 201; 351– 367; 368– 391.

Hultborn, H., Lindstrom, S. and Wigstrom, H. (1979) On the function of recurrent inhibition in the spinal cord, *Exp. Brain Res.,* **37**, 399–403.

Jankowska, E. and Lindstrom, S. (1971) Morphology of interneurones mediating Ia reciprocal inhibition of motoneurones in the spinal cord of the cat, *J. Physiol.,* **226**, 805–823.

Jankowska, E., Jukes, M. G. M., Lund, S. and Lundberg, A. (1967) The effect of DOPA on the spinal cord. 5. Reciprocal organisation of pathways transmitting excitatory action to alpha motoneurones of flexors and extensors, *Acta Physiol. Scand.,* **70**, 369–388.

Laporte, Y. and Lloyd, D. P. C. (1952) Nature and the significance of the reflex connections established by large afferent fibres of muscular origins, *Am J. Physiol.*, **169,** 609–621.

Lloyd, D. P. C. (1943) Neuron patterns controlling transmission of ipsilateral hind limb reflexes in cat. *J. Neurophysiol.*, **6,** 111–120; 293–315; 317–326.

Lundberg, A., Malmgren, K. and Schonberg, E. B. (1987) Reflex pathways from group II muscle afferents. Papers 1, 2 and 3. *Exp. Brain Res.*, **65,** 271–306.

Matthews, P. B. C. (1969) Evidence that the secondary as well as the primary endings of the muscle spindles may be responsible for the tonic stretch reflex of the decerebrate cat. *J. Physiol.*, **204,** 365–393.

Renshaw, B. (1940) Activity in the simplest spinal reflex pathway. *J. Neurophysiol.*, **3,** 373–387.

Investigating reflex pathways and their function

6

6.1 *The tendon jerk*

The best known example of a spinal reflex is a tendon jerk. No one knows when this was first discovered, but it was first brought to the attention of scientists in 1875 by Erb and Westphal. Initially the tendon jerk was regarded as the direct response of muscle to percussion. However, when Sherrington demonstrated that the quadriceps jerk in the cat could be abolished by dorsal root section, the reflex nature of the response was confirmed. Later measurements of the conduction velocity of the afferent fibres gave values (in the cat) of up to 120 m s^{-1}, suggesting involvement of group Ia fibres. Indeed, it is now accepted that although the tendon hammer stimulus may stretch muscles by only 50 μm, the rapid onset of stretch makes this a particularly powerful stimulus for the primary spindle endings.

Traditionally, the tendon jerk is regarded as monosynaptic; however, other pathways also may contribute to the response. Microneurographic recordings in the nerves of the leg following a tendon tap to the ankle show that in man the muscle spindle afferent volley may be dispersed by 20 to 30 ms. This dispersion is due to (1) double firing of some receptors; (2) dispersion in the conduction velocity of the Ia afferents (from about 40–60 m s^{-1}); and (3) differences in the times at which individual receptors discharge their first impulse. This dispersed afferent volley evokes a dispersed wave of depolarization in spinal motoneurones, the duration of which can be estimated by analysing the firing pattern of single motor units. The time at which single units discharge after a tendon tap is not fixed; there is a jitter of 5–6 ms from one trial to the next. Since motor units usually are made to fire during the rising phase of the composite EPSP, this means that the EPSP rise time also is of the order of 5–6 ms. Impulses arriving at any time within this 6 ms period can contribute to excitation of the motoneurone. Thus it is possible that the first-arriving volley at the spinal cord could produce both monosynaptic and polysynaptic excitation within this period.

This is not proof that such polysynaptic inputs do contribute to the tendon jerk reflex. However, since the possibility exists, it is a warning not to attribute all pathologies of the tendon jerk to changes in the spinal monosynaptic reflex arc. It is possible that changes in polysynaptic pathways may be responsible for enhancement or depression of the tendon jerk in disease states in man.

A final point is, why is the EPSP rise time so much shorter than the temporal dispersion of the afferent volley? One reason is that later-arriving EPSPs may be superimposed upon the decay phase of EPSPs produced by earlier parts of the volley, and may therefore be missed. Another possibility is that the excitation is actively turned off by spinal inhibitory mechanisms involving, for example, Ib inhibition from the homonymous muscle.

6.2 *The H-reflex*

A reflex analogous to that of the tendon jerk also may be evoked in some muscles by electrical stimulation of nerve trunks. In 1926 Paul Hoffman found that low intensity stimulation of the tibial nerve in the popliteal fossa could produce a reflex contraction of the triceps surae muscles without direct activation of the muscle via the α-motoneurones. The reflex had a latency of some 30 ms and was believed to be an electrically elicited analogue of the tendon jerk. The reflex, now known as the H-reflex in his honour, occurs because the group Ia fibres are larger than, and have a lower threshold to electrical stimulation, than the α-motoneurones. Therefore, at very low stimulus intensities, Ia afferents may be the first fibres to be activated. The relatively pure and synchronous Ia afferent volley then produces reflex activation of triceps surae motoneurones. However, as the stimulus strength is turned up, the axons of α-motoneurones are stimulated and a direct muscle response (M-wave) is elicited. At high stimulus intensities, the H-reflex disappears. This is due to two factors: (1) antidromic firing of motor fibres renders the motoneurones refractory to the reflex input; and (2) the antidromic motor volley collides with the orthodromic reflex volley set up by the Ia input. H-reflexes can be obtained in many muscles. Figure 6.1 shows the behaviour of H-reflexes in the wrist and finger flexor muscles.

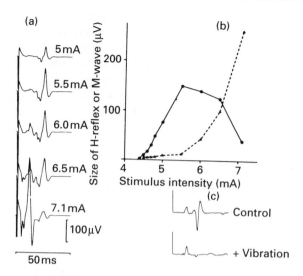

Figure 6.1 Relationship between the strength of an electrical stimulus to the median nerve at the elbow and the size of the H-reflex and direct M-wave in the wrist and finger flexor muscles of humans. In (a), an H-reflex with a latency of some 20 ms appears at low intensities. At 6.0 mA, an earlier response appears. This is the M-wave, a muscle response caused by stimulation of efferent motor fibres in the mixed nerve. As the M-wave increases in size, the H-reflex gets smaller. This is shown graphically in (b), where the peak-to-peak size of the H-reflex (●—●) and M-waves (▲----▲) are plotted on the same axes. (c) shows the effect on the H-reflex of vibrating the flexor tendons at the wrist. Vibration was applied 0.5 s before the sweep and continued throughout. The H-reflex is abolished, although the M-wave remains unchanged. (From Day *et al.*, 1984, Figure 1; with permission.)

As with the tendon jerk, there are several caveats against assuming that the H-reflex is a purely monosynaptic reflex. Although there is less dispersion in the afferent volley set up by electrical stimulation of nerve (no repetitive firing and all fibres activated at the same time), the spread of conduction velocities in the Ia afferent fibres still ensures that the volley arriving at the spinal cord is not completely synchronous. Measurements of the EPSP rise time show it to be of the order of 2 ms or so, giving time for the fastest conducting volleys to evoke di- or even trisynaptic excitation of the motoneurones in addition to the conventional monosynaptic response (Burke, Gandevia and McKeon, 1983).

In the past, it was assumed that both the tendon jerk and the H-reflex were monosynaptic. Because of this it was thought that comparison of the amplitude of the two responses could provide information as to the level of activity in the γ system. The rationale was that, with little or no γ drive, the spindle sensitivity would be low, and its response to a tendon tap would be small. In contrast, the H-reflex would be unaffected by the level of γ drive.

As shown above, however, the central timing of H- and tendon reflexes is very different, allowing for the possibility of much more spinal processing of the tendon jerk than the H-reflex. Because of this, it is not possible to assume that both responses use the same pathway, and hence it is not correct to compare the amplitudes directly. In addition, some doubt has recently been cast over the assumption that the level of γ drive affects spindle output following a tendon tap. In some cases, the spindles respond maximally whether or not γ activity is present.

Spindle after-effects and their influence on spinal reflexes

As described in Chapter 4, the response of a muscle spindle to stretch depends upon the past history of the muscle. A series of rapid muscle stretches, or a brief strong contraction, causes the actin–myosin bonds to break and then to reform. Imagine that a muscle can be held at one of three different lengths, long, intermediate and short. If the actin–myosin bonds are allowed to reform at a long muscle length, return of the muscle to an intermediate length will produce slack in the muscle spindle because of the stiffness of the bonds in the striated region. Conversely, if the bonds reform at a short length, the spindle may even be slightly stretched when it is returned to the intermediate position. Thus, stretch given after conditioning at a long length results in a smaller response from the spindle than when the actin–myosin bonds are allowed to reform at a short length. Changes in spindle sensitivity are reflected in the amplitude of the tendon jerk. In these experiments, a muscle is contracted or subjected to a rapid series of stretches and then held for several seconds at either a long or short conditioning length. It is then returned to an intermediate length and the tendon jerk elicited. The response is much larger when the conditioning is given at a short length than at a long length. In both cats and humans, recordings of spindle afferent volleys have confirmed that stretches from an intermediate position given after conditioning at a short length produce a larger volley than after conditioning at a long length (Edin and Vallbo, 1988; Gregory *et al.*, 1990).

6.3 *Long latency stretch reflexes*

Since complex, polysynaptic stretch reflex pathways are not facilitated in the relaxed state, and because the afferent volley produced by the tap is so short lasting, the tendon jerk is produced primarily through activity in the monosynaptic pathway. Stretch of an actively contracting

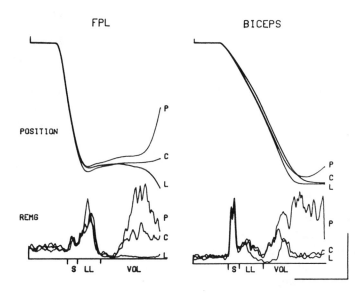

Figure 6.2 Reflex and voluntary EMG responses to muscle stretch in flexor pollicis longus (FPL) and biceps brachii (biceps). Top traces represent position of the thumb or elbow; bottom traces are average (of 16) rectified EMG responses. Stretches were given by a torque motor every 5–6.5 s and in separate sets of trials the subject was instructed either to do nothing (C), resist the stretch voluntarily as rapidly as possible (P) or to let go on perceiving the stretch (L). The two components (short-latency or tendon jerk component, S; long-latency, LL) of the stretch reflex EMG response were little affected by these instructions. However, the voluntary response (VOL, beginning about 100 ms after onset of stretch) is changed considerably. This affects the trajectory of the final portion of the position traces. Note how the long-latency reflex component is much larger than the short-latency reflex in the distal FPL muscle. Horizontal calibration 100 ms; vertical calibration 30° or 150 µV. In FPL, the short-latency component begins about 22 ms after stretch, and the long-latency after some 40 ms. (From Marsden, Rothwell and Day, 1983, Figure 2; with permission.)

muscle, however, produces a more complex response: the tendon jerk response is often followed, at about twice its latency, by a **long latency stretch reflex,** which may last 100 ms or more (Figure 6.2). This response may be followed by even later EMG activity, but this is difficult to distinguish from the voluntary reaction of the subject to the initial muscle stretch. Under these conditions, voluntary reaction time can be as short as 90–100 ms.

The origin of the long latency components of the stretch reflex (sometimes called the M2 response, as compared with the M1 tendon jerk component) is still very much a matter of debate. Several mechanisms probably contribute, perhaps even to different extents in different muscles. Three main factors are involved.

(1) The segmented EMG response could be due to a segmented muscle spindle afferent input.
(2) The long latency reflex could be mediated by more slowly conducting afferents (e.g. group II secondary spindle afferents), than those responsible for the tendon jerk.
(3) The same fast conducting (group Ia) afferents could be involved in both the tendon jerk and long latency response, but the latter could be delayed because the impulses traverse a longer pathway. In particular, it has often been suggested that this pathway involves a transcortical route, up the dorsal columns to the sensorimotor cortex and back down the corticospinal tract to the spinal cord.

Evidence can be adduced for and against each explanation.

(1) Evidence supporting segmented spindle afferent volleys comes from observations using microneurography during muscle stretch in the human. Such segmentation occurs because of oscillations in the muscle–tendon compliance and gives rise to muscle spindle bursts at intervals of 20–30 ms. These are appropriate to produce long latency responses through repeated operation of the tendon jerk pathway. This cannot be a unique explanation of the long latency reflex since very rapid muscle stretches, which are complete within 10–20 ms, may under certain circumstances evoke a long latency reflex, presumably without giving rise to a segmented afferent volley. In addition, segmentation of the afferent volley does not explain the results of pathological studies described below.
(2) Evidence for a role of slowly conducting afferents (group II fibres) comes from experiments using muscle vibration. A 200 ms stretch of a muscle may evoke a long latency stretch reflex, but vibration of the same muscle for 200 ms or so only produces a short latency (spinal) response. Arguing that vibration stimulates primary endings far more than secondary endings, Matthews (1984) concluded that stretch, by stimulating both primaries and secondaries well, evoked a long latency stretch reflex because of additional input in group II fibres.

    There are two lines of evidence against a unique role for group II afferents in production of the long latency component of the stretch reflex. First, if it were true that slowly conducting afferents were wholly responsible for the long latency response, then those muscles nearest the spinal cord would be expected to have the shortest latency M2 responses. In an ingenious series of observations, Marsden, Merton and Morton (1973) showed that this was not the case. Stretch reflex responses were recorded in both the long thumb flexor and the infraspinatus muscle. The latency of the tendon jerk was about 22 ms for the thumb and about 10 ms for infraspinatus, yet the latency of the long latency response was approximately the same in each case, at about 40 ms. If a slow conducting afferent pathway were responsible, the M2 response in infraspinatus, a muscle very close to the spinal cord, would have been at a much shorter latency.

The second piece of evidence against the group II theory comes from experiments involving cooling the arm. Cooling slows the conduction velocity of small diameter afferents more than that in large diameter fibres, so that the onset of the long latency component of the reflex should be delayed more than the onset of the tendon jerk response. In fact, cooling the arm delays both components of the reflex by the same amount, suggesting that large diameter afferents are involved in each case (Matthews, 1989).

(3) Evidence in support of the transcortical hypothesis of long latency reflexes comes mainly from clinical observations. Patients with lesions at any point on the postulated transcortical reflex root (dorsal columns, sensorimotor cortex or corticospinal tract) all have reduced or absent long latency reflexes even though the tendon jerk component of the response is present and frequently enlarged.

Such evidence although compelling, is not direct proof of the transcortical hypothesis and there are still arguments as to whether the long latency pathway may involve polysynaptic spinal routes which are subject to descending influence from supraspinal structures. Removal, for example, of a hypothetical facilitatory influence on this pathway could explain the absence of long latency responses in the patients referred to above.

In conclusion, it is possible that all these pathways can contribute to the long latency stretch reflex. Indeed, it is becoming clear that different pathways may contribute to different extents in different muscles. A striking example of this is shown in Figure 6.3. Because of an injury to the high cervical region at birth, this patient had been left in the peculiar position of having the movements of both hands controlled by only one (the left) cerebral hemisphere. Thus, transcranial stimulation (see Chapter 9) of the left motor cortex produced short latency EMG responses in both hands. Stimulation of the right side produced no response in either hand. When stretch was given to the long flexor of the right thumb, a long latency stretch reflex was evoked as usual. This was accompanied by a reflex in the left (unstretched) thumb flexor at the same latency. It appears that in this case, there is a strong transcortical contribution to the long latency stretch reflex of the flexor pollicis longus, such that stretch input from the right side reaches the left hemisphere, which then drives the thumb flexor of both sides to

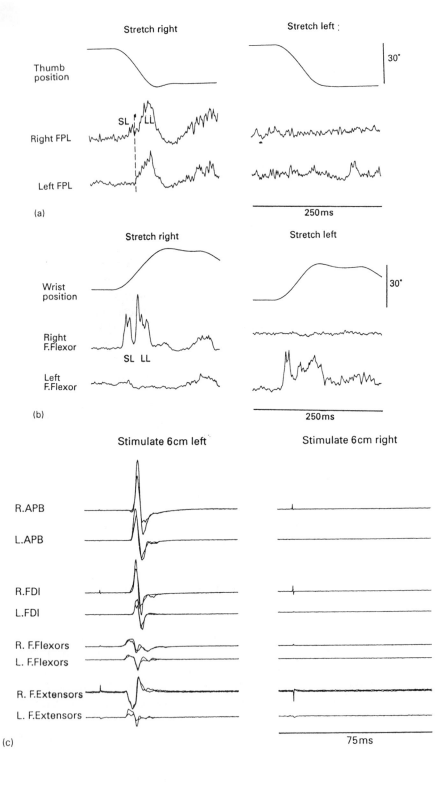

Stretch right

Stretch left :

Thumb position

30°

Right FPL

SL LL

Left FPL

(a)

250 ms

Stretch right

Stretch left

Wrist position

30°

Right F.Flexor

SL LL

Left F.Flexor

(b)

250 ms

Stimulate 6 cm left

Stimulate 6 cm right

R.APB

L.APB

R.FDI

L.FDI

R. F.Flexors

L. F.Flexors

R. F.Extensors

L. F.Extensors

(c)

75 ms

contract together. In contrast, when the stretch was given to the right flexor muscles of the wrist, the long latency component of the stretch reflex was only seen on the right. As in any normal subject, there was no response on the left. Presumably, in this muscle, the transcortical contribution to the long latency stretch reflex was small so that no contralateral responses appeared.

6.4 *Vibration reflexes*

A final reflex from muscle spindles that may be mentioned here is the tonic vibration reflex (TVR). Vibration of a muscle belly or its tendon at 50–150 Hz produces a slowly developing reflex contraction which is sustained throughout the period of vibration (Figure 6.4) and which subsides slowly when it is stopped. The reflex is seen best if the attention of the subject is distracted. Unlike the stretch reflex, it is possible voluntarily to prevent the reflex occurring if visual feedback is given. Muscle spindle Ia afferents are particularly well excited by such stimulation although in man, in contrast to the cat, many secondary spindle endings also are activated. The reflex pathway probably involves both the monosynaptic reflex arc as well as longer polysynaptic pathways, which would account for the slow onset and decay of the response. Some descending facilitation of these pathways must be necessary for the reflex to occur since no TVR may be recorded in the lower limbs of paraplegics with complete spinal cord transection. Vibration of purely cutaneous receptors may evoke a similar, although smaller, muscle contraction.

In addition to its excitatory effects on motoneurones, muscle vibration also has a concurrent inhibitory effect on the monosynaptic reflex

---

Figure 6.3 Left, stretch reflexes elicited by stretching the long flexor of the thumb (a) or the wrist flexor muscles (b) in a patient with mirror movements. In this patient, stretch of the right thumb produced a small short latency (spinal) stretch reflex in the right, and a bilateral long latency EMG response. When the left thumb was stretched, no stretch reflex was evident in either side. When the wrist flexor muscles were stretched, the responses were strictly ipsilateral. Stretch of the right flexor muscle produced short and long latency reflexes in the right side but not in the left. Stretch to the left muscle produced long and short latency reflexes in the left muscle but not in the right. (c) Shows the effect of transcranial magnetic stimulation of the left or right motor cortices. Recordings are made from the left and right abductor pollicis brevis (APB), first dorsal interosseous (FDI) and the flexors and extensors of the wrist and fingers. Note that stimulation of the left hemisphere produced bilateral responses in all the muscles illustrated. Stimulation of the right hemisphere produced no responses in either side. The conclusion is that the left hemisphere was connected to the distal arm muscles of both sides of the body, whereas the right hemisphere was not connected to either. (Rothwell *et al.*, 1991).

(Figure 6.4). Thus, vibration of the triceps surae inhibits the ankle jerk or H-reflex while simultaneously producing a TVR in the muscle. This is not due to occlusion of the afferent volley in the H-reflex or tendon jerk by the constant high level of muscle afferent discharge produced

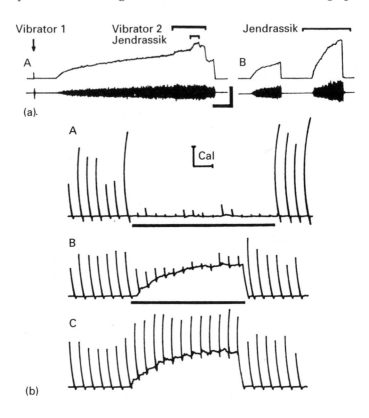

Figure 6.4 (a)Tonic vibration reflex (TVR) in quadriceps muscle of a normal human. A vibrator was applied to the ligamentum patellae and after a short delay produced reflex activation of quadriceps. The effect is increased by performance of the Jendrassik manoeuvre or by applying a second vibrator to the muscle belly. Upper trace, muscle force; lower trace, quadriceps EMC. Time calibration is I s, vertical calibration 5 kg or 1 mV. (From Burke, Andrews and Lance, 1972, Figure 2; with permission.) (b) Suppression of quadriceps tendon jerks by continuous muscle vibration in a normal human. In A and B, the thick horizontal line indicates the period of vibration. Records are of the force of quadriceps contraction, and show the responses to repeated tendon taps given every 5 s. In A, vibration evokes no tonic contraction, yet still suppresses the tendon jerks. In B, a TVR develops and again suppresses the tendon jerks. In C, the subject made a voluntary contraction similar to the reflex contraction of B. Voluntary contraction does not suppress tendon jerks. Time calibration is 10 s, force calibration 0.4 kg for A and 0.6 kg for B and C. (From DeGail, Lance and Nielson, 1966, Figure 1; with permission.)

by vibration (the 'busy-line' phenomenon). Recordings of gross neural discharge using near-nerve electrodes inserted through the skin show that the afferent volley from a tendon tap or H-reflex is preserved during vibration. Since the α-motoneurones are excited by vibration, the effect on the monosynaptic reflex has usually been ascribed to presynaptic inhibition of muscle spindle Ia afferent terminals. In experiments performed on anaesthetized cats, such presynaptic inhibition in the spinal cord can be recorded as a slow electrical potential from the dorsal roots (see dorsal root potential in Chapter 5). However, a second factor may also play a part in vibratory suppression of the H-reflex. Activation of Ia presynaptic terminals by continuous vibratory input may lead to **post-activation depression** of transmitter release. Thus an H-reflex stimulus given during or shortly after vibration may produce less synaptic activity than when given in the resting state; the EMG response would then be smaller than expected (Crone and Nielsen, 1989).

Experiments with muscle vibration, therefore, illustrate that Ia fibres have polysynaptic as well as monosynaptic excitatory connections with spinal α-motoneurones. In addition, the fibres produce a presynaptic inhibition of their own terminals (Lance, Burke and Andrews, 1973, review vibration).

6.5 *The servo hypothesis and* α–γ *coactivation*

The servo theory

Since the discovery of the stretch reflex, there has been no generally agreed hypothesis to describe its role in the control of movement. The most influential has been the servo hypothesis put forward by Merton in 1953 (Eldred, Granit and Merton, 1953). In its simplest form, the theory envisaged that slow movements could be initiated and controlled by the activity of the γ-efferents to the muscle spindles (Figure 6.5). Rather like power-assisted steering in a motor car, an increase in activity in the fusimotor (γ) system would produce contraction of the spindle poles and stretch the sensory receptors. The increased afferent discharge would, via the stretch reflex, produce contraction of the main extrafusal muscle until the stretch on the spindle was removed. The extrafusal contraction would 'follow' that of the spindles.

This method of control would be rather slow because of the extra delay involved in traversing the servo loop (20 ms in the arm, 30 ms in the leg). However, the advantage of this system would be that the muscle servo would follow the spindle contraction even if the external load were to change. For example, if the load on the main muscle were

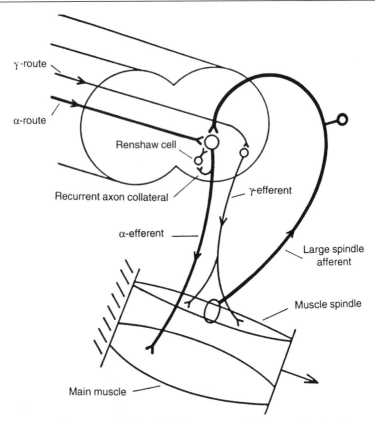

Figure 6.5 Diagram of Merton's 'follow-up servo' theory of muscular contraction. Two routes were available from the brain to control the activation of spinal α-motoneurones: projections directly to the α-motoneurones (α-route) and projections to the γ-motoneurones (γ-route). The α-route was thought to be reserved for 'urgent' muscle activation. The γ-route could produce contraction more slowly by activating muscle spindles and exciting α-motoneurones reflexly via the spindle afferent fibres. (From Hammond, Merton and Sutton, 1956, Figure 1; with permission.)

increased during the contraction, so that the muscle was stretched, there would be an increased mismatch between extra- and intrafusal lengths. More power would be called up via the servo loop to overcome the increased mismatch and the obstacle would be overcome. This type of contraction was said to be via the γ-route. Direct activation of muscles via the α-route, which would bypass the delay in the servo loop, was reserved for rapid, urgent muscle contraction.

Unfortunately, this attractively simple theory is probably incorrect. The most direct evidence against the simple version of the servo theory has come from recordings of spindle afferent discharge during movements in conscious man. The sequence of events preceding onset of movements in the servo theory would be: (1) γ-efferent firing; (2) contraction of spindle poles; (3) stretch of sensory endings; (4) discharge in spindle afferent fibres; and (5) discharge of α-motoneurones. However, percutaneous recording of single spindle afferents during slow isometric finger movements showed that this sequence was not followed. Spindle discharge always began **after** the start of muscle activity, rather than **before** it (Figure 6.6). Slow voluntary movements are not initiated by the γ-route.

Another argument against the theory is that in order for such a servo to work effectively, the 'gain' of the reflex must be fairly high. That is, the amount of extrafusal muscle contraction called up by a given mismatch of spindle and muscle length must be sufficient to produce movement and overcome any minor obstacle *en route*. In normal humans, it seems that the gain of the stretch reflex generally is quite low. Thus, in Figure 6.2, the subject was trying to hold his thumb or elbow in a given position against a small load. According to the servo theory, the muscle power to maintain this position would be supplied by a very small mismatch between extra- and intrafusal muscle length driving the contraction via the reflex arc. When the force against the wrist suddenly was doubled, the reflex contraction in the flexor muscle was not sufficient to restore the wrist position. In fact, in this case, very little compensation was achieved by the reflex correction, which would be a

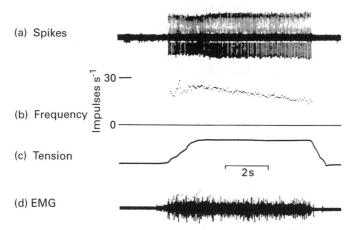

Figure 6.6 Discharge of a single muscle spindle afferent in the finger-flexor muscles (a, direct record; b, instantaneous frequencygram of same unit) during a weak voluntary isometric flexion of the index finger. The recording was made with a tungsten microelectrode, of about 10 μm tip diameter, inserted into the median nerve above the elbow. Note how the afferent only begins to fire after the onset of EMG activity (d) in the muscle. (From Vallbo, 1970, Figure 6; with permission.)

poor performance to expect from a servo system. True compensation began with the onset of voluntary activity. It should be noted, however, that although the gain of the stretch reflex compensation for relatively large disturbances, such as those shown in Figure 6.2, is low, this system may be much more sensitive to small perturbations. This is discussed in detail below.

α–γ coactivation

Although spindle activation does not precede extrafusal contraction, Figure 6.6 suggests that the behaviour of spindles is not entirely passive during voluntary movements. If it was, then spindles lying in series with the extrafusal muscle would be unloaded by the extrafusal contraction, causing a decrease, rather than an increase in spindle firing. (Even in isometric contractions, there is some spindle unloading due to stretch being taken up by the tendon compliance.)

The increase in spindle discharge seen in Figure 6.6 is evidence that there is an active contraction of the spindles which more than compensates for the extrafusal unloading. Both α and γ-motoneurones must have been active in the movement. This concept usually is referred to as α–γ coactivation, meaning that the command to contract a muscle does not go solely to the α-motoneurones, or solely to the γ-motoneurones (as suggested by the follow-up servo theory), but to both sets of motoneurones at the same time.

One of the advantages of α–γ coactivation sometimes is said to be that gamma discharge during the contraction prevents spindle unloading, and enables spindles to respond to disturbances even during active muscle shortening. However, if the gain of the stretch reflex is low, and compensation for unexpected disturbances to movement is made mainly through voluntary intervention, then it is reasonable to ask why α–γ coactivation is of any advantage. After all, what is the point of detecting displacements, if there is no rapid way to compensate for them?

The answer seems to be that the stretch reflex gain is much higher for small disturbances of movement, around the limit of perception, than for larger displacements. An example is shown in Figure 6.7. The subject was flexing the top joint of his thumb against a small steady resistance offered by a motor. This produced upwards deviation of the trace on the left of the figure. (This control trace is not shown for clarity, but would lie in between the lines marked +1% and −1%.) In random trials, the resistance of the motor was increased or decreased for the remainder of the movement either by a large amount, or by only 1% of the standing force. When large disturbances were given, the thumb movement never regained its initial trajectory. The stretch reflex was

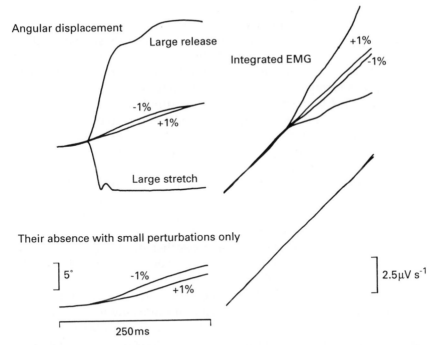

Figure 6.7 Reflex compensation for small, but not large disturbances to voluntary movements. The two traces on the left show thumb movements made by a subject flexing the distal joint of the thumb against a constant torque of 0.1 N m. In control trials (not illustrated) this movement produced a steady upwards deflection of the trace for the period of the sweep. In random trials, the torque offered by the motor was increased or decreased by a large or a small (± 1 %) amount, 50 ms after the start of the recording sweep. This torque change was too small to be detected by the subject. When larger disturbances were included, they were not corrected by any reflex mechanisms. However, when small perturbations (± 1 %) were given then these were compensated completely and the trajectory of the thumb regained its target before the end of the sweep. Corresponding changes could be seen in the EMG response from the thumb muscle (right traces). The display here is of integrated, rectified, EMG activity. In control trials the activity was constant, which means that when integrated it would produce a constant sloping line at about 45° (not illustrated). In trials in which the thumb was stretched, the integrated EMG trace deviates upwards from the control trace indicating an increase in EMG activity at the time when the deviation begins. This occurs about 50 ms after the displacement given to the thumb. Conversely, when the thumb was released, the integrated EMG activity declines. The bottom two traces show that the complete correction for small perturbations which were seen in the upper two traces is not present when only small perturbations are presented during the movement. The disturbances go uncorrected, and there is no change in the ongoing level of EMG activity during the task. (From Marsden, Rothwell and Day, 1983, Figure 10; with permission.)

ineffective in compensating for the displacement. However, with the small ±1% disturbances, the thumb movement had regained its initial trajectory by the end of the sweep. Since the subjects could hardly feel the perturbation, the implication is that this correction was reflex in origin. In fact, close inspection of the EMG records showed that the small disturbances lead to an increase or decrease of the EMG activity (as appropriate) after a latency of some 40–50 ms. Thus the gain of the stretch reflex is much higher for small than for large disturbances. This is probably due to the relatively high sensitivity of spindle endings to small displacements (see Chapter 4). The result is that if a movement is progressing approximately on course, then reflex mechanisms can correct small errors. If large errors ensue, the system is overwhelmed, and voluntary corrective mechanisms are engaged. Finally, it should be mentioned that full correction for small displacements only occurs if they are intermixed (as in normal movement) with large disturbances. If small disturbances are given alone, then they go uncorrected. Why this occurs is unknown.

The fact that spindles normally increase their discharge during movement (although this is not true for fast movement as described below) has another interesting consequence. Although this discharge does not 'drive' the contraction via the stretch reflex circuitry, as envisaged by the servo-theory, it does **support** voluntary movement. That is, stretch reflex feedback continues throughout a task, and assists the voluntary descending command to produce muscle contraction. If this assistance is removed, then increased voluntary effort is needed to produce a given size of contraction. Thus, the maximum voluntary contraction strength of the ankle dorsiflexor muscles is decreased during partial local anaesthetic block of the nerve which supplies them. Such a block is sufficient to interrupt conduction in small γ fibres much more than that in larger α fibres and hence prevents α–γ coactivation. Spindle discharge is no longer available to support the contraction and voluntary strength is reduced. Electrical stimulation of the peripheral nerve shows that the small effect of anaesthetic on the α fibres themselves is insufficient to explain the effect on voluntary contraction.

Although it is agreed that α–γ coactivation does occur, the main current area of controversy is whether α–γ coactivation is the rule or whether there are occasions in which α activity and γ activity can be controlled independently. In the cat, there are indeed sites within the brainstem, stimulation of which can produce pure γ activity, but the question remains as to whether the brain ever activates these centres independently of those producing α activity during normal movement. The very existence of a separate γ system, in addition to the shared skeletofusimotor β system, which is all that is seen in amphibia, argues for the possibility of independent control.

Unfortunately, it is not easy to resolve the question of independence versus coactivation because it is very difficult to record directly from the small γ-efferent fibres during movement. Because of this, the main evidence for the existence of α–γ coactivation comes from observations like those above in which the activity of fusimotor neurones is deduced from the spindle afferent discharge. α–γ Coactivation is presumed to occur if the spindle discharge during active movements is different from that seen in passive movements of the same kind.

α–γ *coactivation in humans*

In human studies, using the technique of microneurography to record directly from muscle spindles in peripheral nerve, there is no good evidence that α- and γ-motoneurones can be activated independently during voluntary contraction of muscle.

In the past, it was often thought that the Jendrassik manoeuvre (forceful contraction of muscles distant from the site of a tendon jerk) increases the amplitude of the tendon reflex by increasing fusimotor drive to spindles in the relaxed target muscle, and thereby making them more sensitive to the tendon tap. However, microneurographic recordings have shown no change in spindle discharge during contraction of distant muscles so long as the target muscle remains completely relaxed. It is now thought that the Jendrassik manoeuvre increases the size of the tendon jerk by facilitating transmission at central synapses rather than by increasing spindle sensitivity.

Why should it be possible to demonstrate α–γ independence in the cat (see below), whereas it is so difficult in humans? One possible factor is that because of the difficulty in maintaining a stable recording, most of the movements that have been examined in humans have been relatively slow and hence the whole range of voluntary movement has not been explored. Perhaps in natural, free, movements, α–γ dissociation may occur as it does in the cat.

An addendum to this story of α–γ coactivation in humans is that although it has not so far been possible to reveal independent activation of the γ system, it is possible to change the balance between α and γ activity. In stereotyped isometric slow voluntary contractions, each spindle ending begins to accelerate its discharge at a particular level of contractile force. This force threshold is reproducible for that ending, but can differ markedly between endings. Threshold differences may be caused (1) by the particular position of the spindle in relation to the active extrafusal muscle fibres, or (2) by a recruitment threshold for γ-motoneurones innervating that ending.

Whatever the reason, given that a contraction is always the same, changes in the threshold at which a spindle is activated give a measure of the level of α–γ balance. Two procedures have been shown to affect this balance: caloric stimulation of the vestibular organs (by passing cold water into the ear) and vibratory stimulation of cutaneous receptors. Both manoeuvres increase the threshold for spindle excitation, and therefore probably decrease the γ drive to the spindles relative to the α activity. Thus different systems (that is, cutaneous afferents and vestibulospinal connections) can activate α and γ systems in different proportions.

Reflex effects onto γ-motoneurones from low threshold cutaneous afferents have often been demonstrated in animal experiments and have recently been described in humans. In standing subjects, stimulation of the cutaneous sural nerve at the ankle with a short train of impulses at about twice perceptual threshold can cause changes in spindle firing rates in the pre-tibial flexor muscles without producing any observable change in their EMG activity. High intensities of stimulation often do cause EMG responses in the pre-tibial muscles, so that the results imply that for cutaneous inputs the threshold for effects onto the γ system is lower than that for the effects onto the α system. Interestingly, the effects are not seen recumbent subjects. It is believed that the responses are 'gated' by the posture assumed by the subjects. (See discussion and further references in Aniss *et al.*, 1990.)

Figure 6.8 Spindle afferent discharge recorded from implanted electrodes in a normal awake cat compared with recordings made from a spindle in the same muscle of an anaesthetized cat. In the anaesthetized preparation, the same length changes were applied to the muscle whilst different combinations of fusimotor stimulation were given in order to match the afferent discharge as closely as possible to that seen in the awake animal. The column in (a) illustrates the original chronic recordings from the awake animal. The upper trace shows the length of the triceps surae muscle, the middle trace shows the instantaneous discharge frequency of a Ia spindle afferent in the muscle, and the lower trace shows the EMG activity of the triceps. In (b) and (c) are shown simulations of the same movement in an anaesthetized preparation. The records are from two different Ia afferents B2, C2 and B3, C3. In column (b) the upper trace shows the stretch applied to the soleus muscle, which was the same as that recorded in the chronic experiments. The middle two traces show the instantaneous frequency discharge of an Ia afferent from the muscle during concomitant stimulation of its fusimotor supply at a rate proportional to the ongoing level of EMG activity in the muscle. With either static or dynamic γ stimulation, the response of the spindle afferent differs in two ways from that seen in the chronic recordings. First the peak of discharge during stretch is smaller, second, in the simulation, the spindle begins to fire at the end of the record at the same time as EMG activity increases. This is not seen in the original chronic recording. In the records of column (c), the fusimotor stimulation was set at a tonic sustained level. The best fit of the spindle discharge to that seen in the chronic state was achieved with a steady 80 Hz dynamic γ stimulation. The conclusion is that in these experiments, the fusimotor drive to the triceps surae muscle spindles does not follow the EMG command to the extrafusal muscle. In other words, there is no α–γ coactivation. (From Hulliger and Prochazka, 1983.)

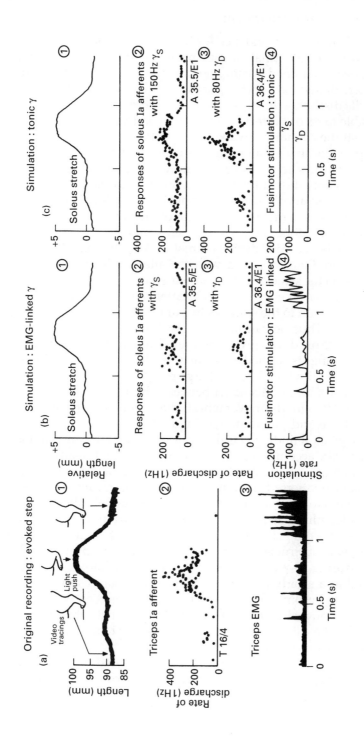

(a) Original recording : evoked step

① Video tracings — Light push

Length (mm)

② Triceps Ia afferent

Rate of discharge (1Hz)

T 16/4

③ Triceps EMG

Time (s)

(b) Simulation : EMG-linked γ

① Soleus stretch

Relative length (mm)

② Responses of soleus Ia afferents with γ_S

A 35.5/E1

③ with γ_D

A 36.4/E1

Rate of discharge (1Hz)

④ Fusimotor stimulation : EMG linked

Stimulation rate (1Hz)

Time (s)

(c) Simulation : tonic γ

① Soleus stretch

② Responses of soleus Ia afferents with 150Hz γ_S

A 35.5/E1

③ with 80Hz γ_D

A 36.4/E1

④ Fusimotor stimulation : tonic

γ_S

γ_D

Time (s)

α–γ *coactivation in the cat*

There is now clear evidence for independent control of both α and γ systems during normal movements in the cat. Two types of movement have been examined in conscious animals, walking and chewing. In the former, deductions about γ activity have been made by examining spindle discharge during movement, whereas in the latter it has proved possible to record directly from both α and γ fibres.

In studies of normal walking, the pattern of spindle firing has been compared to that seen in spindles from an anaesthetized cat subjected to exactly the same passive joint movements. This type of experiment is remarkably difficult to perform. Stable recording electrodes have to be implanted into the dorsal roots of intact cats, together with miniature length gauges to monitor the length of the muscles under study. When this has been done, single spindle afferents, muscle length and muscle EMG can be recorded during normal walking, jumping and so on. These records are then compared with the afferent discharge recorded from spindles in the same muscle of an anaesthetized cat.

Under anaesthetic, the experimenter can control the level of fusimotor stimulation to the spindles by stimulating appropriate ventral root filaments. Thus he can find the precise level of fusimotor discharge required to produce a given spindle afferent response when a particular length change is applied to the muscle. This fusimotor discharge is then compared with the EMG discharge recorded from extrafusal muscle in the intact animal. Mismatch between the EMG and fusimotor input indicates that for the movement studied, there must have been separate control of α and γ motor fibres (Figure 6.8).

The general finding was that most unobstructed and familiar movements were accompanied by a relatively low and steady level of either dynamic or static fusimotor activity, which was independent of the phasic activation of extrafusal muscle. This meant that the spindles behaved like passive stretch receptors throughout the movement. For example, during active shortening of the extrafusal muscle the spindle discharge would silence (unlike the records of Figure 6.6); if the shortening was obstructed, there was no increase in spindle discharge (as seen during α–γ coactivation). Although fusimotor activity usually stayed at a steady level, there were circumstances in which it could change to become phasically active. This was seen when the cat performed unusual or novel tasks. It may be that when new tasks are learned, the cat makes use of the servo-assistance which a phasically active fusimotor system can provide.

α–γ Independence also has been documented during rhythmic jaw movements in the cat. Direct recording from fusimotor neurones revealed two classes of discharge. One type of neurone showed a sustained increase in firing rate during chewing, whereas the other class discharged phasically together with the extrafusal EMG. The two types probably were γ_d and γ_s, respectively. Thus, unlike the limb movements described above, the phasically modulated units (γ_s) provided an α–γ linkage during normal movements. This was not strong enough to prevent unloading of the spindles during rapid extrafusal shortening, but was sufficient to increase spindle discharge if the shortening movements were obstructed. Such experiments suggest that α and γ motoneurones can be independently controlled in the cat, and the type of control varies with the type of movement that is being made (Barnes and Gladden, 1985; Hulliger, 1984; Hulliger *et al.*, 1989, for further details of the work covered in this section).

6.6 *Investigation of activity in other reflex pathways using the H-reflex*

All of the spinal reflex pathways described in the previous chapter can now be analysed to some extent and in limited muscle groups, in humans. The advantage of working with humans is the ease with which it is possible to investigate how transmission in such pathways changes during the performance of different voluntary tasks. In addition, knowledge of the pathological changes in neurological disease helps us to understand some of the clinical symptoms of disordered motor control (reviews by Pierrot-Deseilligny and Mazières, 1984; Fournier and Pierrot-Deseilligny, 1989; Schieppati, 1987).

Reciprocal Ia inhibition

Tanaka (1974) was the first to document reciprocal Ia inhibition in humans. The technique is relatively simple: H-reflexes are elicited in one muscle, and low intensity electric stimuli, designed to stimulate only the largest afferent fibres (the muscle spindle Ia afferents), are applied to the nerve supplying the antagonist muscle. If the antagonist stimulation is given at the appropriate time, it is possible to inhibit the test H-reflex in the agonist. Timing considerations show that the spinal latency for this effect is appropriate for disynaptic inhibition in the spinal cord.

The best place to demonstrate reciprocal inhibition in humans is between the extensor and flexor muscles around the wrist joint where flexor H-reflexes are readily suppressed by 50% or more by low intensity

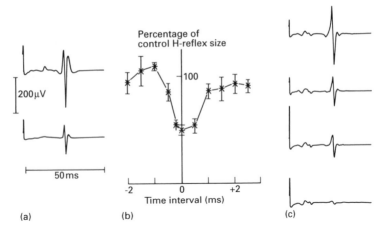

Figure 6.9 Reciprocal inhibition between the extensor and flexor muscles in the forearm. (a) The effect on the H-reflex of the flexor muscles produced by weak isometric extension at the wrist. The upper trace is a control H-reflex. This is suppressed by voluntary extension of the wrist. (b) The time course of flexor H-reflex inhibition produced by radial nerve stimulation. The test H-reflexes were elicited in the flexor muscles by stimulation of the medial nerve in the cubital fossa at different times relative to a conditioning shock applied to the radial nerve in the spiral groove at time zero. The flexor H-reflex is inhibited to about half its value when the two stimuli are applied at approximately the same time. Under these conditions, the afferent input from the radial nerve stimulus arrives at the spinal cord earlier than that from the median nerve stimulus and can cause disynaptic inhibition of the flexor H-reflex. (c) Facilitation of disynaptic Ia inhibition by a descending voluntary command. These responses were recorded after complete anaesthetic block of the radial nerve so that the wrist extensors could no longer be activated. The upper trace shows the average H-reflex after median nerve stimulation given alone. The second trace shows the effect on the H-reflex of attempted voluntary wrist extension (but without any observable movement). The response is inhibited. The third trace shows the effect in the relaxed state of giving a radial nerve conditioning stimulus at T = 0 ms. The final trace shows the combined effect of willed wrist extension and radial nerve shock. The effect of both manoeuvres together was greater than expected from the sum of each on their own. This indicates a voluntary facilitation of the Ia disynaptic inhibitory pathway. (From Day *et al.*, 1981.)

stimuli to the radial (antagonist) nerve. An example is shown in Figure 6.9. In the leg, inhibition from pre-tibial flexors onto the soleus muscle in the calf can be seen, but is much less powerful.

The excitability of the Ia inhibitory pathway changes during voluntary movement. Thus, in the leg, inhibition at rest from tibialis anterior to soleus is small. However, if the antagonist nerve stimulus (the peroneal nerve) is given during, or even slightly before a small voluntary contraction of the tibialis anterior, then the amount of inhibition onto the soleus H-reflex is increased. The increased excitability of the Ia inhibitory pathway is thought to be due to facilitation by the descending command of Ia inhibitory interneurones in the spinal cord which project onto soleus motoneurones. This is compatible with the concept of α–γ coactivation in reciprocal inhibition discussed in Chapter 5.

[It should be noted that this apparently simple demonstration of increased inhibition from tibialis anterior onto the soleus motoneurones during dorsiflexion of the foot is complicated by two confounding factors. First, dorsiflexion of the foot itself decreases the amplitude of soleus H-reflexes. This change in size of the test response complicates the issue of whether a given peroneal nerve stimulus produces more reciprocal inhibition during dorsiflexion than at rest. The way around this problem is to test inhibition at rest with a range of H-reflex sizes in the soleus muscle. Thus, one can compare the amount of inhibition during dorsiflexion with the appropriate size of test H-reflex. Another problem is that dorsiflexion is accompanied by a sustained increase in firing from the muscle spindles in tibialis anterior due to α–γ coactivation. This can depress transmission at Ia synapses within the spinal cord because of transmitter depletion. The effect would be to reduce the effectiveness of Ia reciprocal inhibition after stimulation of the peroneal nerve, and thus counteract the increase in excitability of Ia inhibitory interneurones produced by a descending voluntary dorsiflexion command (Nielsen *et al.*, 1992).]

Ib inhibition

In humans, the combination of test H-reflex and conditioning shock can be used to demonstrate Ib inhibition. Ib effects are not limited to nearby antagonist or synergist muscles like the reciprocal Ia inhibition described above. Ib inhibition and facilitation can be demonstrated between muscles acting at different joints within a limb. Typical Ib effects with a short central delay (1–1.5 ms) have been demonstrated from afferents in the medial gastrocnemius nerve to soleus, quadriceps and hamstrings (inhibitory effects). Facilitatory effects are seen from afferents in the common peroneal nerve to quadriceps and from the radial nerve in the arm to the flexor carpi radialis. A typical experiment would be to evoke test H-reflexes in soleus and condition these by stimulation of the medial gastrocnemius nerve (Figure 6.10). The conditioning stimulus produces an initial small facilitation of the H-reflex, which is presumed to be due to the small number of heteronymous Ia projections from medial gastrocnemius to soleus. This is followed by a short period of inhibition which has a slightly higher threshold than the preceding facilitation and hence is thought to be Ib in origin. Similar effects are seen between medial gastrocnemius and quadriceps. In this case though, both the initial Ia excitation and later Ib depression are much more conspicuous than in soleus.

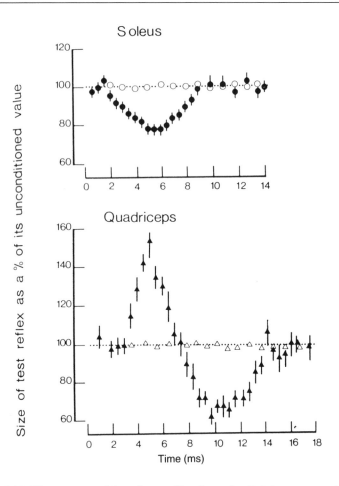

Figure 6.10 Time course of the effects of low intensity (0.8 times motor threshold) stimulation of the medial gastrocnemius nerve on the size of H-reflexes evoked in the soleus or quadriceps muscles. Filled symbols represent the variation in amplitude of the different test reflexes at different times after the medial gastrocnemius nerve stimulation. The vertical bars represent 1 standard error of the mean above and below the mean values. The open symbols show that a pure cutaneous nerve stimulus (sural nerve at the ankle) did not significantly modify the test reflexes. Note the initial small facilitation in soleus at about 1.5 ms and the subsequent much longer depression. The same pattern of facilitation (maximum at 5 ms) and depression is seen in the quadriceps H-reflex (with different timings because of the different peripheral conduction delays), although the effects are much larger than in the soleus. (From Pierrot-Deseilligny *et al.*, 1981, Figure 1; with permission.)

Ib inhibition is modulated both by voluntary contraction and by cutaneous input. Moderate to strong contraction of the test muscle depresses the amount of Ib inhibition that it receives. This is probably

caused by a reduction in the excitability of the Ib inhibitory interneurones in the spinal cord by the descending voluntary command.
Such an effect would obviously be of use in strong contractions: if Ib
inhibition was not removed, it would be like trying to drive a car fast
with the brake on. The situation is different during weak contractions.
Ib inhibition is hardly depressed at all. Under these circumstances, it
may help with fine control of the muscle force.

Cutaneous modulation of transmission in the Ib pathway is a
particularly good example of the degree to which apparently simple
reflex pathways are continuously regulated in the intact normal state.
Most experiments have been performed on the Ib inhibition between
gastrocnemius and quadriceps muscles. Stimulation of the anterior part
of the foot sole depresses Ib inhibition to quadriceps when the subject
is at rest, although it has no effect on the inhibition from gastrocnemius
to soleus. This effect may be brought about by convergence of cutaneous
inputs onto specific groups of Ib interneurones in the spinal cord: those
projecting to quadriceps being excited by cutaneous input, while those
to triceps surae are unaffected.

When tested during contraction of the triceps surae, the cutaneous
facilitation of Ib interneurones to quadriceps is **reversed**: cutaneous
stimulation increases Ib inhibition. Reversal of the cutaneous effect is
thought to be caused by descending control of interneurones in the
cutaneous reflex pathway. By this means, the weighting of cutaneous
effects onto the Ib interneurone may be reversed by voluntary contraction. Exploratory movements of the foot could provide a possible role
for this circuitry. These movements are produced by activity in quadriceps and triceps surae. If the big toe or anterior part of the foot comes
into contact with an object, the cutaneous input would facilitate Ib
inhibition of quadriceps and thereby depress its contraction. The effect
would be to reduce the effect of stubbing one's toe during a voluntary
exploratory movement.

The explanation is not so far-fetched as it may seem. For example, if
one is walking around the house at night, trying to avoid children's toys
scattered about on the floor, there are powerful reflexes which inhibit
one's exploratory movements when an unexpected obstacle is encountered. In daylight, however, one is free to point and press one's toes
against any desired object or to kick a ball with full force. The difference
may be that the excitability of spinal circuits has been reset in a totally
different way. (Pierrot-Deseilligny and Maziers, 1984 give further details on possible Ib actions in humans.)

These recent findings on the adaptability of Ib circuits and their
control by other inputs are in contrast to an older, general theory of Ib
function known as the **regulation of stiffness** hypothesis (Nichols and
Houk, 1976). In essence, this hypothesis proposes that Ib inputs, which

sense muscle force, and muscle spindle inputs, which are sensitive to muscle length, act together to produce reflex responses to mechanical disturbance of a limb. Thus, neither muscle force nor muscle length alone is the regulated variable. Instead, the combined measure, stiffness (equals force/length) is the important factor. Isolated muscle has a rather non-linear stiffness, there being a greater drop in tension for a given decrease in length than the rise seen for an equivalent increase in length. In an intact animal with functioning reflexes, these changes become more symmetrical and hence it can be said that muscle stiffness has become more linear. However, the general applicability of theory has been hard to demonstrate. As seen above, other more variable roles of the Ib pathways have been found.

Renshaw inhibition

The technique used to demonstrate Renshaw inhibition in man is too complex to detail here. It is discussed fully in the papers by Katz and Pierrot-Deseilligny (1982; 1984) and by Hultborn and Pierrot-Deseilligny (1979). Like Ia reciprocal inhibition, it is possible to describe how Renshaw inhibition changes during voluntary contraction. Compared to rest, a weak tonic contraction increases the amount of recurrent inhibition, whilst during a strong contraction, the amount of inhibition is decreased. As discussed above, this is probably caused by changes in excitability of Renshaw cells produced by the descending voluntary command (see Chapter 5).

Presynaptic inhibition

Measurement of presynaptic inhibition in man has forced a re-evaluation of its importance in motor control. In the cat, there is a seemingly diffuse pattern of presynaptic inhibition. For example, in the Ia system, inhibition is found on the terminals of all ipsilateral flexor and extensor muscles from ipsilateral flexors. Because of this, presynaptic mechanisms were often dismissed when considering the fine control of movement. In fact, it turns out that this can be a remarkably specific mechanism, and is modulated quite precisely during different types of movement.

There are three ways of activating presynaptic inhibition in man. The simplest is to use short-lasting vibration to produce presynaptic inhibition of Ia afferents and hence reduce the size of test H-reflexes in a muscle. In the past, it was usual to apply the vibrator to the **same** muscle as used to test the H-reflex. Unfortunately, under these conditions, the results are difficult to interpret since vibration produces activity in Ia afferent fibres used in the H-reflex. The activity produced by vibration can cause **post-activation depression** of transmitter release at the Ia-motoneurone synapse, and reduce the effectiveness of the H-reflex volley without the need to invoke presynaptic inhibition. A better method is to apply the vibration to a different muscle. Presynaptic inhibition can still occur, but will not be complicated by possible post-activation depression. Caution must still be exercised, however, since vibration readily spreads through the bone and other tissues in a limb. Thus, only 2–3 cycles of vibration are used, at very low amplitude, to prevent activation of antagonist muscle spindles.

A second situation in which presynaptic inhibition can be observed is whilst testing for reciprocal inhibition between extensors and flexors in the forearm. As discussed above, disynaptic reciprocal inhibition has a short central latency, and produces depression of the test H-reflex at short intervals. This period of inhibition lasts only a few milliseconds. However, if the interval between conditioning and test shocks is increased, a second, long-lasting period of inhibition occurs. This is thought to be due to presynaptic inhibition of the flexor Ia afferent terminals by the extensor conditioning volley. Its longer latency is compatible with a longer central pathway in the spinal cord. Its long duration is typical of a presynaptic inhibitory time course.

A third method is more difficult to use, but has the advantage that more specific inputs can be used than with vibration, and can be used on muscles in the arm as well as the leg. Full details of the method, and proof of the presynaptic nature of the effect are discussed in the paper by Hultborn *et al.* (1987a). In essence, it relies on measuring the amount of H-reflex facilitation in one muscle provided by a monosynaptic heteronymous Ia volley from another. For example, the soleus H-reflex can be facilitated by stimulation of Ia afferents from quadriceps. This is because the quadriceps volley produces subliminal monosynaptic Ia facilitation of soleus motoneurones. Presynaptic inhibition of quadriceps Ia afferents is therefore detected by a reduction in the amount of heteronymous Ia facilitation.

Using one or a combination of these and other methods, presynaptic inhibition has been shown to be carefully controlled during voluntary movement. For example, during a ramp and hold isometric contraction of soleus, presynaptic inhibition is reduced before the contraction begins and is maintained low during the dynamic phase of movement. This produces facilitation of the soleus H-reflex which begins just before the start of contraction and lasts during the ramp. Presumably this has the effect of increasing the gain of the monosynaptic stretch reflex, which could assist the voluntary command to activate the muscles (remember, this is an isometric contraction of soleus, during which spindle afferent discharge increases due to α–γ coactivation (see above)).

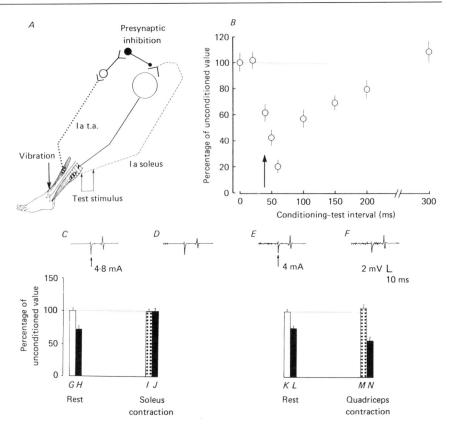

Figure 6.11 Changes in presynaptic inhibition of the soleus H-reflex at the onset of isolated voluntary contractions of the soleus and quadriceps muscles. The experimental arrangement is sketched in *A*. Three cycles of vibration were applied to the tendon of the tibialis anterior muscle at different times before H-reflexes were elicited in the soleus muscle. The vibration inhibits the soleus H-reflex with the time course shown in *B*. Most of this inhibition is due to presynaptic effects, although at very short conditioning-test intervals there may be contamination by group I postsynaptic effects onto soleus motoneurones, and at intervals longer than 50 ms there may be cutaneous effects. The remaining experiments were done with a conditioning-test interval of 40 ms indicated by the arrow at which presynaptic inhibition is the predominant effect. The bar graphs in the lower half of the figure show how the amount of presynaptic inhibition at 40 ms is changed at the onset of an isolated contraction of soleus or quadriceps muscle. The open bars (*G* and *K*: see also raw data record in *C*) show the size of the control H-reflex (= 100 %) at rest. Vibration of tibialis anterior causes presynaptic inhibition of the H-reflex to about 75% of its control value (*H* and *L*: see raw data trace in *D*). At the onset of soleus contraction, the intensity of the test stimulus has been reduced so that the control H-reflex in soleus is approximately the same size as at rest (*I*: see raw data trace in *E*). This H-reflex is no longer inhibited by vibration (*J*: see raw data trace in *F*). The converse happens at the onset of a quadriceps muscle contraction. The amount of presynaptic inhibition of the soleus H-reflex (*M* and *N*) is increased. (From Hultborn *et al.*, 1987b, Figure 1; with permission.)

At the end of the ramp phase, presynaptic inhibition returns to baseline levels and is maintained that way during the tonic contraction. At the end of the tonic phase, during voluntary relaxation, presynaptic inhibition is increased. It is presumed that a reduction in the monosynaptic stretch reflex will aid muscle relaxation.

Such experiments in humans also show that there is separate control over presynaptic inhibition to different muscles within a limb. Selective contraction of one muscle in a limb is accompanied by an increase in presynaptic inhibition to other muscles whilst presynaptic inhibition to the agonist is, as we have seen, decreased. This may be a focusing mechanism which helps stop unwanted contraction of distant muscles. The result also shows that there must be separate sets of presynaptic inhibitory interneurones which project to different muscles, and that these can be addressed separately by a descending voluntary command (see Figure 6.11).

The experiments above have concentrated on isometric contraction. Presynaptic inhibition is controlled quite differently in some isotonic contractions. Much attention has been directed to what happens during standing, walking and running. Analysis of reflex excitability in such conditions is difficult because of the continually changing amount of EMG activity (which itself will affect reflex excitability by changing the background facilitation of motoneurones), and technical difficulties in ensuring that exactly the same size of conditioning and test stimuli are given during different parts of the movement. Despite this, it is now believed that there is increased presynaptic inhibition of Ia afferents in the soleus H-reflex during walking and (more) during running. In other words, during movements in which the voluntary command produces high levels of EMG activity, it is at the same time acting to reduce the gain of monosynaptic feedback in the soleus muscle by presynaptic mechanisms.

(Further reading on presynaptic inhibition in humans can be found in the following: Hultborn *et al.*, 1987; Schieppati, 1987; Stein and Capaday, 1988; Meunier and Pierrot-Deseilligny, 1989; Edamura, Young and Stern, 1991.)

Propriospinal neurones

As described in Chapter 7, the C3/C4 propriospinal neurones seem to form an important relay for descending motor commands related to visually-guided reaching movements in the cat. It is now possible to demonstrate the existence of such a system in humans. The techniques rely on showing that low intensity stimulation of a mixed nerve not only produces a monosynaptic Ia excitation of motoneurones which is responsible for the H-reflex, but also produces a much smaller excitation which occurs 3–4 ms later. The idea that this input comes from a C3/C4 propriospinal-like system is supported by several lines of evidence. First, the extra latency over and above the H-reflex is appropriate for the longer conduction pathway. Second, activity in this pathway is greatly facilitated during voluntary contraction of the target muscle, as expected from the strong monosynaptic corticospinal projections to propriospinal neurones which have been shown in the cat. Third, this late excitation receives both facilitatory and inhibitory inputs from low threshold cutaneous and muscle afferents in the arm. The pattern of this input is quite specific for different groups of propriospinal neurones projecting to different sets of muscles within the arm. This reinforces the suggestion made in the cat that such afferent input may be involved in selecting groups of propriospinal neurones to perform particular synergic movements. Thus, despite the very strong monosynaptic corticospinal inputs to arm muscles that can be demonstrated in man, important inputs may also be transmitted by propriospinal pathways analogous to those seen in the cat (Pierrot-Deseilligny, 1989; Gracies *et al.* (1991) for example, give more details).

6.7 *Cutaneous reflexes*

In contrast to the enormous literature on the reflex effects of muscle afferents, cutaneous reflexes have been very little studied. Their actions depend on the type or modality of the stimulus, its location on the body surface and, because most cutaneous reflexes are polysynaptic, they depend very much on descending supraspinal and other types of afferent inputs.

In the decerebrate cat, stimuli which would be regarded as painful usually result in flexion of the limb to withdraw it from the stimulus, accompanied by extension of the contralateral limb. This is known as ipsilateral flexion with crossed extension, which would presumably be of use in the intact animal, allowing the contralateral leg to take the weight of the animal after ipsilateral withdrawal. These flexion reflexes are considerably enhanced in the spinal preparation and may be triggered off by less noxious stimuli. In a high decerebrate animal, with an intact red nucleus, more complex cutaneous reflexes may be elicited, involving non-noxious input from mechanoreceptors. An example of this is the placing reaction which can be evoked by stroking the dorsum of a paw against the edge of a table while suspending the cat above it. The leg will flex and then extend to lift its paw onto the surface. This may then be followed by the positive supporting reaction induced by pressure sensors in the ventral surface of the paw, leading to co-contraction of flexor and extensor muscles of the leg and the formation of a rigid supportive pillar.

One of the most complex cutaneous reflexes is the scratch reflex of the spinal dog. Very light, moving stimulation of the hairs, mimicking the passage of an insect, can give rise to a vigorous scratching by the ipsilateral foot. The scratching is rhythmic, continues until the stimulus is removed and is directed quite specifically towards the site of stimulation. Despite their range, very little is known about the pathways which mediate these cutaneous effects.

Flexion reflexes in humans

Relatively high intensity electrical stimulation of peripheral nerve, or natural painful stimuli such as pinch or pin-prick can give rise to flexion reflexes in humans. The best place to demonstrate these reflexes electrophysiologically is in the relaxed leg of reclining subjects. Stimulation of the sole of the foot with a short (10–20 ms) train of impulses produces a reflex contraction in tibialis anterior (and/or hamstrings) which consists of two components: a short latency response (minimum 50–60 ms), and a longer latency response at 100 ms or more. The late response produces the more forceful contraction and actually withdraws the foot from stimulus (Figure 6.12). The function of the early part is not clear. In normal subjects, the early response usually has a lower threshold than the late response (Shahani and Young, 1971).

(a)

Figure 6.12 (a) Schematic drawing of the reflex pattern of movement obtained by painful cutaneous stimulation of the ball of the great toe (left) and buttock (right). Filled areas denote the trunk muscles involved in the reflexes. (From Kugelberg, Eklund and Grimby, 1960, Figure 1; with permission.)

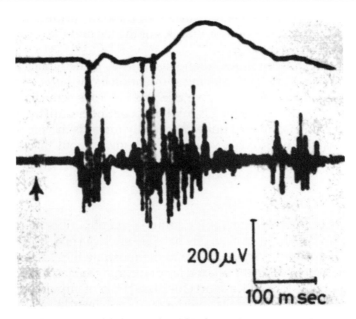

Figure 6.12 Continued. (b) A combined EMG and force record of a typical flexor reflex in a normal subject. The lower line is the EMG recorded from the tibialis anterior in response to a short train of stimuli (arrow) applied to the sole of the foot. The other line is the force exerted by the dorsum of the foot on a force transducer. The main part of the force response is produced by the late component of the flexor reflex.

These responses are rather variable and decrease in size if the stimuli are given too frequently. However, their unpredictability has one useful implication. Since either the early or the late component can appear in isolation after giving the same stimulus, we must conclude that they are mediated by two different reflex pathways. Likewise, since the responses can be seen in patients with complete spinal transection (see below), they must be spinal in origin. It is now believed that they are analogous to the early and late components of the flexion reflex described in spinal cats after stimulation of the flexor reflex afferents (see Chapter 5). Like the responses in spinal cats, flexion reflexes in man are thought to be produced by afferents conducting at velocities within the group II range (40 ms^{-1} or so). The responses have too short a latency to involve activity in unmyelinated, C fibres.

Flexion reflexes usually are investigated in relaxed, reclining subjects because the pattern of responses changes with posture. In fact, these reflexes do not always involve the flexion. The rule of thumb is that the reflex elicited by painful stimulation of a single point on the skin produces reflex muscle activity which is appropriate to withdraw that area from the stimulus (Figure 6.12). If the stimulus is, for example, over

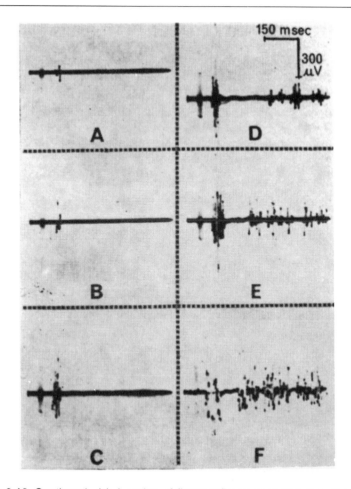

Figure 6.12 Continued. (c) A series of flexor reflexes recorded from the tibialis anterior of a normal subject after different intensities of cutaneous trains applied to the sole of the ipsilateral foot. The stimulus voltage was increased from A through to F (15, 20, 25, 30, 35 and 40 V). Note that the early component of the response has the lower threshold. The later component has a higher threshold and its onset becomes earlier at higher intensities. (From Shahani and Young, 1971, Figures 1 and 3.)

the front of the thigh, the legs and hips will be extended whereas if it is on the back of the thigh, the legs will be flexed.

In humans, particularly interesting variations on this theme occur after stimulation of the sole of the foot. Normally, this results in flexion at the knee, plantar flexion at the ankle and flexion of the toes. In clinical neurology, this **plantar response** usually is produced by firm stroking of the lateral aspect of the sole with a blunt object. It serves to withdraw the sole of the foot from the site of stimulation. However, the response changes after damage to the pyramidal tract. Physiological flexion is produced at all joints: the knee flexes and both the ankle and toes dorsiflex. This is known as Babinski's sign ('the upgoing big toe'). It is a pure flexion reflex quite distinct from the subtle response in the normal human. The normal response is believed to be adapted from the true flexion reflex to allow withdrawal of the foot sole, while maintaining contact of the toes with the ground. With bipedal gait, this adaptation is of obvious advantage in postural stability. In fact, it is only when painful stimuli are applied directly to the ball of the toe that a full flexion reflex is elicited in normal humans.

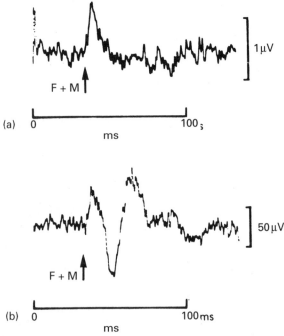

Figure 6.13 Cutaneous reflex responses from the left (a) and right (b) first dorsal interosseous muscle following electrical stimulation of the digital nerves of the index finger. The records are from a patient with a cerebral angioma in the region of the right motor cortex. The patient was required to make a steady isotonic contraction of the muscle while 512 stimuli were given at 3 Hz to the index finger. The traces represent the average modulation of EMG activity that the stimulus produced. On the right, intact, side (b), the response consists of an initial excitation, followed by a depression and later excitation. These are the E1, I1 and E2 phases of the reflex. The onset of E1 (35 ms) occurs at a latency compatible with operation of a spinal reflex arc. This can be shown by measuring the F + M latency in the same muscle (arrow). E2, which begins at about 60 ms and I1 are not present on the left side. E2 is thought to be due to activity in a transcortical reflex loop. I1 is believed to be a spinal reflex, dependent upon the presence of tonic descending input from the brain. Damage to the motor cortex prevents the appearance of both components. (From Jenner and Stephens, 1982, Figure 5; with permission.)

Low threshold cutaneous reflexes in humans

Tactile reflexes are difficult to elicit in relaxed muscles. When detected in the electromyogram, they usually have a shorter latency and a lower threshold than the responses described above. They are more easily seen as fluctuations in the level of activity in tonically contracting muscles. The muscles of the hand and forearm have been extensively studied in this way. Parts of the response are thought to involve activity in trans-cortical reflex pathways. The procedure is to ask the subject to produce a sustained contraction of one or more muscles and then to give small electrical stimuli to the cutaneous nerves of the fingers. An average response of one of the hand muscles is shown in Figure 6.13. There are three main phases: a short latency excitation (E1) followed by inhibition (I1) and later excitation (E2).

The E1 and I1 components are believed to be mediated by spinal circuits, whereas the E2 response is thought to traverse a transcortical reflex pathway. Thus the E2 response is small or absent or delayed in patients with lesions of the motor cortex, dorsal columns or pyramidal tract. Motor cortical damage also reduces the size of the I1 component. Because of this it has been suggested that the I1 response depends on tonic facilitation of spinal interneurones from the brain. In confirmation of this, the I1 response is present at normal latency in patients with intact, but slowly conducting descending pathways, although their E2 response is delayed. The E2 component also shows an interesting development with age. It is completely absent at birth, and only begins to appear in the second year of life. This parallels the development and myelination of the corticospinal tract during the same period, so that maturation of the reflex may depend on maturation of its anatomical pathway.

A second feature of these cutaneous effects is their functional distribution to different fractions of the motoneurone pool. At the level of the single motor unit, it has been shown that electrical stimulation of cutaneous nerves can produce overall inhibition of first-recruited units and excitation of later-recruited units. The implications of this result for the firing order of motor units have been discussed in Chapter 4.

The role of cutaneous inputs in natural movements of the hand

The flexion reflex and the electrically-elicited cutaneous reflexes described above provide only very limited examples of the effectiveness of cutaneous inputs in the control of movement. The role of cutaneous afferents under more natural conditions has been investigated in a series of elegant experiments by Johansson and Westling. Subjects have to lift an object from the table using a precision grip between the thumb and the index finger. They hold it in the air for several seconds and then replace it on the table.

Two forces are exerted by the subject on the object (1) the grip force between the thumb and finger which acts perpendicular to the surface, and (2) the vertical lifting force, or 'load' force which acts against gravity. In order for the object to be lifted from the table, the finger and thumb must grasp it with sufficient force that the friction between the skin and the surface prevents the object from slipping as the weight is taken. The higher the grip force, the greater this friction will be.

One way of making sure that the friction between the fingers and the object is sufficient to prevent slipping is that all objects are gripped with the maximum force. Under these conditions the object will slip only if it is very smooth or very heavy. However, in most circumstances, this is clearly not an appropriate strategy. Maximum grip force could easily destroy a delicate object that we are manipulating. Under normal conditions, the grip force is scaled to the weight of the object, and the frictional quality of the object's surface. Thus, if the object is slippery, the grip force is larger than when the object has a rough surface. Similarly, if the object is heavy, the grip force is larger than when the object is light.

Figure 6.14 illustrates the behaviour of a normal subject lifting a known weight. The fingers touch the object and the grip force begins to increase. Shortly after, the vertical lifting force (load force) also begins

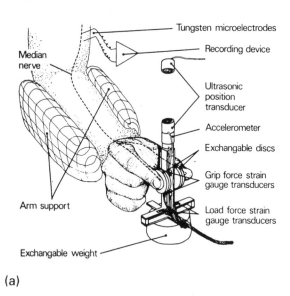

(a)

Figure 6.14 (a) Experimental apparatus used in investigation of the cutaneous influences on control of precision grip.

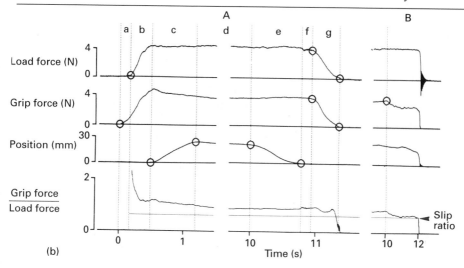

Figure 6.14 Continued. (b) A, principal phases of a lifting trial. The traces show the load force (the vertical lifting force on the object), the grip force, the vertical position between the object and the surface, and the ratio between the grip and the load force. The subject was lifting a 400 g weight with a sandpaper surface. The periods indicated are: **a**, preload phase (increase of grip force from initial contact until object is securely grasped); **b**, loading phase (parallel isometric increase of both grip and load force without removal of object from the table); **c**, transitional phase (object leaves the table and is elevated in the air); **d**, static phase (the object is held stationary in space); **e**, replacement phase; **f**, delay; **g**, unloading phase (parallel isometric decrease of the forces of grip and load). Note the interrupted time scale during phase **d**. On the right (B) is shown a segment of record in which the slip ratio of grip to load force is estimated for this particular combination of object and experimental subject. The subject is asked to hold the object in the air decrease his slip force until slips occur. The dashed vertical line indicates the start of a slow voluntary grip force decrease. The final slip of the object occurs some 2 s later. The grip/load force ratio at which this slip occurs is the slip ratio. (From Johansson and Westling, 1984, Figure 2, Westling and Johansson, 1987, Figure 1; with permission.)

to rise. When the load force equals the weight of the object, then the object begins to move up from the table into the air. It is held for 10 seconds or so and then replaced on the table. On reaching the table, the grip and load force both decline in parallel. Throughout most of the movement, the ratio of grip to load force is approximately constant and slightly above the minimum ratio (the **slip ratio** for this particular object and its surface) at which slips would occur. For a given frictional surface, the grip/load force ratio is kept constant when lifting different weights. The heavier the weight the larger the grip force, but, since the load force also is larger, the ratio between the grip and load forces remains the same. In contrast, if the surface of the object is changed, and made, for example, more slippery, then the grip/load force ratio is changed (in this case it would be increased; we would need to grip harder to support the same weight).

Adaptation of grip and load force is made using cutaneous input. This can be confirmed by anaesthetizing the fingers. When this is done the grip force is not changed when the surface of an object is changed. During lifts of a novel object, the subject makes a guess at the correct ratio of grip to load force. This is based on either the visual characteristics of the object, or (if vision is not allowed) on the basis of a previous lift made by the subject. During lifts in which the frictional characteristics of the surface are changed, there is an automatic adjustment in the grip force when afferent input from the fingers is intact, which occurs only 0.1–0.2 seconds after the initial touch. Afferent recordings from single cutaneous fibres in the digits show that this adjustment is probably an initial burst of discharge in SAI and FAI units. The magnitude of this burst (especially in the FAI unit) is strongly influenced by the nature of the surface: the more slippery the surface, the greater the discharge (Figure 6.15).

If this initial adjustment is insufficient for establishing an adequate safety margin, the object will slip (even if only by a very small amount). Slips which occur during the loading phase, when the object is not actually lifted from the table, are rarely noticed by the subject. Nevertheless, they elicit bursts of discharge in FAI, FAII and SAI afferents.

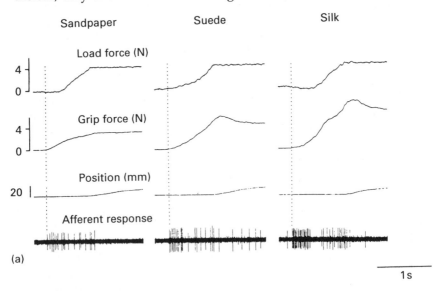

(a)

1 s

Figure 6.15 (a) Examples of initial afferent responses in an FAI unit and the influences of the surface structure. Single trials with sandpaper, suede and silk. The afferent fires much more when the surface is silk than when it is sandpaper. This firing occurs even before the onset of load force in the 'pre-load' period.

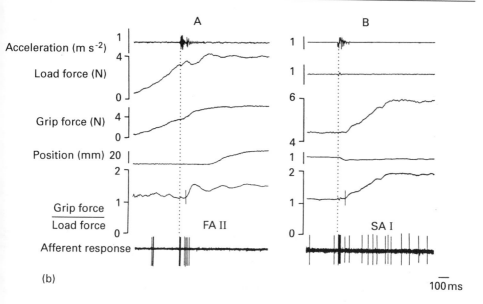

Figure 6.15 Continued. (b) Examples of afferent slip responses and upgradings of the grip force to load force ratio elicited by small but distinct slips. The slips have been detected by an accelerometer on the object (upper traces). In A, the slip occurred during the loading phase, with the object still in contact with the table. The afferent fibre was an FAII unit. The discharge created a permanent increase in the grip/load force ratio. In B, the slip occurred during the static phase of the task, when the object was suspended above the table. The afferent responses from an SAI unit, again produced a sustained increase in the grip/load force ratio. Note that in A, the slip adjusted both the grip force (increase) and the load force (decrease) in the first 100 ms after the slip. In contrast, in B, the load force remained constant, whilst the grip force was increased.

This discharge upgrades the grip/load force ratio with a latency of about 75 ms to onset of force change, or about 65 ms to onset of EMG changes (Figure 6.15). For comparison the most rapid voluntary reaction to a slip has a latency of about 150 ms. The conclusion is that cutaneous afferent discharge can exert powerful and appropriate reflex changes to the commands involved in a coordinated grip. The reflex latency is compatible with the long-loop, transcortical reflexes discussed above, but as yet there is no evidence to favour this pathway in preference to a polysynaptic spinal route. There is no evidence that muscle spindle activity changes during these slips so that the responses can safely be assumed to be cutaneous in origin.

There are two points of interest in these reflex corrections. First, even though the afferent discharge is a phasic burst, the new grip/load force ratio is maintained for a considerable time after the initial slip. The transient change in input has not given rise to a transient reflex correction,

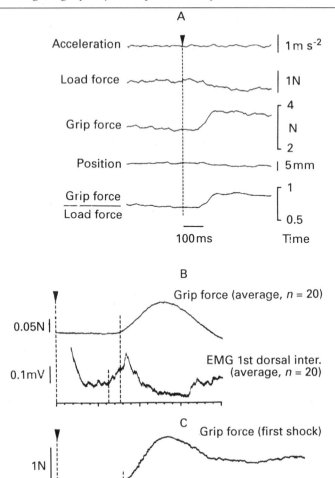

Figure ⌐.15 Continued. (c) Examples of grip force responses to single electrical stimuli applied to the pad of the index finger whilst the subject held an object in the air. A shows the response on the first trial. The grip force is increased by almost 1 N some 80 ms after the stimulus and is maintained at the new level until the end of the sweep. B shows EMG and force responses on an expanded time scale for the next 20 trials. In these traces, the electrical stimulus comes at the start of the sweep and the dotted lines indicate EMG and force onset at 65 and 78 ms, respectively. However, the grip response is very small (0.1 N) and short-lasting in comparison with that seen in the first trial. C shows the grip force response on the first trial on the same timescale as B. (From Johansson and Westling, 1987, Figures, 2, 9 and 5; with permission.)

it has updated the whole command for movement. It is as if a single slip can adjust the 'frictional memory' which is being used to guide the movement. The second feature of these reflex corrections is that they influence the activity of many different muscles. The grip force upgrades activity in both finger and thumb flexors. Indeed, if slips occur in the loading phase of the movement, before the object has left the table, there is even a change in the activity of elbow flexor and radial deviator muscles which slows down the rate at which the load force on the object increases. Slips have no effect on load force if the object is suspended in the air. Any adjustment in load force at this point would mean that the object would move up or down in the air.

Upgrading of the grip force on an object can also be induced by weak electrical stimulation of cutaneous afferents in the region of skin which is in contact with the object being lifted. Figure 6.15 shows an example of this effect. In contrast to the responses to natural slips, the responses to electrical stimulation adapt very rapidly after repeated presentation. As shown in the figure, the response on the first trial is more than 10 times larger than the average response over the next 20 trials. Somehow, the electrically-elicited input is recognized as inappropriate and the response on subsequent trials is scaled down. Perhaps a similar effect explains the small size of reflex response in the task-irrelevant electrically-elicited cutaneous reflexes described above. The lesson from this is that although it is possible to investigate reflex **pathways** with electrical stimulation, it is only possible to investigate their **function** under natural conditions.

Changes in the frictional characteristics of an object's surface are detected during the loading phase of contraction, before the object leaves the support surface. However, if changes are made to the weight of the object without affecting the surface characteristics, then there are no clues as to the change until the loading phase is complete and the object has left the table. For example, if a subject has practised lifting an 800 g weight, then an unseen change to a 200 g weight will produce no change in the afferent discharge during the loading phase. Only when the object moves off the table too early and too fast is there any clue that its weight has changed. Under these conditions, the sensory signals relating to lift-off disrupt the muscle activation pattern with a latency of 80–110 ms and correct for the excessive forces applied. Interestingly, it can also be shown that **absence** of expected input can update the commands to move. This occurs when a practised load of, say, 200 g is unexpectedly replaced by an 800 g load. Again, all initial inputs are the same when lifting the new weight. It is only when the object fails to move at the expected time that the programme is updated: grip and load force continue to increase after a short hiatus at the time of expected lift-off.

In summary, cutaneous inputs are essential for the control of grip force. Unlike most reflexes which have been studied, phasic afferent discharges do not produce a transient correction of muscle commands. An afferent burst is capable of permanently updating the commands to move. One explanation of this is that the cutaneous input can adjust the 'memory' of the object being manipulated, and it is this memory which drives the final output to muscle (Johansson, 1991, for summary).

6.8 *Pathophysiology of spinal reflexes*

Most of the conditions in which disorders of spinal reflexes can be recorded are due to abnormal activity in descending pathways which control interneuronal excitability in the spinal cord. In some cases there is overactivity in these pathways; in others underactivity. The latter sometimes results in reorganization of synaptic connections at the spinal cord. Perhaps most frequently there is a failure to modulate interneuronal excitability during the course of voluntary movement. This may cause much greater disability to the patient when he tries to move than when he is at rest. Diseases which attack the spinal interneuronal circuitry itself are rare, but do exist.

The decerebrate preparation

A clear example of the effect of descending activity on the excitability of spinal mechanisms can be seen in the decerebrate cat. The classical decerebrate preparation is produced by sectioning the brainstem between the superior and inferior colliculi (intercollicular decerebration). This produces such intense stiffness of the body muscles, particularly in the anti-gravity extensor muscles, that the animals may be able to stand unassisted. Nevertheless, because they lack postural reflexes, they are extremely unstable.

Sherrington was the first to show that this increase in muscle tone was due to a reflex mechanism. Sectioning the dorsal roots abolished the contraction. In addition, since activity in a muscle remained even after section of all cutaneous nerves, the reflex contraction must have been due to muscle afferent activity. The increased tone of such preparations is known as decerebrate **rigidity**. Unfortunately, as we shall see below, decerebrate rigidity in animals does not correspond to clinically defined rigidity in man. It is a confusion of terms which it is now too late to change.

What is the mechanism responsible for increased stretch reflex activity in the intercollicular decerebrate? The usual explanation is that it is due to overactivity of the fusimotor system leading to increased sensitivity of muscle spindle afferents. This is supported by the presence of high tonic spindle discharge in the decerebrate cat, and by a relatively high sensitivity of the receptors to stretch. In addition, Matthews and Rushworth (1958) showed that if procaine, a local anaesthetic, is applied to the muscle nerve, rigidity disappears before the large fibres are blocked. Small fibres are affected first by local anaesthetic and so it is supposed that blocking fusimotor efferents may reduce the sensitivity of spindles to stretch and decrease their background discharge.

Excess fusimotor activity is certainly one factor which contributes towards increased excitability of the stretch reflex arc. However, other factors also must be involved. One reason for this is the observation that vibration of a muscle, producing a large tonic Ia input to the cord, does not produce a tonic muscle contraction in spinal animals (nor in spinal man, see below). Thus, a large Ia input alone does not necessarily produce stretch reflex rigidity. It is thought that other spinal reflex pathways also have to be facilitated in order for decerebrate rigidity to occur. Details are not known; they may include facilitation of polysynaptic Ia connections to motoneurones, or decreased presynaptic inhibition of the terminals of Ia afferents.

The decerebrate preparation has particularly large stretch reflexes because of the unopposed action of centres in the brainstem which act on spinal reflex mechanisms after decerebration. The intercollicular section which produces decerebrate rigidity is thought to lead to an imbalance between descending facilitatory and inhibitory systems which project to flexor and extensor systems in the spinal cord. Section at the intercollicular level tips the balance towards facilitation of extensor systems, with a large portion of the excitability directed towards the fusimotor system. The effects are different if sectioned at other levels. No rigidity occurs if the section is made above the level of the colliculi, to include part of the red nucleus, or if the section is too caudal, below the level of the vestibular nuclei.

A second type of hypertonus can be observed in the cat if ablation of spinal parts of the anterior cerebellum is carried out in addition to intercollicular decerebration. This produces an animal with extreme hypertonus, which cannot be abolished by dorsal root section. These animals are said to show α-**rigidity** to emphasize the direct driving of α-motoneurones independent of stretch reflex pathways. The anterior lobe of the cerebellum has a inhibitory projection to the neurones of the lateral vestibular nucleus of Dieters. Removal of the inhibition is responsible for a direct vestibulospinal excitation of extensor α-motoneurones (see Chapter 10).

Spasticity in humans

Spasticity is seen after lesions of the upper motoneurone, or in the chronic stage of spinal transection. This section is devoted to spasticity typically seen after capsular hemiplegia. Spinal section is dealt with below. Spasticity in hemiplegia is characterized by increased tendon jerks, the clasp-knife phenomenon and an increase in muscle tone which is greater in flexors than extensors in the arm, and greater in the extensors than the flexors in the leg.

Muscle tone in spasticity

Muscle tone is evaluated in relaxed muscle. An increased tone (hypertonus) means that there is an increased resistance to passive (usually rather slow) movements of the limb made by the examiner. Two mechanisms contribute to the resistance felt during passive manipulation of limb segments: the inherent viscoelastic properties of the muscle itself, and the tension set up in the muscle by reflex contraction caused by muscle stretch. In a normal muscle at complete rest, the stretch reflex component of tone is not present. This is partly because the very low level of background fusimotor discharge in relaxed human muscle ensures that the afferent response to such relatively slow movements is small. In addition, the central level of α-motoneurone excitability is very low. Together these factors explain the lack of reflex muscle activity during manipulation. In fact, muscle tone is the same in relaxed normal individuals as it is in totally anaesthetized patients, in whom fusimotor discharges are absent.

The increased tone in relaxed spastic muscle is caused to a large extent by increased stretch reflexes which oppose movement. Examples of the response to different velocities of biceps stretch in relaxed muscle and the spastic arm of a hemiparetic subject are shown in Figure 6.16. In the normal, no reflex EMG is seen until the stretch velocity is greater than $240°\,s^{-1}$. A short burst of activity at these velocities may be analogous to the tendon jerk. Its rather long latency is probably caused by the time taken for mechanical displacement applied at the elbow actually to produce muscle stretch and cause biceps spindles to discharge. In a spastic subject, large, long duration EMG responses are seen even at the lowest velocity of stretch. These responses are thought to be produced by activity in mono- and polysynaptic stretch reflex pathways. The largest increases in stretch reflexes are seen in the flexors of the upper limb and extensors of the lower limb, corresponding to the distribution of clinical hypertonia.

Interestingly, the situation is quite different if muscle tone is evaluated in the active state. This is not usually performed clinically because

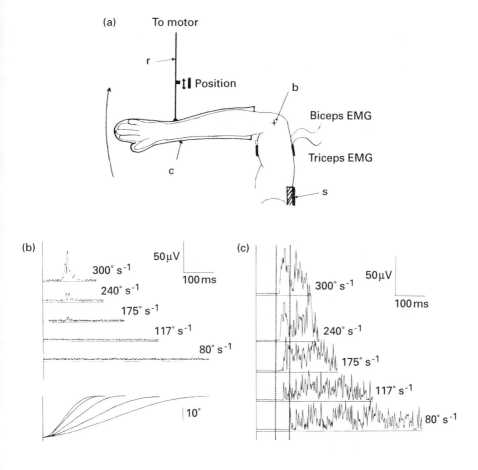

Figure 6.16 The EMG responses of the biceps muscle of a normal (b) and a spastic (c) human subject to a 30° extending displacement applied to the relaxed elbow (see experimental arrangement in (a)). Each trace represents a rectified averaged EMG record obtained from ten displacements at the velocity specified beside it. The length of the trace corresponds to the duration of the displacement. The position records of the displacement are shown in the bottom left. The vertical dotted lines in c indicate the range of timings seen in normals. (From Thilmann *et al.*, 1991a, Figure 1; with permission.)

of difficulties in estimating the resistance to passive movement under these conditions. However, using torque motors it is now relatively easy to evaluate the resistance to movement in actively contracting muscles. In the active state, the resistance of a limb to imposed movement is very similar in both patients and normals. It is as if the enhanced tone of spastic muscle at rest is due to failure to switch off the stretch reflex mechanisms during relaxation. The fact that tone is only increased at rest means that disorders of tone in spasticity cannot contribute to the difficulties that these patients have in making voluntary movements.

The increased responses to stretch at rest may be caused either by increased spindle sensitivity, producing a larger discharge for a given stretch, or by changes in the excitability of the central reflex pathway (for example by changes in presynaptic inhibition, or motoneuronal excitability). It seems that in spasticity the latter is responsible since the microneurographic studies made of spindle behaviour so far in humans have shown no evidence of receptor hypersensitivity. Similarly, experimental models of spasticity in monkeys have been made by lesioning motor cortical areas. In these animals, resting tone is increased without any change in spindle sensitivity (Burke, 1983, for a discussion).

The final question in spasticity concerns the peculiar distribution of increased tone to extensors of the leg and flexors of the arm. Reasons for this are not clear. At the present time it is usually said that these muscles represent the 'anti-gravity' muscles in humans. Extending the leg raises the trunk from the ground, and flexing the arm raises it against gravity. (Note that the toe flexor muscles, because of their anti-gravity action in raising the body are usually defined as 'physiological extensors', together with the other anatomical, extensors in the calf.) In the cat, the anti-gravity muscles (which because of its quadripedal stance are all extensors) appear to be controlled differently from other muscles. For example, stimulation of certain areas in the brainstem causes limb extension, whereas other areas when stimulated cause flexion. Similarly, decerebrate rigidity is primarily directed towards extensors. Perhaps a similar differential control exists in man, so that central lesions can release anti-gravity mechanisms separately from others.

Changes in muscle tone produced by changes in the mechanical properties of muscle

In the past, the discussion of muscle tone has focused very closely on the contribution of reflex mechanisms to the increased stiffness of muscle. However, it is becoming clear that other, purely mechanical, factors may also contribute to the increased tone of spastic muscles. Disuse of weak muscles because of lack of voluntary power leads to changes within the muscle, including fibrosis and probably even changes in the properties of the muscle fibres themselves (see Chapter 3). Such changes make the muscle stiff to move, even in the absence of any muscle activity. There may also be changes in the stiffness of the tendon. The contribution of these factors to the clinical impression of increased tone in spasticity is, however, very much a matter of controversy (e.g. Dietz, 1992; Thilmann *et al.*, 1991).

Tendon jerks in spasticity

Tendon jerks are what Sherrington referred to as a 'fractional manifestation' of the stretch reflex. What is meant by this is that activity in mono-, oligo- and polysynaptic pathways contributes to the stretch reflex as evoked by a relatively slow stretch of muscle (as used, for example, to assess muscle tone). However, a very brief stretch, as produced by a tendon tap, evokes a relatively pure mono- (or oligo-, see section on tendon jerks in normal subjects above) synaptic activation of spinal motoneurones. Because of its clinical importance, it is worth considering this 'fractional manifestation' in more detail.

Tendon jerks are much more easily elicited in spastic patients than in normal subjects. Three differences are noted clinically: (1) the threshold for eliciting jerks is reduced so that, for example, ankle or knee jerks can sometimes be produced even by tapping the tendon lightly with the fingers, (2) the maximum size of the response is increased, and (3) tendon jerks can be obtained in many more muscles than usual.

Like increased muscle tone, it was once thought that the increased tendon jerks in spasticity were due to hyperexcitability of the muscle spindles to stretch. In particular, an increased response to the phasic stretch produced by a tendon tap would be caused by increased γ_d fusimotor action. However, since spindle behaviour appears normal in spasticity, this cannot be the correct explanation. The conclusion is that increased tendon jerks in spasticity are caused by increased excitability of central synapses involved in the reflex arc. The question remains as to which synapses are changed and how they are changed. At present, it is thought that both pre- and postsynaptic factors are involved.

Spasticity is caused by interruption of descending input to the cord from the brain. The physical destruction of axons causes degeneration of their terminals on the spinal motoneurones and interneurones. When this occurs, two things happen in the spinal cord. Over a short time interval, a proportion of the remaining intraspinal synapses, which had until the lesion been inactive, become active. These so-called 'silent synapses' appear to have no function in the intact animal, but following the lesion are called into action by some unknown mechanism. Second, over a longer time period, the remaining spinal synapses sprout new terminals which take over the spaces left vacant by degenerating terminals of descending axons. The nett effect of both these phenomena is to increase the effectiveness of spinal cord connections. In the case of the pathway involved in the tendon jerk, this includes the Ia-motoneuronal synapses and perhaps the di- and trisynaptic relays from afferents to motoneurones that could be involved in the tendon jerk.

Synaptic efficiency may also be increased by presynaptic mechanisms, although less is known about these. Vibration inhibition of the H-reflex provides one piece of evidence in favour of this hypothesis. In normal man, vibration of a muscle produces inhibition of homonymous H-reflexes by a premotoneuronal mechanism (see above). In spasticity, this effect is reduced, as if the amount of presynaptic inhibition of Ia afferent terminals is decreased. If so, then it may be that removal of **tonic** levels of presynaptic inhibition can increase the effectiveness of the Ia-motoneuronal synapse and increase the size of the tendon jerk.

The pathological changes in tendon jerks following lesions in the upper motoneurone in spasticity are an interesting mirror to the normal development of the tendon jerk in children. At birth, tendon jerks can be elicited in many more muscles than in the adult. In addition, taps to one muscle often cause short latency (probably monosynaptic) responses in many nearby muscles, including antagonists as well as synergists. This is known as reflex irradiation. Over the first two to four years of life, the tendon jerk pattern develops into that seen in adults. Since muscle spindle sensitivity is thought to be similar in adults and children, the development of the tendon jerk is thought to reflect changes in spinal cord circuitry such as the reorganization of both the projection pattern and the effectiveness of Ia synapses (O'Sullivan, Eyre and Miller, 1991).

Clasp-knife phenomenon

The clasp-knife phenomenon is best seen in the knee extensor muscles of spastic patients. As the knee is flexed, resistance builds up gradually, and then as a certain point is reached, the resistance suddenly melts away and the remainder of the stretching movement can be completed with little force. The mechanism of this effect has been debated for some time.

It was originally suggested that the inhibitory actions of Ib afferents might mediate the clasp-knife reaction. It was thought that these fibres had a high threshold to muscle tension and hence could account for the abrupt onset of stretch reflex inhibition, which is characteristic of the clasp-knife effect. The phenomenon was believed to be a protective reaction to prevent production of dangerously high tensions in muscle. However, with the discovery that tendon organs are very sensitive to actively produced muscle tension, this theory could no longer hold: tendon organs do not fire **only** when dangerously high muscle tensions are produced. Also, it was found that the underlying inhibition responsible for the clasp-knife reaction far outlasts the actual reduction of muscle tension, whereas Ib activity declines in parallel with muscle tension. Ib effects are not now believed to be the sole factor in the phenomenon, although they may contribute to its initiation.

Closer examination of the clasp-knife phenomenon has suggested other mechanisms. Figure 6.17 is one of the few published records of a clasp-knife phenomenon in spastic man. What appears to happen is that

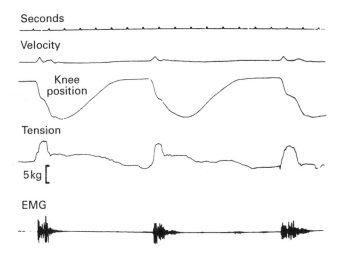

Figure 6.17 Spastic stretch reflex and the clasp-knife phenomenon. Quadriceps EMG (bottom trace) was elicited by stretching the muscle from the extended position of the knee joint to 90° flexion. Velocity of stretch is shown (second trace), with joint position (third trace) and muscle tension (fourth trace) (estimated by force exerted by the examiner on a transducer at the ankle). On flexing the knee, tension and EMG build up rapidly, and then subside almost completely before flexion is completed. This sudden decrease in resistance to movement is known as the clasp-knife phenomenon. (From Burke, Gillies and Lance, 1970, Figure 1; with permission.)

on flexing the knee from the extended position there is a slow, velocity-sensitive build-up of tension in the quadriceps muscle which opposes movement. As the reflex tension increases, it begins to oppose the movement imposed by the examiner, and sometimes is so powerful as to stop the movement entirely. The reduction of stretch velocity reduces the velocity-sensitive input to the reflex, and reflex tension declines. The examiner, noticing the decrease in speed of movement, then resumes a rapid knee flexion. However, at this angle of the knee, the movement is no longer opposed by reflex muscle contraction. It is the absence of reflex contraction at this stage that constitutes the clinical impression of the clasp-knife phenomenon. The melting away of resistance in the clinical situation therefore is due to two factors: (1) reduction in stretch velocity by the opposing muscle contraction; and (2) a decrease in sensitivity of the quadriceps stretch reflex.

The latter effect has been described in some detail by Burke and colleagues who showed that the velocity-sensitive stretch reflex in quadriceps is smaller if stretch is imposed when the knee joint is flexed, than it is when the knee is extended. They refer to this as a position-dependent inhibition of quadriceps stretch reflex, and argue that this is all that is necessary to produce a clinical clasp-knife phenomenon. The reasoning is that quadriceps stretch reflex activity decreases throughout the range of joint movement, but that this is only noticed by the examiner when he tries to increase the rate of knee flexion following a reflex-induced slowing. The novelty of the explanation is that the clasp-knife phenomenon is no longer regarded as a sudden onset of active inhibition. Instead, it is possible to get the same clinical impression from a gradual build-up of reflex inhibition.

Receptors responsible for this inhibitory activity have been suggested to be group II afferents from muscle spindle secondary endings. All group II afferents once were thought to have inhibitory effects on extensor muscles (that is, they belonged to the flexor reflex category of afferent fibres), and it was postulated that with increasing muscle stretch, there would be a length-dependent inhibition of the extensor muscles. However, this explanation in its simplest form fell from favour when it was discovered that group II spindle afferents from extensor muscles could have, like the group Ia fibres, autogenetic excitatory actions (see Chapter 4).

More likely candidates are the group II and III non-spindle afferents. Indeed, some afferents of the group III class have recently been discovered to respond to muscle stretch only when the tension reaches a high level. Such afferents are ideal fibres to sharpen the point at which inhibition becomes noticeable as the clasp-knife phenomenon. Thus the clasp-knife phenomenon is no longer regarded as a pure Ib effect, but is thought to be a complex phenomenon resulting from the actions of many different classes of afferent fibre. The reason that the reaction is not present in rigidity is usually ascribed to a lack of facilitation or active inhibition of the requisite spinal interneuronal pathways (Rymer, Houk and Crago, 1979).

Other spinal reflex pathways in spasticity
Abnormalities have been shown in three other spinal reflex pathways in spasticity: Ia reciprocal inhibition, Ib autogenic inhibition and Renshaw inhibition. The effects observed are thought to be caused by changes in the excitability of spinal interneurones following lesions of descending supraspinal pathways (Pierrot-Deseilligny and Mazières, 1984).

There is an imbalance in the distribution of reciprocal inhibition in the calf. In spasticity, no inhibition can be demonstrated from tibialis anterior onto the soleus, although inhibition from soleus to tibialis anterior is strong. The soleus as a leg extensor muscle shows increased tone in spasticity. This means that any attempt to contract the anterior tibial muscles is doubly disadvantaged, both by the strong resistance to stretch of the hypertonic soleus muscles, and by the lack of any reciprocal inhibition from tibialis anterior which might have been available to cancel this out.

Ib inhibition from medial gastrocnemius to soleus is also lacking in spastic hemiplegia. Thus, if the discharge from Ib afferents in the autogenetic inhibitory pathway has any important role in regulating muscle contraction or changing the gain of the stretch reflex, then this would be missing in spasticity. Again, this would increase the consequences of the increased resting tone of the extensor muscles. It should also be noted that if the results obtained in the calf can be extrapolated to the quadriceps muscle, then the decrease in Ib inhibition in spasticity is a further argument against the possible role of Ib effects in the clasp-knife response.

Renshaw inhibition also has been explored in spastic patients with lesions of the upper motoneurone. Unlike reciprocal Ia inhibition, no abnormalities have been seen at rest. However, as might be expected because of the lesion in descending pathways, the control of Renshaw cell activity during voluntary contraction is severely diminished. Renshaw cell activity remains high even during voluntary activation of muscles.

Parkinson's disease

Rigidity

Patients with Parkinson's disease have increased muscle tone which is evenly distributed between flexor and extensor muscles of both limbs. The increased resistance of the limbs to passive movement imposed by an examiner is due to increased stretch reflex activity produced by muscle stretch. An example is shown in Figure 6.18. The more rigid the patient, the larger the reflex response produced by passive movement.

In the past, there have been many attempts to relate the increased reflex response to activity in a particular stretch reflex pathway. Since the tendon jerk is probably little changed in Parkinson's disease, several workers investigated the long latency stretch reflex which is produced by relatively rapid muscle stretch. These responses are increased in patients compared with normal, and it is probable that some of this increase is due to hyperexcitability in the transcortical component of the response. A possible mechanism would be reduced tonic inhibition of such transcortical responses by underactivity in the supplementary motor area caused by reduced basal ganglia output (see Chapter 11). However, despite the attraction of the theory, the size of long latency stretch reflex in Parkinson's disease is not well correlated to clinical rigidity. The conclusion is that, when manipulated rather slowly by an examiner, many pathways contribute actively to the overall reflex response. Some pathways, like the monosynaptic pathway may function relatively normally. Others, like the long latency ones may be hyperactive. The

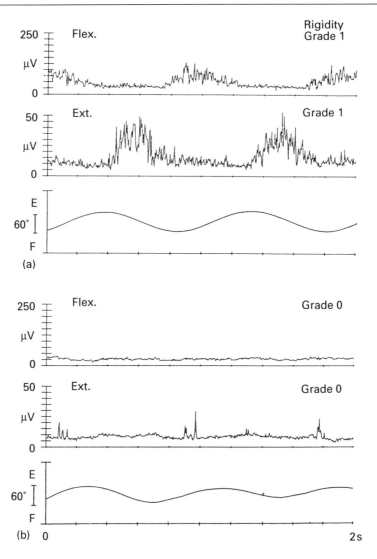

Figure 6.18 Rectified, surface electromyograms from the wrist flexor (flex.) and extensor (ext.) muscles during manually applied alternating extension (E) and flexion (F) movements of the wrist in a patient with Parkinson's disease and moderate rigidity. EMG and displacement records are simultaneous averages from three movement sequences. The rigidity grade assessed by the clinical examiner during the movements illustrated are indicated to the right of each record. The value against a flexor EMG record denotes rigidity during extension movements and that against the extensor EMG records denotes rigidity during flexion movements. (a) Records obtained prior to medication; (b), records obtained following medication when rigidity had almost disappeared. (From Meara and Cody, 1992, Figure 3; with permission.)

net result, though, is due to the operation of all together, and this is what the examiner feels.

As with spasticity, it is thought that muscle spindle function is normal in Parkinson's disease, so that the increased stretch reflexes must represent abnormal central processing of stretch input. Nevertheless, the normal tendon jerks, the even distribution of increased tone, and lack of a clasp-knife reaction mean that there must be quite separate mechanisms involved in producing rigidity and spasticity. Unfortunately, the precise cause of rigidity has yet to be established.

Activity in spinal reflex pathways in Parkinson's disease
Although Parkinson's disease is a disease of the basal ganglia (see Chapter 11), it is now clear that it is accompanied by changes in the excitability of some spinal reflex pathways. These changes are due not to lesions of supraspinal descending pathways, but to abnormalities in the activity of these systems.

At rest, the excitability of the Ia reciprocal inhibitory pathway is increased between both the forearm extensor and flexor muscles and between the tibialis anterior and soleus muscle. In contrast, the excitability of the Ib autogenic inhibition between medial gastrocnemius and soleus is decreased (or even replaced by facilitation). These changes are related to the level of rigidity seen in the patients, and Delwaide (Delwaide, Pepin and Maertens de Noordhout, 1991) has made a specific suggestion that they are caused by increased activity in the dorsal reticulospinal system, which in cats can facilitate Ia inhibitory interneurones and inhibit Ib inhibitory interneurones. Renshaw inhibition is normal.

Flexor reflexes also are changed in Parkinson's disease. The early component of the flexor response is much more readily obtained than in normal subjects, and the late component almost always is absent. Remembering that monoaminergic precursors (such as L-dopa) given to spinal cats can facilitate the long latency component of the flexor reflex, and inhibit the early component, then the opposite finding in Parkinson's disease may be related to the changes in dopamine function in the spinal cord late in the condition.

Dystonia

Dystonia is a disease of the basal ganglia discussed in detail in Chapter 11. It is characterized by abnormal co-contraction of antagonist muscles during action, or even at rest. Despite the symptoms, there is some debate as to whether there is a significant change in the level of disynaptic reciprocal Ia inhibition between antagonist muscles at rest. More impressive, though is a clear reduction in the amount of presynaptic inhibition of Ia fibres, at least as tested between the large diameter afferents from extensor muscles in the forearm onto the Ia fibres from the flexor. Such a disorder of presynaptic mechanisms has been attributed to changes caused by the disease in the level of activity in descending supraspinal systems. In view of the careful control of presynaptic inhibition during different tasks, which has been revealed in recent work on man, a lack of presynaptic effects might well cause some of the clinical symptoms of excessive muscle activity in dystonia (Nakashima *et al.*, 1989).

Effects of spinal cord transection

In all mammals, acute transection of the spinal cord produces an initial period of areflexia, during which all reflexes have extremely high thresholds. The duration of this areflexia varies, being only an hour or so in the cat, to days or weeks in humans. It is caused by the sudden removal of facilitatory descending inputs to the cord from supraspinal structures, and is followed by the gradual reappearance and eventual hypersensitivity of some reflexes.

In humans, acute transection is immediately followed by flaccid paralysis and areflexia. The first signs of recovery are the automatic reflex emptying of bladder and rectum, followed by the appearance of cutaneous reflexes in some muscles. These invariably result in flexion withdrawal of the limb and may be accompanied by emptying of the bladder and rectum (the **mass** reflex). Later on, tendon jerks appear and, after some months, the enhanced flexor reflexes may lead to a permanent posture of flexion of the hip, knee and ankle, resulting in paraplegia in flexion. The stretch reflexes are enhanced in this state, but the predominant final effect of spinal transection is to enhance transmission in flexor reflex pathways.

There have been surprisingly few electrophysiological studies of flexor reflexes in patients with spinal transection (Shahani and Young, 1971; Roby-Brami and Bussel, 1987, 1992). The early (less than 100 ms) and late (greater than 100 ms) reflex seen in the legs of normal man after short trains of nerve stimuli behave in the following way after spinal transection. First, the late component has a lower threshold than the early component. Indeed, the early component is sometimes difficult to observe in some patients. Second, the latency of the late component is increased, if the duration of the stimulus is prolonged. Third, the late reflex is inhibited by a preceding flexor reflex stimulus to the opposite leg (Figure 6.19). These three findings are strikingly similar to the flexion reflexes described above in spinal cats. They show that not only are the pathways from flexor reflex afferents virtually identical in the two species, but also that some part of the interleg coordination which may

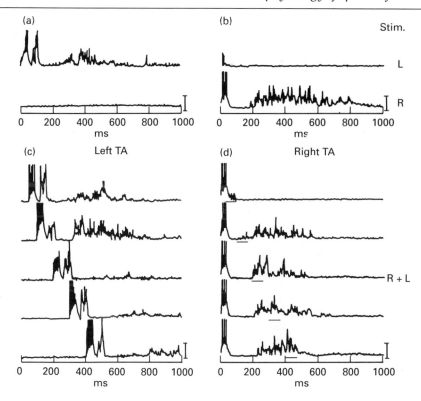

Figure 6.19 Flexion reflexes in the tibialis anterior muscles on the right and left side (R, L) of a paraplegic patient with a complete spinal cord transection at the T8 level. A and B show the average of five rectified EMG traces recorded simultaneously from each side after stimulation of the left (a) or right (b) sural nerve with a 30 ms train of stimuli. The responses are strictly unilateral, and on the left consist of an early and late component, whereas on the right there is only a late component. The records in (c) and (d) show the results of applying bilateral stimulation and illustrate the interaction between the reflexes in the two legs. Stimuli were always given to the right sural nerve first (d), and the left sural nerve was stimulated 50–400 ms later (c). Responses in (c) and (d) are from the ipsilateral muscle only. Note that the response in the right tibialis anterior is abolished if the left sural nerve is stimulated 50 ms later (first row of records). Thereafter, when the left sural nerve is stimulated 200, 300 or 400 ms after the right sural nerve, then the late response on the left was virtually abolished. The early response on the left was still present, and the responses on the right were unaffected. The bars underneath the EMG records on the right side indicate the time of the stimulus to the left sural nerve. (From Roby-Brami and Bussel, 1992.)

underlie primitive locomotor generators is present in the isolated human spinal cord. Clearly this is of some importance to those involved in rehabilitation of spinal patients.

6.9 *Movement without reflexes: deafferentation*

Given the wide variety and adaptability of the reflex pathways described in this chapter, it is sometimes easy to imagine that these

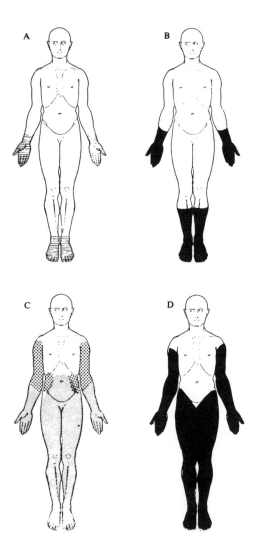

Figure 6.20 (a) Schematic illustration of the distribution of sensory loss in a typical 'deafferented' patient with a severe sensory peripheral neuropathy. A, absent vibration sense (cross-hatched); B, impaired temperature sensation (solid); C, absent (stippled) and reduced (dots) pinprick appreciation; D, absent light touch (heavy stippling).

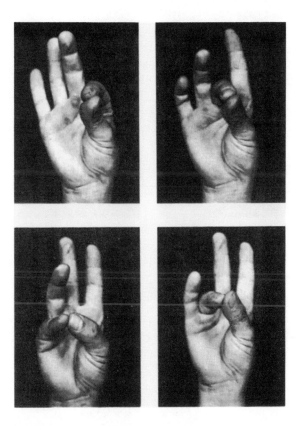

Figure 6.20 Continued. (b) Sequential series (clockwise) of photographs showing the patient's ability to perform individual finger movements with the right hand with his eyes closed. Reading from top to bottom, he opposed his thumb to each finger in turn.

responses are **essential** for movement. However, this is not the case. Surprisingly complex movements can be made in the absence of any sensation (and hence any reflexes) at all. Several studies have been carried out on adult patients who have been effectively deafferented by severe sensory peripheral neuropathies. Patients may be left with virtually normal muscle strength, despite complete loss of sensation, particularly in the distal extremities. A typical pattern of sensory loss is shown in Figure 6.20a.

The question is what sort of movements can these patients make? Figure 6.20b shows finger movements made by the patient with the devastating sensory loss shown above. When this patient closed his eyes he could move his thumb to touch each finger in turn. This was despite the fact that he could not feel his fingers, nor did he notice if the examiner

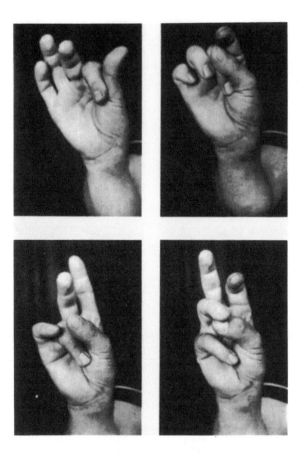

Figure 6.20 Continued. (c) As in (b), after he had been performing the sequence for 30 s without visual feedback. The movements break down. (From Rothwell *et al.*, 1982, Figures 1 and 4; with permission.)

interfered with the movement. Thus, the whole set of instructions to all the hand and forearm muscles involved in the task could be sent out by the brain in the absence of any feedback. He could also outline figures and numbers in the air without vision. Nevertheless, the movements were not entirely normal. Normal subjects can continue almost indefinitely to oppose fingers and thumb in turn. The performance of this patient degraded considerably over 2–3 trials (Figure 6.20c), resulting in the final decomposition of the movement. Presumably, undetected errors crept into the movement and were never compensated. In this case, feedback was necessary to fine-tune the central commands, but was not necessary to run them in the first instance.

There are other situations when feedback is more important. Typical

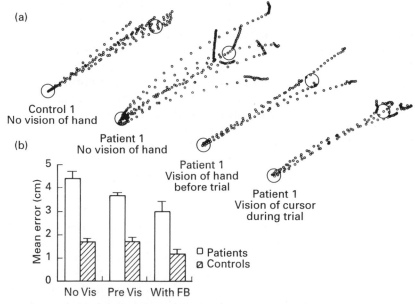

(a)

Control 1
No vision of hand

Patient 1
No vision of hand

(b)

Patient 1
Vision of hand
before trial

Patient 1
Vision of cursor
during trial

□ Patients
⊠ Controls

No Vis   Pre Vis   With FB

Mean error (cm)

Figure 6.21 The effect of vision of the hand before or during movement on the accuracy of movements made by deafferented patients. Subjects sat in front of a computer graphics tablet and had to use their whole arm to move a stylus and produce motion of a cursor on a computer screen before them. The movement shown here was from a central start position to a target in the upper right direction about 12 cm away. Under normal conditions, the subjects' arm was covered, and they viewed only the screen. The cursor was usually blanked for the duration of the movement. (a) Typical trajectories of movement between the two targets (large circles) from a normal subject, and a deafferented patient. The small circles indicate the position of the cursor at 20 ms intervals during movement. The normal subject makes movements which are fairly accurate in both direction and extent even when no visual feedback of target or arm position is allowed. The patient's movements are inaccurate in comparison, and show slow drifts at the end of the movement. However, the patient's movements were improved considerably when she was allowed to look at the position of her hand before the start of each movement (vision of hand before trial). The arm was then covered, and the movement itself made without any visual feedback. Further improvement was seen if the position of the cursor on the computer screen was switched on during the course of the movement (vision of cursor). (b) The mean error (distance from target to termination of movement) for this movement in the three different conditions (no vision of either arm or cursor, NoVis; vision of the arm before movement, PreVis: vision of both arm and cursor, WithFB), for three normal subjects and three deafferented patients. Note that the movements of normals are uninfluenced by prevision of their arm, but those of the patients are improved considerably. (From Ghez *et al.*, 1990, Figure 7; with permission.)

examples are manipulation of objects in the environment, when it is not possible to prespecify the exact motor commands in advance of a movement. Fastening buttons, picking coins from a table, or writing are all movements in which afferent input is essential in order that the muscle activity can be correctly scaled to the movement that is required. Deafferented patients cannot make such movements except with the greatest difficulty, and even then under continuous visual guidance. They grip pens very tightly so as to stop them falling out of their grasp when they write. The fine adjustments of grip which normal subjects make continuously as they move the pen over the page are absent.

The examples above have concentrated on the effect of afferent input in updating motor commands as a movement progresses. This is typical of the feedback role usually assigned to reflexes. However, afferent input is also important in helping to specify motor commands even before a movement has begun. Deafferented patients, particularly if they close their eyes, have no precise information on the relative position of the various segments of their limbs. Thus, arm movements of deafferented patients made in the dark are less accurate than normal not just because they cannot update their muscle commands throughout the movement to compensate for any small inaccuracies that might have occurred in the initial commands, but also because the initial command itself was innaccurate due to lack of information about the initial state of the limb. This leads to the surprising finding that arm movements made in the dark can be improved if patients are allowed to look at the position of their arm just before they have to move it (Figure 6.21). The conclusion is that afferent input is used to update continuously the internal 'model' that we have of the body, and that this is lacking in deafferented patients. This 'updating' takes us out of the realm of reflexes into the brain itself.

Bibliography

Review articles and books

Barnes, W. J. P. and Gladden, M. H. (eds) (1985) *Feedback and Motor Control in Vertebrates and Invertebrates,* Croom Helm, London.

Dietz, V. (1992) Human neuronal control of automatic functional movements: interaction between central programmes and afferent input, *Physiol. Rev.,* **72,** 33–69.

Hulliger, M. (1984) The mammalian muscle spindle and its central control. *Rev. Physiol. Biochem. Pharmacol.,* **101,** 1–110.

Marsden, C. D., Rothwell, J. C., Day, B. L. (1983) Long-latency automatic responses to muscle stretch in man: origin and function. *Adv. Neurol.,* **39,** 509–539.

Pierrot-Deseilligny, E. and Mazières, L. (1984) Circuits réflexes de la moelle épinière chez l'homme, Parts 1 and 2, *Rev. Neurol. (Paris),* **140,** 605–614; 681–694.

Schieppati, M. (1987) The Hoffmann reflex: a means of assessing spinal reflex excitability and its descending control in man. *Prog. Neurobiol.*, **28,** 345–376.

Original papers

Aniss, A. M., Diener, H.-C., Hore, J., Burke, D. and Gandevia, S. C. (1990) Reflex activation of muscle spindles in human pretibial muscles during standing, *J. Neurophysiol.*, **64,** 671–679, 661–670.

Burke, D. (1983) Critical examination of the case for and against fusimotor involvement in disorders and muscle tone. *Adv. Neurol.*, **39,** 133–150.

Burke, D., Gillies, J. D. and Lance, J. W. (1970) The quadriceps stretch reflex in human spasticity. *J. Neurol. Neurosurg. Psychiatr.*, **33,** 216–223.

Burke, D., Andrews, C. J. and Lance, J. W. (1972) The tonic vibration reflex in spasticity, Parkinson's disease and normal man. *J. Neurol. Neurosurg. Psychiatr.*, **35,** 477–486.

Burke, D., Gandevia, S. C. and McKeon, B. (1983) The afferent volleys responsible for spinal proprioceptive reflexes in man. *J. Physiol.*, **339,** 535–552.

Crone, C. and Nielsen, J. (1989) Methodological implications of the post-activation and depression of the soleus H-reflex in man. *Exp. Brain Res.*, **78,** 28–32.

Day, B. L., Marsden, C. D., Obeso, J. A. and Rothwell, J. C. (1981) Peripheral and central mechanisms of reciprocal inhibition in the human forearm, *J. Physiol.*, **317,** 39P.

Day, B. L., Marsden, C. D., Obeso, J. A. *et al.* (1984) Reciprocal inhibition between the muscles of human forearm. *J. Physiol.*, **349,** 549–554.

DeGail, P., Lance, J. W. and Nielson, P. D. (1966) Differential effects on tonic and phasic reflex mechanisms produced by vibration of muscles in man. *J. Neurol. Neurosurg. Psychiatr.* **29,** 1–11.

Delwaide, P. J., Pepin, J. L. and Maertens de Noordhout, A. (1991) Short-latency autogenic inhibition in patients with parkinsonian rigidity. *Ann. Neurol.*, **30,** 83–89.

Edamura, M., Yang, J. F. and Stein, R. B. (1991) Factors that determine the magnitude and time course of human H-reflexes in locomotion. *J. Neurosci.*, **11,** 420–427.

Edin, B. B. and Vallbo, A. B. (1988) Stretch sensitisation of human muscle spindles, *J. Physiol.*, **400,** 101–111.

Eldred, E., Granit, R. and Merton, P. A. (1953) Supraspinal control of the muscle spindles and its significance. *J. Physiol.*, **122,** 498–523.

Fournier, E. and Pierrot-Deseilligny, P. (1989) Changes in transmission in some reflex pathways during movement in humans. *NIPS*, **4,** 29–32.

Ghez, C., Gordon, J. and Gilardi, M. F. (1990) Roles of proprioceptive input in the programming of arm trajectories. *Cold Spring Harb. Symp.*, **55,** 837–847.

Gracies, J. M., Meunier, S., Pierrot-Deseilligny, E. and Simonetta, M.

(1991) Pattern of propriospinal-like excitation to different species of human upper limb motoneurones. *J. Physiol.*, **434**, 151–167.

Gregory, J. E., Mark, R. F., Morgan, D. L., Patak, A., Polus, B. and Proske, U. (1990) Effects of muscle history on the stretch reflex in cat and man. *J. Physiol.*, **424**, 93–107.

Hammond, P. H., Merton, P. A. and Sutton, G. C. (1956) Nervous gradation of muscular contraction. *Brit. Med. Bull.*, **12**, 214–218.

Hulliger, M. and Prochazka, A. (1983) A new stimulation method to deduce fusimotor activity from afferent discharge recorded in freely moving cats. *J. Neurosci. Meth.*, **8**, 197–204.

Hulliger, M., Durmuller, N., Prochazka, A. and Trend, P. (1989) Flexible fusimotor control of muscle spindle feedback during a variety of natural movement. *Prog. Brain Res.*, **80**, 87–100.

Hultborn, H. and Pierrot-Deseilligny, E. (1979) Changes in recurrent inhibition during voluntary contractions in man studied by an H-reflex technique. *J. Physiol.*, **297**, 229–251.

Hultborn, H., Meunier, S., Morin, C. and Pierrot-Deseilligny, E. (1987a) Assessing changes in presynaptic inhibition of Ia fibres: a study in man and the cat. *J. Physiol.*, **389**, 729–756.

Hultborn, H., Meunier, S., Pierrot-Deseilligny, E. and Shindo, M. (1987b) Changes in presynaptic inhibition of Ia fibres at the onset of voluntary contraction in man. *J. Physiol.*, **389**, 757–772.

Jenner, J. R. and Stephens, J. A. (1982) Cutaneous reflex responses and their central nervous pathways studied in man. *J. Physiol.*, **333**, 405–419.

Johansson, R. S. (1991) How is grasping modified by somatosensory input? in D. R. Humphrey and H.-J. Freund, *Motor Control: Concepts and Issues* (eds), John Wiley, London, pp. 331–355.

Johansson, R. S. and Westling, G. (1984) Roles of glabrous skin receptors and sensorimotor memory in automatic control of precision grip when lifting rougher or more slippery objects. *Exp. Brain Res.*, **56**, 550–564.

Johansson, R. S. and Westling, G. (1987) Signals in tactile afferents from the fingers eliciting adaptive motor responses during precision grip. *Exp. Brain Res.*, **66**, 141–154.

Katz, R. and Pierrot-Deseilligny, E. (1982) Recurrent inhibition of α-motoneurones in patients with upper motoneurone lesions. *Brain*, **105**, 103–124.

Katz, R. and Pierrot-Deseilligny, E. (1984) Facilitation of soleus-coupled Renshaw cells during voluntary contraction of pre-tibial flexor muscles in man. *J. Physiol.*, **355**, 587–603.

Kugelberg, E., Eklund, K. and Grimby, L. (1960) An electromyographic study of the nociceptive reflexes of the lower limb. Mechanism of the plantar responses. *Brain*, **83**, 394–410.

Lance, J. W., Burke, D. and Andrews, C. J. (1973) The reflex effects of muscle vibration, in J. E. Desmedt (ed.), *New Developments in Electromyography and Clinical Neurophysiology*, Vol. 3, Karger, Basel, pp. 444–462.

Marsden, C. D., Merton, P. A. and Morton, H. B. (1973) Is the human stretch reflex cortical rather than spinal? *Lancet*, **i**, 759–761.

Matthews, P. B. C. (1984) Evidence from the use of vibration that the human long-latency stretch reflex depends upon spindle secondary afferents. *J. Physiol.*, **348,** 383–416.

Matthews P. B. C. (1989) Long-latency stretch reflexes of two instrinsic muscles of the human hand analysed by cooling the arm. *J. Physiol.*, **419,** 519–538.

Matthews, P. B. C. and Rushworth, G. (1958) The discharge from muscle spindles as an indicator of the gamma efferent paralysis by procaine. *J. Physiol.*, **140,** 421–426.

Meara, R. J. and Cody, F. W. J. (1992) Relationship between electromyographic activity and clinically-assessed rigidity studied at the wrist joint in Parkinson's disease. *Brain*, **115,** 1167–1180.

Meunier, S. and Pierrot-Deseilligny, E. (1989) Gating of the afferent volley of the monosynaptic stretch reflex during movement in man. *J. Physiol.* **419,** 753–763.

Nakashima, K., Rothwell, J. C., Day, B. L. *et al.* (1989) Reciprocal inhibition between forearm muscles in patients with writer's cramp and other occupational cramps, symptomatic hemidystonia and hemiparesis due to stroke. *Brain*, **112,** 681–697.

Nichols, T. R. and Houk, J. C. (1976) The improvement in linearity and the regulation of stiffness that results from the action to stretch reflex. *J. Neurophysiol.*, **39,** 119–142.

Nielsen, J., Kagamihara, Y., Crone, C. and Hultborn, H. (1992) Central facilitation of Ia inhibition during tonic ankle dorsiflexion revealed after blockade of peripheral feedback. *Exp. Brain Res.*, **88,** 651–656.

O'Sullivan, M. C., Eyre, J. A. and Miller, S. (1991) Radiation of phasic stretch reflex in biceps brachi to muscles of the arm in man and its restriction during development. *J. Physiol.*, **439,** 529–543.

Pierrot-Deseilligny, E. (1989) Peripheral and descending control of neurones mediating non-monosynaptic Ia excitation to motoneurones: a presumed propriospinal system in man. *Prog. Brain Res.*, **80,** 305–314.

Pierrot-Deseilligny, E., Morin, C., Bergego, C., and Tankov, N. (1981) Pattern of group I fibre projections from ankle flexor and extensor muscles in man. *Exp. Brain Res.*, **42,** 337–350.

Roby-Brami, A. and Bussel, B. (1987) Long-latency spinal reflex in man after flexor reflex afferent stimulation, *Brain*, **110,** 707–725.

Roby-Brami, A. and Bussel, B. (1992) Inhibitory effects on flexor reflexes in patients with a complete spinal cord lesion. *Exp. Brain Res.*, **90,** 201–208.

Rothwell, J. C., Traub, M. M., Day, B. L. *et al.* (1982) Manual motor performance in a deafferented man. *Brain*, **105,** 515–542.

Rothwell, J. C., Colebatch, J. G., Britton, T. C. *et al.* (1991) Physiological studies in a patient with mirror movements and agenesis of the corpus callosum. *J. Physiol.*, **438,** 34P.

Rymer, W. Z., Houk, J. C. and Crago, P. E. (1979) Mechanisms of the clasp-knife reflex studied in an animal model. *Exp. Brain Res.*, **37,** 93–113.

Shahani, B. T. and Young, R. R. (1971) Human flexor reflexes. *J. Neurol.*

Neurosurg. Psychiatr., **34,** 616–627.

Stein, R. B. and Capaday, C. (1988) The modulation of human reflexes during functional motor tasks. *Trends Neurosci.*, **11,** 328–332.

Tanaka, R. (1974) Reciprocal Ia inhibition during voluntary movements in man. *Exp. Brain Res.*, **21,** 529–540.

Thilmann, A. F., Fellows, S. J. and Garms E. (1991a) The mechanism of spastic muscle hypertonus. *Brain,* **114,** 233–244.

Thilmann, A. F., Fellows, S. J. and Ross, H. F. (1991b) Biomechanical changes at the ankle joint after stroke. *J. Neurol. Neurosurg. Psychiatr.,* **54,** 134–139.

Vallbo, A. B. (1970) Slowly adapting muscle receptors in man. *Acta Physiol. Scand.,* **78,** 315–333; Vol. 80, 552–566.

Westling, G. and Johansson, G. (1987) Responses in glabrous skin mechanoreceptors during precision grip in humans. *Exp. Brain Res.,* **66,** 128–140.

Ascending and descending pathways of the spinal cord

7

7.1 *Ascending pathways*

There are a large number of ascending pathways, all of which have important direct projections to areas of the brain concerned with movement. In this chapter, a short summary of the types of sensory fibre which contribute to each tract, and the cells of origin of the tracts, will be given. The data is mostly from anatomical studies in the cat (and monkey); comparable human studies have not yet been performed.

Afferent input to the spinal cord (Figure 7.1)

The site at which sensory fibres terminate within the dorsal horn of the spinal cord is related to the diameter of the afferent fibre. Large diameter fibres enter more medially than small fibres within the dorsal root entry zone, and descend deeper into the grey matter before forming synapses. Many fibres branch on entry to the cord and give off an ascending and descending axon which terminate in nearby segments. Rexed's lamina I and the dorsal part of lamina II receive input from the smallest myelinated (Aδ) fibres innervating mechanosensitive nociceptors in the skin. Lamina II receives input from unmyelinated nociceptive (C) fibres, and the deeper laminae III–VI receive input from larger diameter (Aα) myelinated fibres. Group I muscle afferent axons send most of their input to lamina V and deeper. Apart from the majority of fibres of the dorsal column pathway, the main ascending tracts arise from cells in specific laminae of the spinal cord, and in some instances from particular subgroups of cells within a lamina.

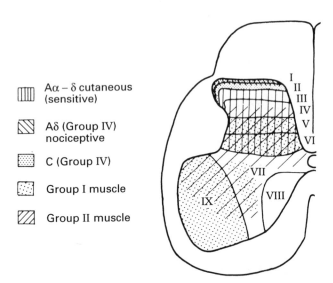

Figure 7.1 Diagrammatic representation of the distribution of cutaneous and muscle afferent fibres to the spinal grey matter. (From Brown, 1981, Figure 10.1; with permission.)

Pathways to thalamus and cerebral cortex

Dorsal column pathway (Figure 7.2).
The axons which travel within the dorsal columns of the spinal cord are a more heterogeneous group than was previously supposed. The 'classical' dorsal column system is made up of large myelinated dorsal root fibres which ascend immediately on entering the cord and terminate in the dorsal column nuclei of the medulla. Their cell bodies lie in the dorsal root ganglion and their axons, the longest in the whole body, are known as first-order (or primary) afferent fibres. At the segment of entry to the dorsal columns, new fibres occupy the most lateral position, but as they ascend, they are displaced medially by fibres entering at successively higher levels. Afferents from the leg travel in the gracile fasciculus, and afferents from the arm in the cuneate fasciculus. This orderly behaviour results in a somatotopic organization of the dorsal column fibres within the gracile and cuneate fasciculi.

In addition to the first-order afferent fibres, the dorsal columns also contain many axons of second-order neurones which also terminate in the dorsal column nuclei. These fibres arise from neurones which have

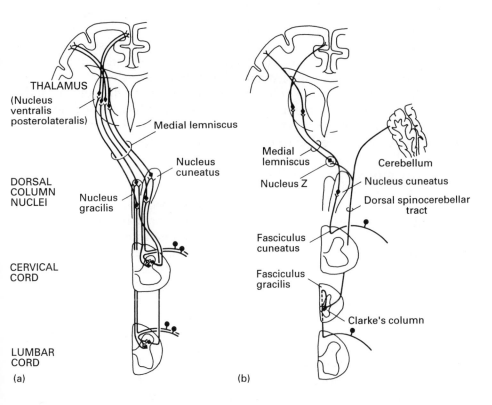

Figure 7.2 Main anatomical features of the dorsal column–medial lemniscal pathway. (a) Main routes taken by fibres carrying impulses from low threshold cutaneous and rapidly adapting deep receptors. Note that many dorsal column fibres are axons of second-order neurones with cell bodies in the dorsal horn. Other second-order axons ascend in the dorsolateral columns to terminate in the dorsal column nuclei. (b) Main routes taken by fibres carrying impulses from slowly adapting receptors in joints and muscles. Note that those from the lower limb only travel in the dorsal columns for a short distance before synapsing in Clarke's column and ascending the dorsolateral columns with the dorsospinocerebellar tract. The axons relay in nucleus z and not in the dorsal column nuclei proper. (From Brodal, 1981, Figure 2.10; with permission.)

their cell bodies in lamina IV of the dorsal horn and receive their afferent input trans-synaptically from first-order dorsal root fibres. They are known as second-order fibres. Some of them ascend in the dorsolateral funiculus, rather than in the dorsal columns and then terminate in the dorsal column nuclei.

Propriospinal neurones also travel in the dorsal columns. These are fibres which connect together segments of the cord without projecting as far as the brain. Those within the dorsal columns are usually branches of first-order afferent fibres. These fibres may be of any diameter, not necessarily being limited to large Aα or Aβ range. The smaller the diameter, the less distance they appear to travel before passing back into the grey matter. Finally, the dorsal columns also appear to contain some descending fibres, arising from the dorsal column nuclei.

The ascending dorsal column fibres terminate in the cuneate and gracile nuclei in the medulla. Anatomical and physiological studies suggest that there are two patterns of termination in the dorsal column nuclei. A 'cluster' region receives input mainly from the distal parts of the body, via first-order afferent fibres, whereas a 'reticular' zone receives more proximal inputs with a large contribution from second-order afferents.

Cells in the 'cluster' region are modality-specific and tend to have smaller receptive fields than those in the 'reticular' zone. The latter have larger receptive fields and often receive inputs from various types of receptor, including painful ones. Input to the 'cluster' region comes mainly from large diameter peripheral afferent fibres: units innervating low threshold mechanoreceptors in skin (hair follicles, Meissner corpuscles, Merkel discs, Ruffini endings and Pacinian corpuscles), joint receptors and muscle receptors.

Exceptions to these rules of termination are afferents from the slowly adapting joint and group I muscle receptors from the lower limbs. Afferents from these receptors only travel in the dorsal columns for a short distance. Above lumbar levels, the fibres travel in the dorsolateral funiculus. Many of them are collaterals of the dorsal spinocerebellar tract, and as well as sending fibres to the spinal parts of the cerebellar cortex, also terminate in nucleus z, just rostral to the gracile nucleus. Nucleus z, then, is an important relay site for transmission of proprioceptive input from the legs to the cerebral cortex.

Afferent fibres from the 'reticular' and 'cluster' zones of gracile and cuneate nuclei and from nucleus z, ascend in the medial lemniscus in the brainstem, cross the midline in the medulla and terminate in the VPL nucleus of the thalamus (see below). The fibres do not give off any collateral branches to other structures *en route*.

Spinocervical pathway (Figure 7.3a)
The spinocervical tract has been recognized only for the past 30 years. It appears to be particularly prominent in the cat, and rather smaller in monkey and in man. The cervical nucleus consists of a column of cells in the first and second cervical segments of spinal cord, just ventrolateral to the dorsal horn. In the cat it consists of a well-defined mass of cells (the lateral cervical nucleus), but in humans there is no such distinct structure. It is assumed that the cells which form the lateral cervical nucleus in the cat are more diffusely scattered in man. It receives input

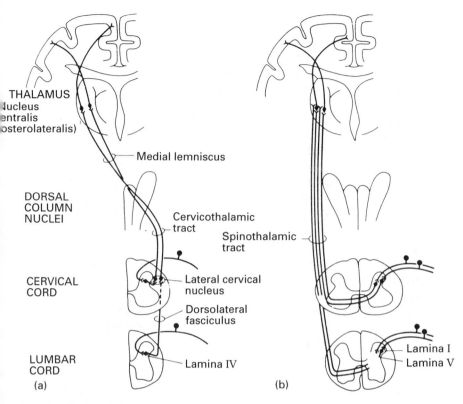

THALAMUS
Nucleus
entralis
osterolateralis)

Medial lemniscus

DORSAL
COLUMN
NUCLEI

Cervicothalamic
tract

Spinothalamic
tract

CERVICAL
CORD

Lateral cervical
nucleus

Dorsolateral
fasciculus

LUMBAR
CORD

Lamina IV

(a)

Lamina I
Lamina V

(b)

Figure 7.3 Simplified diagrams which show the main anatomical features of the spinocervicothalamic (a) and the spinothalamic (b) tracts. The termination of spinothalamic fibres in other parts of the thalamus than the VPL is not shown. (From Brodal, 1981, with permission.)

ipsilaterally from all levels of spinal cord, from second order afferents whose cell bodies lie mainly in lamina IV of the dorsal horn, and whose axons ascend in the dorsolateral funiculus. In this position, they are intermixed with axons of many other ascending (dorsal spinocerebellar) and descending (corticospinal, rubrospinal and raphe-spinal) tracts, making it impossible to determine their role using lesioning techniques.

The main source of input to the spinocervical tract in the cat comes from hair follicle receptors, although there are cells in the cervical nucleus which also respond to noxious stimuli. Such noxious input is under the control of descending systems, which probably act presynaptically on the input to the cells of origin in the dorsal horn. It is not known which descending tracts mediate this effect.

The efferent axons of the cervical nucleus cross in the high cervical cord and travel with lemniscal fibres to the VPL nucleus of the thalamus without synapsing in the dorsal column nuclei.

Spinothalamic tract (Figure 7.3b)
The spinothalamic tract arises from neurones in dorsal and intermediate laminae of the spinal grey matter whose axons cross the cord at segmental level and ascend in the contralateral ventrolateral funiculus. Fibres join the tract medially, but are pushed lateral as they ascend the cord by fibres entering the tract at higher levels. In man, the spinothalamic tract is large: larger than in the monkey and much larger than in the cat. In the monkey, the cells of origin lie in layers I and V (and adjacent laminae IV and VI) in the cervical cord. At lumbar levels, a substantial number are found in layers VII and VIII.

The afferent axons from cells in lamina I tend to be of smaller diameter than those of deeper layers. They respond to intense mechanical or painful heat stimuli, as expected from the Aδ terminations in layer I. Afferents from layer V are larger and are activated by both light tactile and also by intense noxious stimuli, and are said to have a 'wide dynamic range'. Spinothalamic fibres from muscle and joint receptors arise from cells in laminae V, VII and VIII. The tract ascends through the medulla and pons occupying a space more dorsal than the lemniscal fibres, and terminates mainly in the VPL nucleus of the thalamus. Its fibres give off collaterals to the reticular formation *en route*.

The spinothalamic tract is sometimes subdivided into two components: the neospinothalamic tract and the paleospinothalamic tract. The neospinothalamic tract is said to convey sensations of touch and deep pressure, and is located in the ventromedial portion of the tract. In contrast, the paleospinothalamic tract is said to convey sensations of pain and temperature, and occupies the dorsolateral portion of the tract. However, there is some debate over this, with some authors claiming that the two types of fibre are completely intermixed within the anatomical spinothalamic tract. Even if this is true, it may be that the two types of fibre terminate in different portions of the thalamus (see below).

Thalamic terminations of somatic afferents
Laminae of white matter subdivide the thalamus into its three major grey masses: the anterior, lateral and medial nuclear groups. More detailed subdivision has been attempted by many authors on anatomical grounds, the most commonly accepted being that of Walker (cited in Hassler, 1959). Walker's classification was developed from a study of monkey and chimpanzee and will be used here (Figure 7.4). The nomenclature for human thalamus was developed from this by Hassler (1959), and is considerably more detailed.

The ventral portion of the lateral cell mass of the thalamus is the main region of termination of somatosensory afferents. It is subdivided into an anterior (VA), lateral (VL) and posterior part, the posterior part itself being further subdivided into the ventralis posterior lateralis (VPL),

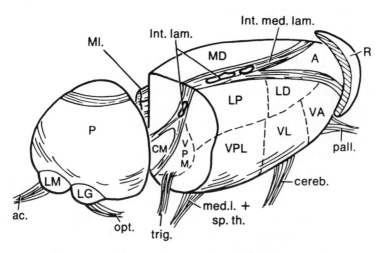

Figure 7.4 Three-dimensional reconstruction of the right human thalamus. The posterior part is separated from the rest by a cut, to display some of the internal structure. Only the rostral tip of the reticular nucleus (R) is shown. Input from trigeminal (trig), medial lemniscal (med. l.) and spinothalamic (Sp. th.) afferents is shown entering the VPM (ventroposterior medialis) and VPL (ventroposterior lateralis) nuclei. Cerebellar input ends in the VL (ventrolateral) and pallidal input in the VA (ventroanterior) nucleus. A, anterior; MD, dorsomedial; MI, midline; CM, centromedian; P, pulvinar; LM, medial geniculate; LG, lateral geniculate. More details of the precise regions of termination are described in the text and in Chapter 8. (From Brodal, 1981; with permission.)

ventralis posterior medialis (VPM) and the ventralis posterior inferior (VPI). All of these nuclei send efferent fibres to the cerebral cortex, with the exception of parts of VA and VPI. There are two other smaller areas of the thalamus where somatosensory afferents have been found to terminate: the nucleus centralis lateralis (CL), a part of the intralaminar nuclear group, and the medial part of the posterior thalamic nucleus (POm). This latter nucleus is a rather ill-defined cell mass at the caudal thalamus lying outside the more classically defined ventroposterior and geniculate nuclei.

The vast majority of fibres in the medial lemniscus (which include afferents from the gracile and cuneate nuclei as well as from nucleus z and the lateral cervical nucleus) terminate in the caudal part of the VPL nucleus (VPL$_c$). Fibres from the gracile nucleus end most laterally and those from the cuneate nucleus more medially. Trigeminal afferents end in the VPM nucleus. Some small diameter lemniscal fibres also are found in the POm region, but their terminals are sparse. Terminals of spinothalamic afferents were once believed to overlap with those of the lemniscal afferents. However, more recent studies have shown that the situation probably is not quite so simple as this. Spinothalamic fibres in the monkey have quite a different pattern of termination in VPL than those of the medial lemniscus. Endings are sparse in the central region and more dense along the borders of VPL (VPL$_o$ ventroposterior lateral pars oralis). In particular, there is a region of dense innervation in the rostral part of VPL in the border region with VL. Smaller regions of spinothalamic termination are in the POm and the CL nuclei. The POm nucleus has been extensively studied as a possible 'pain' relay for the small diameter fibre component of the spinothalamic tract (the paleo-spinothalamic tract).

There have been many studies of the physiological properties of the cells in the VPL nucleus. As a rule, cells respond only to one modality of sensory input and are grouped together according to modality in continuous layers of tissue. Cells in the rostral and caudal parts of VPL are responsive to input from deep receptors (muscle spindles, tendon organs, joint receptors), whereas those in the central part of the nucleus receive cutaneous inputs. Because of the superimposed (mediolateral) topographical arrangement of the afferent fibres, such an organization means that a small group of cells in any one part of the VPL nucleus responds to stimulation of one particular kind coming from one partic-ular part of the body.

In deeply anaesthetized animals, the receptive fields of VPL neurones are very small but more recent investigations, using conscious animals, have tended to suggest that receptive field sizes in freely moving animals are larger. In particular, cells responding to noxious stimuli have been found in VPL. Such cells had previously been seen only in recordings from the posterior nuclear group (POm) of anaesthetized animals, and the finding of similar cells in VPL of conscious prepara-tions has tended to focus attention away from POm as a possible candidate for specific transmission of painful sensations.

Pathways to the cerebellum

Dorsal spinocerebellar and cuneocerebellar tracts (see Figure 10.8 in Chapter 10).
The thick myelinated axons of the of the dorsal spinocerebellar tract arise from the cells of Clarke's column (which is also known as the nucleus dorsalis). Clarke's column lies at the base of the dorsal horn in spinal segments T1 to L2 in man (to L3/L4 in cat) and receives monosynaptic input from group Ia and Ib muscle afferents and joint receptors. There is also an input from cutaneous, touch and pressure receptors and a smaller input from group II spindle afferents. As Clarke's column finishes at the second lumbar segment, afferent input from lower levels reaches it via the dorsal columns of the lower lumbar

cord. The axons of cells in Clarke's column cross into the ipsilateral lateral column where they ascend dorsolaterally to the lateral corticospinal tract. Physiological studies have shown that their receptive fields are small and that many of the fibres carry modality-specific information.

There are two subgroups of neurones in the dorsal spinocerebellar tract. One group receives proprioceptive input from group Ia, Ib or II muscle afferents, the other receives input from cutaneous and high threshold muscle afferents. Within each group cells may receive convergent input from the various submodalities, but there are very few cells which receive convergence from both groups of inputs. Proprioceptive inputs sometimes are particularly strong. A single primary afferent fibre may give off 50 synapses onto the proximal dendrites of a dorsal spinocerebellar tract cell. Corresponding EPSPs are large (5 mV), and the coupling between input and output strong. The fibres course through the inferior cerebellar peduncles and end as mossy fibres chiefly in the hindlimb region (see Chapter 10) of the intermediate zone of the ipsilateral cerebellar cortex.

The dorsal spinocerebellar tract does not convey any information from the forelimbs. Clarke's column only begins at the level of the first thoracic segment: its cervical cord analogue is the external (or accessory) cuneate nucleus, which lies in the medulla, lateral and rostral to the cuneate nucleus. This forelimb analogue of the dorsal spinocerebellar tract is known as the cuneocerebellar tract. The inputs which synapse in the external cuneate originate from muscle afferents. Those which synapse in the main cuneate nucleus and then join the projection to cerebellum are of cutaneous origin. The bulk of the efferent fibres end in the forelimb regions of the ipsilateral cerebellar cortex, although a smaller number of fibres recently have been found to project to the rostral part of the VPL nucleus of the contralateral thalamus. This unexpected projection therefore is a further pathway in which proprioceptive input can reach the main somatosensory relay nucleus of the thalamus.

Ventral and rostral spinocerebellar tracts
Like the dorsal and cuneocerebellar tracts, the ventral and rostral spinocerebellar tracts convey impulses from hind and forelimbs, respectively. The cells of origin of the ventral spinocerebellar tract are believed to be the 'spinal border cells' situated in the dorsolateral part of the ventral horn, and some cells situated laterally in Rexed's laminae V, VI and VII. Their axons are smaller in diameter than those of the dorsal spinocerebellar tract, and the majority cross the midline to ascend in the contralateral lateral fasciculus. Physiological studies indicate that the input to the ventral spinocerebellar tract is distinct from that of the

dorsal pathway. Although there is a small contribution from group Ia and Ib muscle afferents, the main input comes, polysynaptically, from flexor reflex afferents. The cells have wide receptive fields and in contrast to the cells of origin of the DSCT, there is convergence between proprioceptive and cutaneous afferents. In addition, there are inputs from descending systems, including the corticospinal, rubrospinal and vestibulospinal tract. In fact, many of the inputs to the spinal α-motoneurones also project in parallel to the ventral spinocerebellar tract cells. The activity in these cells is, for example, modulated during locomotion in deafferented cats, while that of DSCT cells is abolished. Thus, the dorsal spinocerebellar fibres are much more dependent upon peripheral input for their discharge than those of the VSCT.

The complex convergence of inputs from many different sources onto the VSCT makes it difficult to envisage precisely what it is signalling back to the cerebellum. To the extent that the motoneurones receive the same input, the VSCT input might be regarded as a form of efference copy signal. Another possibility, which emphasizes the fact that these neurones receive input both from spinal inhibitory interneurones as well as the afferent fibres which excite them, is that they monitor the excitability of spinal interneuronal circuits. The ventral spinocerebellar tract axons traverse the medulla and most of the pons, entering the cerebellum via the superior peduncle. Their terminations are mainly in the contralateral hindlimb areas of cerebellar cortex, with a sparse termination also in ipsilateral cortex.

The forelimb equivalent is the rostral spinocerebellar tract. Its cells of origin are probably located in lamina VII of the cervical cord, and receive flexor reflex afferent (FRA) input from the forelimb. The axons ascend ipsilaterally, unlike those of the ventral spinocerebellar pathway, and pass partly in the brachium conjunctivum and partly in the restiform body to end as mossy fibres in cerebellar cortex. FRA afferents produce mainly excitatory actions on the cells of the rostral spinocerebellar tract, whereas their action is predominantly inhibitory onto the ventral pathway.

Spino-olivary-cerebellar tracts
The axons of all the cells of the inferior olive go to the contralateral cerebellum, to end in the cerebellar nuclei and as climbing fibres in the cerebellar cortex. The inferior olive is subdivided into a principal nucleus and a dorsal and medial accessory nucleus. Spinal inputs pass only to the two accessory olivary nuclei. There are two main pathways by which spinal afferents reach the olive: a direct route via the spino-olivary tract, which ascends contralaterally in the ventral funiculus of the cord, and an indirect route via a relay in the contralateral dorsal column nuclei. These two routes correspond to the ventral funiculus spino-

olivary-cerebellar tract (VF-SOCT) and the dorsal funiculus spino-olivary-cerebellar tract (DF-SOCT) as defined physiologically by Oscarsson (Oscarsson and Sjkolund, 1977).

Ascending pathways to other structures

Spinoreticular tract
Contrary to earlier beliefs, there is a substantial afferent pathway to the reticular formation from the spinal cord which is not formed from collaterals of other pathways. The medial lemniscus does not give off any collaterals to the reticular formation, nor are the spinothalamic collaterals very dense. The main input comes from fibres which ascend in the ventrolateral fasciculus, intermingled with spinothalamic afferents. In the brainstem the axons deviate from the spinothalamic tract and run dorsally and medially, giving off branches which drop perpendicularly to the axis of the brainstem, to contact the vertically oriented dendrites of the reticular formation cells. The projection is mainly bilateral and many fibres have terminations throughout the length of the reticular formation. Two particularly dense regions of spinal afferent terminals are in the medulla (the nuclei reticularis pontis caudalis and oralis) and the pons (nucleus reticularis gigantocellularis).

Of particular interest in motor control is the somatotopically organized spinal input to the lateral reticular nucleus, which lies lateral to the inferior olive. Physiological studies show that the input is polysynaptic and mainly from flexor reflex afferents with extremely large receptive fields, sometimes covering all four limbs. The efferents from the nucleus travel to the cerebellar nuclei and also end as mossy fibres in the cerebellar cortex.

Spinovestibular and spinotectal pathways.
The number of spinovestibular fibres is small in comparison to the number of vestibulospinal tract fibres. They are thought to be collaterals of spinocerebellar tract fibres, and end in parts of the vestibular nuclei which receive no direct input from vestibular fibres. Some of the areas of termination project to the cerebellum, or VPL thalamus. Spinal afferents also project to the deep layers of the superior colliculus, although their function is unknown.

7.2 *Descending motor pathways*

Several areas of the brain can directly influence the activity of the spinal cord via their descending fibre connections. The main fibre tracts are

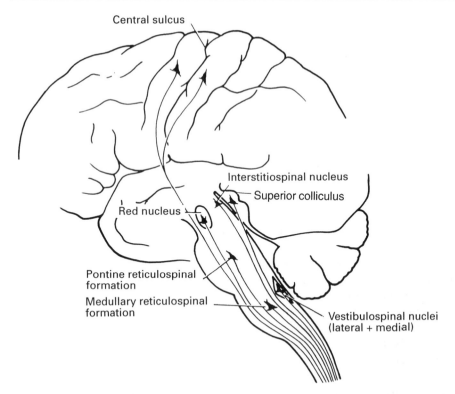

Figure 7.5 Summary of fibre systems descending to the spinal cord from the brain. Shown here are the corticospinal, rubrospinal, interstitiospinal, tectospinal, pontine and medullary reticulospinal and vestibulospinal pathways. (From Brodal, 1981, Figure 4.1; with permission.)

shown in Figure 7.5, and comprise the corticospinal, rubrospinal, interstitiospinal, tectospinal, pontine and medullary reticulospinal and vestibulospinal tracts. Although each fibre tract will be described separately, it must be remembered that, under normal physiological conditions, they will never act alone. This is because each region of origin receives input from many other areas of brain (as well as afferent input from sense organs). The cerebral cortex, for example, does not influence the spinal cord solely by way of the corticospinal tract, but also via its interconnections with the red nucleus, reticular formation, tectum and interstitial nucleus (Nathan and Smith, 1955, 1982, review descending tracts in man).

At one time it was believed that all the descending tracts were involved only with efferent, motor commands. However, over the past 30 years it has been shown that these tracts also have important actions on afferent systems, and on reflex pathways within the spinal cord.

They may regulate transmission of sensory information from spinal cord to brain and influence the excitability of spinal interneurones, as well as providing a motor output. It is as if the engine of a motor car, besides powering the driving wheels, also adjusted the damping on the suspension or the sensitivity of the steering mechanism at the same time.

Pyramidal tract

The pyramidal tract was one of the first large fibre bundles to be recognized as a particular tract in the brain. By definition, it consists of those fibres with axons running in the medullary pyramids. The vast majority of these fibres come from the cerebral cortex and continue to the spinal cord. These fibres are known as the **corticospinal** tract. Other fibres leave the tract in the pyramids or even rostral to the pyramids, and innervate the cranial nerve motor nuclei. This portion of the fibres is known as the **corticobulbar** tract. Although it is not strictly true, because some corticobulbar fibres do not travel as far as the pyramids, the term **pyramidal** tract usually is taken to be a common name for the corticospinal and corticobulbar fibres.

Pyramidal tract fibres originate in both motor and sensory regions of cerebral cortex. Degeneration studies and anatomical methods using retrograde axonal transport suggest that in the monkey, 60% of fibres in the pyramidal tract come from frontal motor areas of cortex. The other 40% of fibres come from primary somatosensory cortex (areas 3, 1, 2) and parietal cortex (areas 5 and 7). The figures for the human are likely to be similar, although the results of surgical excision indicate that as many as 60% of the fibres may arise in area 4. Unlike the pyramidal tract of the cat, no fibres have been found to arise from cells in prefrontal or secondary somatosensory cortex. Pyramidal tract fibres from precentral cortex have a motor function in the spinal cord, whereas those from the postcentral cortex appear to be involved in the regulation of afferent inputs into the cord.

About one million fibres are found in the pyramidal tract in humans. This is more than in any other animal. There are about 800 000 in the chimpanzee, 400 000 in the macaque monkey and 186 000 fibres in the cat. In relation to weight, man has no special superiority in numbers, but it is probably more appropriate to relate the figures to the number of muscles to be controlled, which is about the same in all these species. Of the pyramidal tract fibres in humans, 6% are unmyelinated: their origins and functions are unknown. The remainder of the fibres are myelinated and have a range of diameters. Some 2% are large diameter (11–20 μm) and conduct impulses at velocities of 50 m s^{-1} or so. They are known to electrophysiologists as **fast** corticospinal fibres. Ninety per cent of axons have diameters in the range of 1–4 μm. These are the **slow** corticospinal fibres, and conduct at speeds of about 14 m s^{-1}.

All pyramidal tract fibres have cell bodies in layer V of cerebral cortex. By chance, the corticofugal cells of layer V in any area of cortex also are known to microscopists, on the basis of their shape, as pyramidal cells. Thus, the pyramidal-shaped cells in layer V of pre- and postcentral cortex give rise coincidentally to the pyramidal tract. There is an ill-defined subclass of giant pyramidal-shaped cells in the motor cortex known as Betz cells. In humans, there are some 34 000 of these Betz cells, and it is the axons of such cells which probably account for the large diameter fibres in the pyramidal tract. The majority of small diameter fibres are made up of the axons of small pyramidal shaped cells of cortical layer V.

On leaving the cerebral cortex, the pyramidal tract axons descend in a small region of the posterior limb of the internal capsule. They then course through the middle portion of the cerebral peduncle, and into the pons. The pyramidal tract fibres split into many parts to pass between the pontine nuclei before rejoining as a distinct bundle in the medullary pyramids. Most of the other fibres which run together with the pyramidal tract in the cerebral peduncles terminate in the pons. The majority of the pyramidal tract fibres decussate in the brainstem and travel down to all levels of the spinal cord in the dorsolateral columns (the lateral corticospinal tract). A small number of uncrossed fibres descend within the ventromedial part of the spinal cord (the ventral corticospinal tract), but terminate at thoracic levels in humans. Finally, there is a proportion of uncrossed fibres which join the crossed fibres of the lateral corticospinal tract. Throughout most of their course (except within the pons, and perhaps even within the medullary pyramid), the corticospinal fibres are thought to preserve an approximate somatotopic arrangement. Fibres to the leg are found lateral to those to the arm and trunk in the cerebral peduncles and dorsolateral spinal cord. In the internal capsule, the fibres to the leg are found most posteriorly.

In the cat, lateral corticospinal fibres terminate mainly in the intermediate zone of the spinal grey matter (Figure 7.6). There are very few direct contacts with spinal motoneurones of lamina IX and activation of motoneurones is achieved via spinal interneurones. Precentral fibres terminate ventral (laminae VI and VII) to those from postcentral cortex (laminae IV and V), consistent with a role of postcentral fibres in regulating afferent input within the dorsal horn of the spinal cord. In contrast to the cat, in the monkey some fibres of the lateral corticospinal tract end directly on spinal motoneurones. Such monosynaptic connections from cerebral cortex to motoneurones are thought to be more important in man, particularly to the muscles of the hand and forearm.

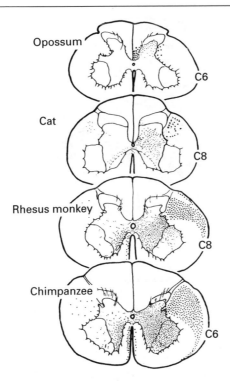

Figure 7.6 Distribution of corticospinal fibres from the left hemisphere to low cervical segments in four different species. Note how the pattern of termination changes from innervation of the dorsal horn, to intermediate zone and, finally, in the rhesus monkey and chimpanzee, to the motor nuclei (lamina IX) of the ventral horn. The locations of the axons in the lateral and ventromedial tracts also are shown. Note the predominantly crossed component of the lateral tract and the uncrossed projection in the ventromedial tract. (From Kuypers, 1981, Figure 12; with permission.)

Within the cervical and lumbar enlargements of the spinal cord, the motoneurones are arranged somatotopically (see Figure 7.11). Those innervating the axial muscles lie most medially and their cell bodies form a ventromedial column in the grey matter, which runs the whole length of the cord. Motoneurones which innervate the extremities are situated in the lateral part of the ventral horn. In the caudal segments of the cervical and lumbar enlargements, the most dorsally located cells in the ventral horn supply the most distal muscles of the limbs. They are nearest the lateral corticospinal tract and receive the densest corticospinal projections. The uncrossed fibres of the ventral corticospinal tract terminate in the ventromedial cell group and innervate the axial muscles. Unlike the lateral tract, the medial corticospinal fibres commonly have a bilateral projection to motoneurones on both sides of the cord.

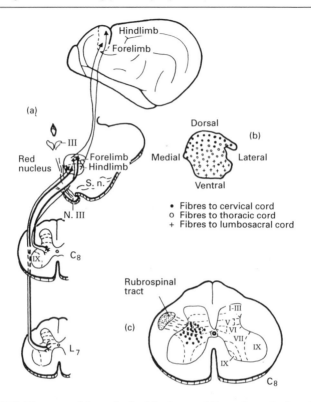

Figure 7.7 Diagram of the principal features of the rubrospinal and corticorubral projections in the cat. (a) Somatotopic projection from motor cortex (anterior sigmoid gyrus in the cat) to the dorsal and ventral regions of the red nucleus. In the cat, it is thought that rubrospinal tract fibres arise from both small and large cells of the rostral and caudal portions of the nucleus. Accordingly, the rubrospinal projection is shown to consist of axons with large and small diameters. In the monkey, this is not the case: only large cells contribute axons to the rubrospinal tract. (b) Topographical arrangement of projection neurones in the red nucleus. (c) Main region of termination of the rubrospinal tract within the intermediate zone of the spinal grey matter. (From Brodal, 1981, Figure 4.6; with permission.)

Corticospinal fibres do not only give excitatory projections to moto-neurones; they also project strongly to spinal inhibitory interneurones such as the Ia interneurone and Renshaw cells. In addition, they also innervate γ-motoneurones as well as the large α-motoneurones.

Rubrospinal tract (Figure 7.7)

Opinions differ considerably over the importance and even the presence of a rubrospinal tract in man (Nathan and Smith, 1982). For convenience, it is best to describe first the tract and its origins in the cat, where it is undoubtedly an important descending projection to spinal cord.

The red nucleus contains cells of various size. Small cells are located in the rostral (parvocellular) portion and large cells in the caudal (magnocellular) part. In the cat, descending spinal projections arise from cells of all sizes in both rostral and caudal parts. As expected, the rubrospinal tract consists of axons with a wide range of diameters, with a maximum conduction velocity (120 m s^{-1}) exceeding that of the pyramidal tract. It decussates near its origin and travels the contralateral brainstem in a ventromedial position (von Monakow's bundle) where it gives off collaterals to many areas, including the interpositus nucleus of the cerebellum and the vestibular nuclei. In the spinal cord, it lies ventral and slightly lateral to the lateral corticospinal tract. The two tracts intermingle and in the cat terminate together on interneurones in laminae V, VI and VIII of the spinal grey. Some of these interneurones are shared between the two tracts so that impulses in one descending system can facilitate transmission in the other. Monosynaptic transmission to spinal α-motoneurones does not occur in the cat, although such connections are seen in the monkey. Electrophysiologically, stimulation of the cat red nucleus tends to produce di- or trisynaptic excitation of contralateral flexor motoneurones and inhibition of extensor motoneurones.

The tract is arranged somatotopically. Cells from the dorsomedial part of the nucleus give rise to axons which travel medially in the cord and innervate cervical segments. Cells from the ventrolateral parts of the nucleus have axons which terminate in lumbar regions. Two of the main inputs to the red nucleus preserve this somatotopy. They come from the ipsilateral cerebral (particularly motor) cortex and from the contralateral interpositus nucleus of the cerebellum. For example, inputs from the arm area of motor cortex and from cerebellar regions receiving afferent input from the forelimb both project to cells in the dorsomedial red nucleus and thence to the cervical cord. Such projections from cortical motor areas obviously give rise to the possibility of a cortico-rubrospinal pathway operating in parallel with the direct corticospinal route. Synapses from cortex are, however, located more distally on the dendrites than those from the interpositus, and probably provide the weaker input. The projection from interpositus nucleus is particularly strong. In the cat, the great majority of its output travels via the superior cerebellar peduncle (brachium conjunctivum) to the red nucleus. Electrical stimulation of the peduncle is a better way of exciting rubral cells than stimulation of the nucleus itself.

In contrast to the situation in humans, the crossed rubrospinal tract of the cat is the major output of the red nucleus. However, there are other outputs and one of these, an ipsilateral connection to the olive, takes particular predominance in humans. The human red nucleus, to a much greater extent than the cat, can be clearly divided into a rostral parvocellular and a caudal magnocellular part. In man and monkey, the most recent work with retrograde labelling techniques shows that the rubrospinal tract takes origin only from the large (and possibly medium-sized) cells of the caudal part. Since there are only 150 to 200 large cells in this region in humans, the rubrospinal tract is probably very small indeed. In contrast, the parvocellular region of the red nucleus is much larger than in lower animals. Its afferent and efferent connections are quite separate from those of the caudal nucleus. Descending axons travel ipsilaterally in the central tegmental tract, giving off projections to the reticular formation and terminate in the olivary nuclei. This tract is much larger in humans than in cats, where it has usually received little attention.

The major input to the rostral part of the red nucleus comes from the ipsilateral dentate nucleus. Thus, since there is a strong projection from the olive (via the inferior cerebellar peduncle) to cerebellar cortex, the red nucleus forms part of a cerebello-rubro-olivary-cerebellar loop. Lesions to the central tegmental tract in humans, which interrupt this loop, give rise to palatal myoclonus. This consists of rapid twitch-like contractions of the soft palate in the pharynx at about 120 min^{-1}. A possible explanation is that they arise from instability in the loop circuit outlined above.

Vestibulospinal tracts

As with the rubrospinal tract, there are little data available on the human vestibulospinal tract. However, what results there are, from pathological investigations, are in agreement with results from experimental animals. Almost all the work has been performed on the cat.

There are two vestibulospinal tracts: medial and lateral. The names derive from the nuclei of origin within the vestibular complex (Figure 7.8). The lateral tract is larger and arises from both large and small cells in the lateral vestibular (or Deiter's) nucleus. The cells are somatotopically arranged within the nucleus. Rostroventral parts project to the neck and forelimb, dorsolateral parts of the hindlimb. The fibres descend ipsilaterally in the ventrolateral columns of the spinal cord, occupying a progressively more medial position from cervical to lumbar segments. Terminations are distributed mainly to laminae VIII and VII. The medial vestibulospinal tract is said to arise only from the medial vestibular nucleus, but large contributions from both the lateral and descending vestibular nuclei have been noted electrophysiologically. Axons travel in the medial longitudinal fasciculus together with some reticulospinal fibres and descend only to upper thoracic levels of the spinal cord in the most medial part of the ventral white matter. The area of termination is

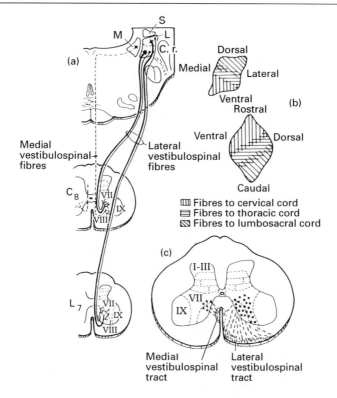

Figure 7.8 Vestibulospinal pathways in the cat. The major tract is the lateral vestibulospinal tract, which arises somatotopically (b) in Deiter's nucleus (the lateral vestibular nucleus (L). Fibres terminate in the medial portion of the ventral horn of the spinal cord (c) at both cervical and lumbar levels. The medial vestibulospinal tract is much smaller and only travels as far as the cervical segments. Its cell bodies lie in the medial vestibular nucleus (M). (From Brodal, 1981, Figure 4.7; with permission.)

the same as that for the lateral vestibulospinal tract. Both tracts produce disynaptic excitation of extensor motoneurones and inhibition of flexor motoneurones.

The input to the vestibular nuclei comes from two main sources: cerebellum and labyrinth. Some fibres also arise in the reticular formation. Input from the labyrinth travels via the vestibular branch of the VIIIth nerve. Fibres from the semicircular canals terminate mainly in the superior vestibular nucleus and the rostral part of the medial and descending nuclei. Fibres from the utricle terminate in the same regions and also in the rostroventral part of Deiter's nucleus. The descending spinal projections from these regions travel in both medial and lateral vestibulospinal tracts and end in the cervical cord. They are the basis for strong reflex connections between labyrinthine receptors and the

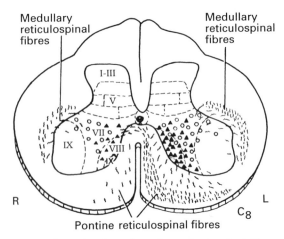

Figure 7.9 Transverse section of a cat spinal cord at the C8 level showing the pattern of termination of reticulospinal tracts within the grey matter and the location of their axons in the white matter. Projections from the left reticular formation are shown. (From Nyberg-Hansen, 1965, Figure 9; with permission.)

neck musculature. Indeed, monosynaptic connections (both excitatory and inhibitory) from the vestibular nuclei to neck motoneurones have been described. Ascending fibres also arise from the rostral part of the medial nucleus and from the superior nucleus and travel rostrally in the medial longitudinal fasciculus to the oculomotor nuclei. They are involved in the labyrinthine control of eye movements.

The cerebellar projections to the vestibular nuclei are extensive and very complex. There are direct ipsilateral pathways from spinal and vestibular regions of cerebellar cortex, as well as ipsi- and bilateral pathways from the fastigial nucleus, to all four vestibular nuclei. In particular, these are the main inputs to the dorsolateral part of Deiter's nucleus, which provides the only vestibulospinal fibres to innervate motoneurones of the lumbar region.

Reticulospinal tracts (Figure 7.9)

The reticulospinal tracts arise from an area in the bulbar reticular formation known as the medial tegmental field. This is the rostral continuation of spinal grey matter in the medial part of the intermediate zone. Several nuclear groups can be distinguished within it and include the paramedian and interfascicular nuclei in the medulla and the gigantocellular, paragigantocellular and nucleus reticularis pontis in the pons.

The fibres which arise from cells in the pontine medial tegmental field descend bilaterally at first through the medullary tegmental field. They continue, mainly ipsilaterally, into the ventral funiculus of the spinal cord. This tract, which runs the length of the cord is known as the medial reticulospinal tract. The fibres of cells in the medullary medial tegmental field descend bilaterally (but mainly ipsilaterally) to the spinal cord. In the brainstem these medullary reticulospinal fibres constitute a vertically orientated band running just lateral to the medial longitudinal fasciculus. The fibres in the dorsal portion of this band enter the spinal ventral fasciculus, running together with the pontine fibres in the medial reticulospinal tract. The fibres from the ventral portion of the band enter the ventrolateral funiculus of the spinal cord, forming the lateral reticulospinal tract.

Unlike other descending tracts, there is no somatotopic organization in either of the reticulospinal projections. Electrophysiological studies reveal that two thirds of fibres terminating at cervical levels also project to the lumbar enlargement. The main areas of termination are in the ventral grey matter (laminae VIII and medial lamina VII), with the fibres of the ventrolateral columns tending to be located dorsal to those of the ventral columns. Some studies also indicate that there may even be direct monosynaptic connections with motoneurones of neck and axial muscles.

Some of the main inputs to the reticular regions which give rise to the descending tracts are the sensorimotor areas of cerebral cortex. This input is bilateral. Two other important inputs come from ascending spinal cord fibres and from the fastigial nucleus of the cerebellum (particularly to the medullary region). There is also another area of the reticular formation which gives rise to long descending pathways. It is located in the ventrolateral (rather than medial) pontine tegmentum close to the rubrospinal tract. It gives rise to a crossed reticulospinal tract which travels in the spinal dorsolateral funiculus with the corticospinal and rubrospinal tracts. It is known sometimes as the pontine component of the rubrospinal tract.

Tectospinal and interstitiospinal tracts

There are two other main descending tracts from brain to spinal cord. Little work has been performed on either, in comparison with the other descending systems, and almost nothing is known about them in humans.

The tectospinal tract comes from cells in the deep layers of the superior colliculus. Its fibres cross at the dorsal tegmental decussation of the midbrain and travel just ventral to the medial longitudinal fasciculus. In the cord they are found in the medial part of the ventral funiculus and travel only as far as cervical segments. Most of the terminations are in laminae VI, VII and VIII of the high cervical segments, and the effects are probably directed via interneurones to motoneurones of the neck muscles. The main input to the superior colliculus is from visual cortical areas, and thus the tract may take part in head and neck orienting reactions to visual stimuli.

The interstitiospinal tract arises from cells in the interstitial nucleus of Cajal in the rostral midbrain. The nucleus lies ventral to the medial longitudinal fasciculus and the efferent axons descend ipsilaterally in this bundle to the ventromedial funiculus of the spinal cord. Terminations are given off to all levels of the cord, mainly in laminae VII and VIII, overlapping with the area of termination of vestibulospinal fibres. The input to the interstitial nucleus comes mainly from the vestibular nuclei and cerebral cortex.

Propriospinal neurones

Propriospinal neurones have a cell body in one spinal segment and axons which terminate in other segments. They are subdivided into two categories: short and long propriospinal neurones. The former have connections only over a limited number of spinal segments (within the lumbar or cervical enlargements), whereas the latter have long axons which connect together distant spinal segments.

One particular group of propriospinal neurones has been investigated in great detail in the cat, and has recently been described in man. It is the C3/C4 propriospinal system. This system is described here because in contrast to other spinal interneuronal systems, the the descending excitation to the C3/C4 propriospinal neurones is much more powerful than is the input from peripheral afferents. Because of this, it can be regarded as an important pathway for transmitting descending motor commands.

The C3/C4 propriospinal neurones have cell bodies in the eponymous segment of spinal cord. They receive monosynaptic excitation from the corticospinal tract, and weaker excitation from most other descending motor pathways except for the vestibulospinal system. In addition there is some input from cutaneous and muscle afferent fibres in the forelimb. The majority of these propriospinal neurones have short axons which terminate in the lower cervical segments directly on spinal motoneurones or Ia inhibitory interneurones. Most project to many motor nuclei often innervating muscles acting over different joints. This wide distribution may therefore allow the system to control certain synergies of muscle activity, and would be in contrast to the more limited direct projection of the corticospinal system to small groups of motor nuclei which act around a single joint.

The activity in the C3/C4 propriospinal neurones can be shaped by inhibitory connections from local interneurones. These interneurones receive input both from supraspinal structures and from forelimb afferents (see Figure 7.10). It is thought that input from both sources could help select the correct combination of propriospinal neurones to produce appropriate synergies of movement. In addition, the inhibitory input from forelimb afferents may also be important in terminating movements.

The C3/C4 propriospinal system is the only major relay system in the spinal cord that has been explored with behavioural experiments. This is because of its favourable anatomical location in which the propriospinal neurones are quite separate to the α-motoneurones which they innervate. In addition, the propriospinal axons descend in a separate part of the cord from the direct descending projections from

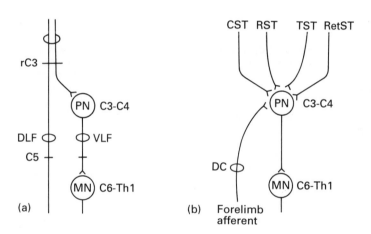

Figure 7.10 Schematic wiring diagrams of some of the neuronal circuits involved in the C3–C4 propriospinal systems. (a) C3–C4 propriospinal neurones (PN) mediate disynaptic excitation from descending motor pathways onto forelimb motoneurones (MN). The propriospinal neurones have axons which descend in the ventrolateral funiculus (VLF), whilst most of the other pathways descend in the dorsolateral funiculus (DLF). The pathways which course in the DLF would also have synapses onto cervical motoneurones which are not illustrated in this diagram. Thus, a lesion of the VLF would interrupt the propriospinal input to lower cervical motoneurones whilst leaving the direct corticospinal and rubrospinal innervation intact. (b) Convergence of monosynaptic excitation to propriospinal neurones from different descending motor systems (corticospinal, CST; rubrospinal, RST; reticulospinal, RetST; tectospinal, TST), and from low threshold primary afferents from the forelimb (via the dorsal columns, DC).

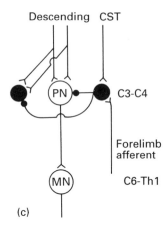

Figure 7.10 Continued. (c) The inhibitory pathways which mediate feedback inhibition from afferents in the forelimb onto the propriospinal neurones. The inhibitory interneurones are monosynaptically activated from forelimb afferents (which also have monosynaptic excitatory projections to propriospinal neurones as shown in b). These inhibitory interneurones can also inhibit other inhibitory interneurones in the same segment. (From Hultborn and Illert, 1991, Figure 3; with permission.)

corticospinal and rubrospinal systems (see Figure 7.10). Anatomical techniques have shown that C3/C4 propriospinal neurones are more active during visually-guided reaching movements of the forelimb than when the cats are running on a treadmill. Lesioning experiments have confirmed this special role in visually-guided reaching movements. Thus, a lesion of the ventrolateral funiculus interrupts the propriospinal input to lower cervical motoneurones, whilst leaving the direct corticospinal and rubrospinal innervation intact. Such a lesion affects the precise aiming of the cat's paw towards a target even though the major descending inputs remain intact. It is therefore believed that the C3/C4 propriospinal system has a crucial role in mediating cortical control of reaching, and that it may allow selection of correct muscle synergies in this complex task. The large number of convergent inputs from both the limb and other descending pathways onto the C3/C4 propriospinal system may shape the motor command *en route* to the motoneurones.

Finally, it should be noted that some C3/C4 propriospinal neurones have long axons which terminate in the lumbar region. These

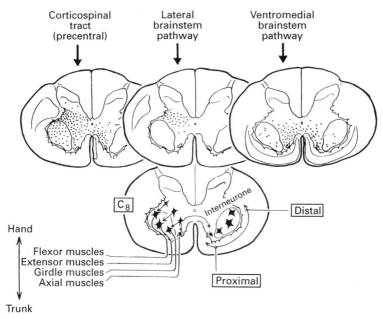

Figure 7.11 Diagram of the approximate patterns of termination of the three major categories of descending pathways: corticospinal and group A (ventromedial) and B (lateral) brainstem pathways. The sections are from the lower cervical cord of a monkey. The bottom section shows the arrangement of motoneurones and interneurones of the intermediate zone. Lateral brainstem pathways terminate in regions of the intermediate zone that connect with motoneurones supplying distal muscles. Ventromedial brainstem pathways terminate on regions which project to proximal and axial motoneurones. The corticospinal tract has terminals directly onto both types of motoneurones as well as interneurones. (From Lawrence and Kuypers, 1968, Figure 3; with permission.)

propriospinal neurones receive monosynaptic excitation from all descending motor tracts including the vestibulospinal system. Because of the input from the vestibular nuclei, it seems likely that these propriospinal neurones might contribute to postural adjustments during reaching with the forelimb. Indeed, these long C3/C4 propriospinal neurones receive collateral innervation from the short axoned C3/C4 propriospinal neurones. (Reviews: Hultborn and Illert, 1991; Pierrot-Deseilligny, 1989.)

7.3 *Summary of descending pathways (Figure 7.11)*

The mass of descending pathways described above may be simplified by dividing them into three groups. This classification was devised by

Kuypers and is important because the anatomical distinction between the pathways also reflects functional differences between their actions on spinal motoneurones (see Chapter 9 for effects of lesions of these pathways). The three groups are the corticospinal tract and the two groups (A and B) of descending brainstem pathways. The group A brainstem pathways comprise the interstitiospinal, tectospinal, lateral and medial vestibulospinal and the reticulospinal tracts from the pontine and medullary medial tegmental areas. All these tracts travel in the ventral and ventromedial funiculi of the spinal cord and terminate to a greater or lesser degree, bilaterally in laminae VII and VIII. This is a region of long propriospinal neurones. The fibres have many collaterals, and seem especially well suited for the synergistic activation of large numbers of muscles. Many elements of these pathways also have direct connections with neck and axial musculature. The group B brainstem pathways consist of the rubrospinal and the crossed reticulospinal tract from the pontine lateral tegmental region. They travel in the contralateral dorsolateral funiculi of the spinal cord, terminating in the intermediate zone in laminae V–VIII. These fibres have relatively few collaterals, ending either directly on motoneurones or on short propriospinal fibres of the spinal cord. Their connections are directed mainly to distal muscles.

The corticospinal (CST) tract in monkey and man can be regarded as having two components which parallel the two descending groups of brainstem pathways. They terminate on both distal extremity muscles (lateral CST) and on axial and proximal muscles (ventral CST). In contrast to the brainstem pathways (especially group A), the corticospinal pathway is exceptionally well organized, with a high degree of topographical representation. It would seem to provide a highly differentiated system superimposed on those from the brainstem and probably contributes the ability to produce fractionated movements in both these groups of muscles.

Investigating descending motor pathways in humans

Corticospinal tract

The technique of transcranial stimulation of the motor cortex is explained in detail in Chapter 9. Single stimuli over the motor areas readily produce EMG responses in contralateral body muscles. Several lines of evidence suggest that at least the initial part of these responses is produced by activity in the large diameter component of the corticospinal tract. First, indirect estimates, made from measurements of EMG latencies in hand and leg muscles, as well as direct measurements made during spinal surgery of the descending volleys produced by motor cortex stimulation, show that the spinal efferent pathway conducts at about 60 m s^{-1} which would be consistent with the large diameter of the some of the corticospinal axons in man. Second, there is very little variability in the latency of EMG responses produced by this form of stimulation. Indeed, within a muscle, single motor units are usually activated with a jitter of 1 ms or less, implying that the connection from the cortex has very few interposed synapses. This, together with the fact that responses are most easily obtained in contralateral small hand muscles would be consistent with activity in the monosynaptic component of the corticospinal tract which is particularly well developed in humans.

Although the largest responses are seen in hand muscles, motor cortex stimulation can evoke activity in virtually all muscles of the body, including structures such as the diaphragm, oesophagus, and external anal sphincter. Using the criteria of minimal latency variability and rapid conduction velocity, inputs to many of these muscles are also likely to be monosynaptic. Such inputs have never been described in monkey, where the monosynaptic component of the corticospinal tract is limited to muscles in the hand and arm. As in other animals, in humans, projections from motor cortex to fusimotor neurones and to spinal interneurones including the Ia and Ib inhibitory interneurones have also been described.

There are substantial changes in corticospinal conduction over the first 2–4 years of life (Figure 7.12). At birth, the corticospinal conduction time may be five times that seen in normal adults. This then decreases slowly to reach adult values at about the age of 4 years. The long conduction time at birth is thought to be due to the fact that the pyramidal tract is incompletely myelinated and hence its conduction velocity is much slower than in the adult. Myelination is only completed over the first few years of life during which time the conduction velocity increases and the conduction time decreases. A similar process occurs in peripheral nerve, where the conduction velocities of motor and sensory axons only reach adult values at about 2 years. Interestingly, the conduction time in the central motor pathway is approximately constant from the age of 4 years to adult, even though the conduction distance from motor cortex to spinal cord increases as a child grows. This means that there is a continuing, gradual increase its conduction velocity throughout childhood (reviewed by Rothwell *et al.*, 1991).

Other descending motor pathways

The corticospinal tract is the only pathway which it is possible to investigate with some certainty in man. However, techniques are

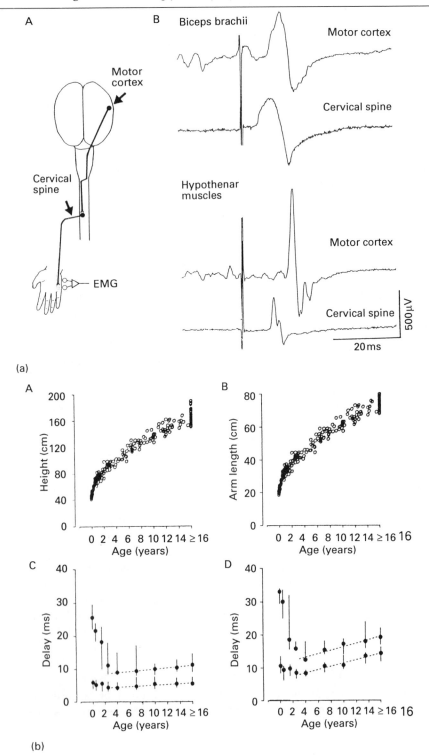

(a)

(b)

available which probably activate descending motor pathways other than those activated by transcranial stimulation of motor cortex.

The startle reaction occurs in normal subjects when they receive unexpected (usually auditory) sensory stimuli (Figure 7.13). EMG recordings from many muscles of the body show that although the onset latency of the response can be variable, there is a very constant order in which the muscles are activated. The pattern of activation gives information on the conduction velocity of the pathway and also the site of origin of the response within the brain. Usually, the first muscle to contract after any loud noise is the orbicularis oculi, which produces a blink of the eyelid. However, this response is not thought to be part of the startle proper since it persists on repeated presentation of the same stimulus. A true startle habituates very rapidly and has virtually disappeared after 2 or 3 stimuli. If the contraction of orbicularis oculi is ignored, then the first muscle to respond in a startle is the sternocleidomastoid, the muscle which turns the head from side to side and flexes the neck. This is followed by activity in arm, trunk and leg muscles. The interval between the sternocleidomastoid activation and other muscles of the body can be two or three times as long as the interval between the same muscles when the motor cortex is stimulated with a transcranial method. If we assume that the conduction velocity of the peripheral pathway from spinal cord to muscle is the same in each case, then the

---

Figure 7.12 (a) Method of estimating corticospinal tract conduction time in man. The motor cortex is stimulated through the scalp using a transcranial magnetic stimulus. This activates the large diameter component of the corticospinal pathways and produces responses in contralateral muscles at short latency. In order to estimate the peripheral conduction delay from spinal cord to muscle, the same magnetic stimulator can be placed over the cervical region of spinal column where it activates peripheral motor axons as they exit the intervertebral foramen. Stimulation here thus gives an estimate of the conduction time from cord to muscle. The central conduction time from cortex to cord (which includes a synapse between the corticospinal fibres and spinal motoneurones, plus a small amount of conduction in peripheral nerve from motoneurone to the intervertebral foramen) is estimated by subtracting the peripheral conduction time from the total conduction time. The EMG records on the right are from the biceps brachii and hypothenar muscles of a normal subject. The pairs of traces show responses in each muscle to stimulation of motor cortex and cervical spine. The stimulus artefact indicates the time of application of the stimulus. (b) How conduction times change during the first years of life. Graphs A and B show how the height of the subject and arm length increase with age. Graphs C and D show, for biceps and hypothenar muscles respectively, how the total conduction time from cortex to muscle (upper points on each graph), and the peripheral conduction time from spinal cord to muscle (lower points on each graph) change with age. The central motor conduction time is the difference between the two sets of points. Note that central motor conduction time is extremely long at birth (approximately 20 ms) and then decreases to adult values by the age of about 4 years. After 4 years, the conduction time is approximately constant. (From Eyre *et al.*, 1991, Figures 1 and 3; with permission.)

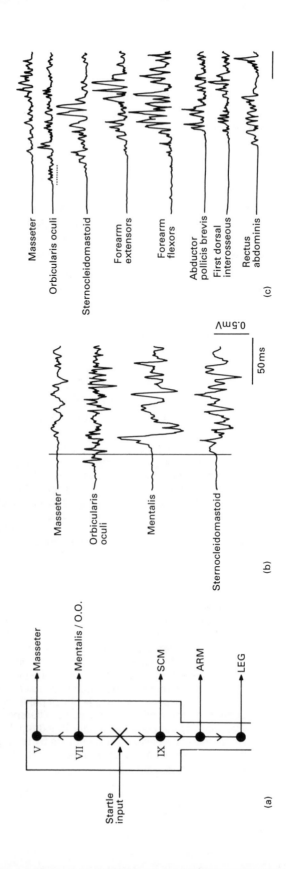

(a)

Masseter

Mentalis / O.O.

SCM

ARM

LEG

V

VII

IX

Startle
input

(b)

Masseter

Orbicularis
oculi

Mentalis

Sternocleidomastoid

0.5 mV

50 ms

(c)

Masseter

Orbicularis oculi

Sternocleidomastoid

Forearm
extensors

Forearm
flexors

Abductor
pollicis brevis

First dorsal
interosseous

Rectus
abdominis

conduction velocity of the central descending tract which is responsible for the startle response must be considerably slower than that of the corticospinal tract. Precisely which tract is being activated is not clear. Nevertheless, analysis of the activation pattern in muscles of the face suggests that the origin of the response lies somewhere within the brainstem, and hence the descending tract may be one of the reticulospinal pathways. Thus, sternocleidomastoid, a muscle innervated by the XIIth cranial nerve leads in the startle; this is closely followed by activity in the mentalis (a muscle innervated by the VIIth cranial nerve), and last by activity in the masseter (a muscle innervated by the Vth cranial nerve). In other words, the startle pattern seems to take origin in the brainstem, and then travel up through the cranial nerve nuclei to innervate the muscles of the face in a caudo-rostral order. These findings in humans are compatible with studies in cats and rats which have also implicated the reticular formation (in particular the nucleus reticularis pontis caudalis) and reticulospinal pathways in the acoustic startle response (Brown *et al.*, 1991; Davis *et al.*, 1982).

A final technique to activate descending motor pathways in man is by galvanic vestibular stimulation. Subjects stand unsupported with their feet together and eyes closed. A small 0.5–1 mA DC (galvanic) electric current is then passed between electrodes placed over the mastoid process just behind each ear. This stimulus is almost unnoticeable by the subject, but causes a clear sway of the body. The reaction

Figure 7.13 (a) Postulated origin of the startle response in man. The figure indicates that the startle centre may lie somewhere in the brainstem and activate cranial nerve innervated muscles in a caudorostral order. Thus, the first muscle activated might be the sternocleidomastoid if the XIth nerve nucleus were nearest to the startle centre. This would be followed by activity in the facial nerve (VII) innervated muscles (mentalis) and then the masseter muscle. In addition, the excitation would spread down the spinal cord and excite muscles of the arm and leg. Panel (b) shows a detail of the pattern of activation of cranial nerve innervated muscles. The records are EMG traces from a single startle response to a loud (120 dB) unexpected noise. The first response occurs in the orbicularis oculi, but, as explained in the text, this is thought not to be part of the true startle itself, but simply a blink reflex to the auditory stimulus. This response persists on subsequent trials whereas responses in other muscles disappear. The true startle begins with contraction of the sternocleidomastoid muscle (solid vertical line), then mentalis (innervated by the VIIth nerve) and then masseter (innervated by the Vth nerve). Panel (c) shows a typical startle from a different subject who had clear responses in arm and trunk muscles. The stimulus was given at the start of the sweep, (horizontal calibration 50 ms). Note the initial blink reflex in the orbicularis oculi muscle (dotted underline). The true startle begins in the sternomastoid, is followed by activity in masseter and then by activity in arm and trunk muscles. The onset latency of responses in the intrinsic muscles of the hand (abductor pollicis brevis, first dorsal interosseous) is much later after the onset of activity in forearm muscles than expected from the conduction distance down the forearm. This is typical of the pattern of activity in the normal startle response. (From Brown *et al.*, 1991, Figures 3 and 4; with permission.)

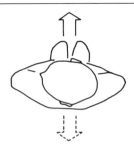

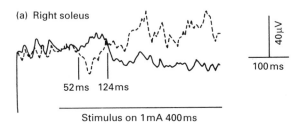

(a) Right soleus

40 μV

100 ms

52 ms 124 ms

Stimulus on 1 mA 400 ms

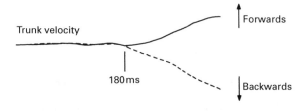

Trunk velocity

Forwards

180 ms

Backwards

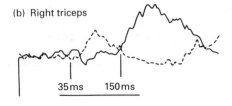

(b) Right triceps

35 ms 150 ms

does not occur in patients with vestibular damage, and it is thought to result from stimulation of the peripheral part of the vestibular nerve.

During normal standing, the stimulus only produces EMG responses in leg muscles (Figure 7.14). These consist of two opposing components: an early small response and a later large response. The later response actually causes the observable body sway; the early response seems to produce no effect on posture. Although responses are usually recorded in leg muscles, if subjects stand on a moving platform and have to use their arms to stabilize their body, then the same galvanic vestibular stimulus will evoke responses in arm muscles. These are similar to, but start earlier than those seen in the leg. As with the startle response, the difference in latency between the onset of activity in arm and leg muscles gives some estimate of the conduction velocity of the spinal efferent pathway. In this case, it seems similar to that of the corticospinal tract. It may represent activity in a vestibulospinal pathway.

There are two points to notice about this vestibular response. First, although the early component occurs first in the arm and is followed by activity in the leg, the late component often occurs first in the leg and is found only later in the arm. The conclusion is that there are two responses produced by separate mechanisms and that the conduction velocity calculation is only adequate for the first component. The second point is that the responses, although produced by a very simple stimulus, are remarkably complex. The direction of body sway induced by the stimulus is always in the direction of the stimulating anode, no matter what the position of the head or trunk. So, if a subject rotates his head 90° to the left, so that his right ear is pointing forwards, then anodal stimulation of the right mastoid will produce a forward sway, whilst anodal stimulation of the left mastoid will give backwards sway. The same stimulus causes activity in different leg muscles depending upon the position of the head. (Britton *et al.*, 1993; Lund and Broberg, 1983; give further details of galvanic vestibular stimulation.)

Figure 7.14 Average EMG responses in the right soleus (a) and right triceps (b) to galvanic stimulation of the vestibular apparatus. The subject stood with his head turned to the left looking over his left shoulder. The right ear faced forwards, and the left ear faced backwards. The EMG responses show the effects of applying a stimulus with the anode on the right ear (solid lines) or on the left ear (dotted lines). The horizontal line under the traces indicates the time for which the stimulus was turned on. In the right soleus muscle there is an initial response which begins at 52 ms followed by a response of opposite polarity at 124 ms. The equivalent responses in the right triceps, which were obtained when the subject was using his right arm to balance whilst standing on a pivoted platform, occur at 35 and 150 ms. The initial part of these EMG responses is very small and produces no visible sway of the body. Only the later parts of the response which persist whilst the stimulus is applied produce changes in body position (see trunk velocity trace in (a)). (From Britton *et al.*, 1993, Figures 1 and 5; with permission.)

Bibliography

Books and review articles

Brodal, A. (1981) *Neurological Anatomy in Relation to Clinical Medicine.* Oxford University Press, Oxford.

Brown, A. G. (1981) *Organisation in the Spinal Cord.* Springer-Verlag, Berlin.

Kuypers, H. G. J. M. (1981) Anatomy of the descending pathways, in V. B. Brooks (ed.) *Handbook of Physiology*, sect. 1, vol. 2, part 1, Williams and Wilkins, Baltimore, pp. 597–666.

Original papers

Britton, T. C., Day, B. L., Brown, P. *et al.* (1993) Postural electromyographic responses in the arm and leg following galvanic vestibular stimulation in man, *Exp. Brain Res.* **94.**

Brown, P., Rothwell, J. C., Thompson, P. D. *et al.* (1991) New observations on the normal auditory startle reflex in man, *Brain*, **114,** 1981–1902.

Davis, M., Gendelman, D. S., Tischler, M. D. and Gendelman, P. M. (1982) A primary acoustic startle circuit: lesion and stimulation studies, *J. Neurosci.*, **2,** 791–805.

Eyre, J. A., Miller, S. and Ramesh, V. (1991) Constancy of central conduction delays during development in man: investigation of motor and somatosensory pathways, *J. Physiol.*, **434,** 441–452.

Hassler, R. (1959) Anatomy of the thalamus, in G. Shaltenbrand and P. Bailey, (eds) *Introduction to Stereotaxis with an Atlas of the Human Brain*, George Threme, Stuttgart.

Hultborn, H. and Illert, M. (1991) How is motor behaviour reflected in the organisation of spinal systems? In D. R. Humphrey and H.-J. Freund (eds) *Motor Control: Concepts and Issues*, John Wiley, London, pp. 49–73

Lawrence, D. G. and Kuypers, H. G. J. M. (1968) The functional organisation of the motor system in the monkey, Parts I and II, *Brain*, **91,** 1–14; 15–36.

Lund, S. and Broberg, C. (1983) Effects of different head positions on postural sway in man induced by a reproducible vestibular error signal, *Acta Physiol. Scand.*, **117,** 307–309.

Nathan, P. W. and Smith, M. C. (1955) Long descending tracts in man. I. Review of present knowledge, *Brain* **78,** 248–303.

Nathan, P. W. and Smith, M. C. (1982) The rubro-spinal and central tegmental tracts in man, *Brain*, **105,** 233–269.

Nyberg-Hansen, R. (1965) Sites and modes of termination of reticulospinal fibres in the cat. An experimental study with silver impregnation methods, *J. Comp. Neurol.*, **124,** 71–100.

Oscarsson, O. and Sjkolund, B. (1977) The ventral spino-olivocerebellar system in the cat. I. Identification of five paths and their termination in the cerebellar anterior lobe, *Exp. Brain Res.*, **28**, 469–486.

Pierrot-Deseilligny, E. (1989) Peripheral and descending control of neurons mediating non-monosynaptic Ia excitation to motor neurones: a presumed propriospinal system in man, *Prog. Brain Res.*, **80**, 305–314.

Rothwell, J. C., Thompson, P. D., Day, B. L. *et al.* (1991) Stimulation of the human motor cortex through the scalp, *Exp. Physiol.*, **76**, 159–200.

Posture

8

The postural control system has to perform three functions: support, stabilization and balance. It must ensure that appropriate muscles are contracted in order to **support** the body against gravity. Without this, the body would be a limp heap on the floor. It must **stabilize** the supporting portions of the body when other parts are being moved. For example, the trunk and the shoulder must remain steady during movements of the arm. Finally, it must ensure that the body is correctly **balanced** on its base of support. In mechanical terms, this means that in a stationary environment, the vertical projection of the centre of gravity should lie within the supporting base. It is not known which centres in the central nervous system perform these various functions. They are likely to be distributed in many separate regions. This chapter is devoted to the physiology of these responses. The pathophysiology is discussed in the chapters which follow.

In order to support, stabilize and balance the body, the postural system needs information about the relative position of the body parts with respect to one another and also information about the position of the body in relation to any external forces which might be acting upon it (e.g. gravity). Three classes of receptor are available for this: somatosensory, visual and vestibular. Somatosensory receptors have been covered in detail in Chapter 4. A description of the vestibular receptors is given in the next section. The visual system is not covered in this book.

8.1 *The vestibular system*

A full description of the vestibular system is given in many textbooks. Only a brief review is given here.

The inner ear consists of a series of cavities in the petrous bone which contain fluid-filled membranous sacs. The vestibular portion of these structures consists of the **otolith** organs, a pair of swellings known as the **utricle** and the **saccule,** and the **semicircular ducts**, three more-

or-less perpendicular tubes of tissue within the bony semicircular canals (Figure 8.1).

The sensory receptor cells, the **hair cells**, lie in specialized regions of the vestibular epithelium. At one end of each semicircular duct, before its connection to the utricle, there is a swelling called the **ampulla** where the epithelium is thickened in the ampullary crest. The tip of the crest contains hair cells, the hairs, or **cilia** of which project upwards towards the **cupula**, a gelatinous diaphragm which covers the ampullary crest and reaches to the roof of the ampulla. The cupula effectively plugs the cavity inside the semicircular duct. When the head experiences angular acceleration, the force exerted by the mass of fluid in the semicircular duct causes the cupula to distort and move the cilia of the hair cells. Because the semicircular ducts are basically round tubes of fluid, it is only angular acceleration which is capable of producing relative motion between fluid and duct, i.e. flexion or extension or rotation of the neck. Pure linear acceleration, as we experience in a moving vehicle, does not have this effect.

The floor of the utricle and the wall of the saccule are also thickened and contain hair cells. These zones are called **maculae** and are covered with a gelatinous substance (the otolith membrane) containing crystals of calcium carbonate known as otoliths (**lithos** is the Greek word for stone). The cilia of the hair cells project into this membrane and are bent whenever there is a relative shearing movement between it and the macular floor. The utricular macula is approximately horizontal when the head is erect. Thus if the head experiences any linear acceleration, or is tilted with respect to gravity, the otolithic membrane moves relative to the floor of the macula and bends the cilia of the hair cells. The macula in the saccule is approximately vertical. Thus, relative shearing movement between the otolithic membrane and the floor of the macula is produced during vertical accelerations of the head.

(It should be noted that although the utricular macula can sense the position of the head with respect to gravity when the head is stationary, this is not necessarily the case when the body is placed in a moving vehicle. If we are standing in an aeroplane which is making a correctly-banked turn, the resultant of gravitational and centrifugal forces acting on the body is at right angles to the floor of the aircraft, not parallel to the line of gravity. It is this resultant force, not gravity, which defines the behavioural vertical for the otolith organs.)

The hair cells

The hair cells of the ampullary crest and maculae are separated from one another by support cells, to which they are joined by tight junctions (Figure 8.1). In mammals, they consist of two types (type I and type II).

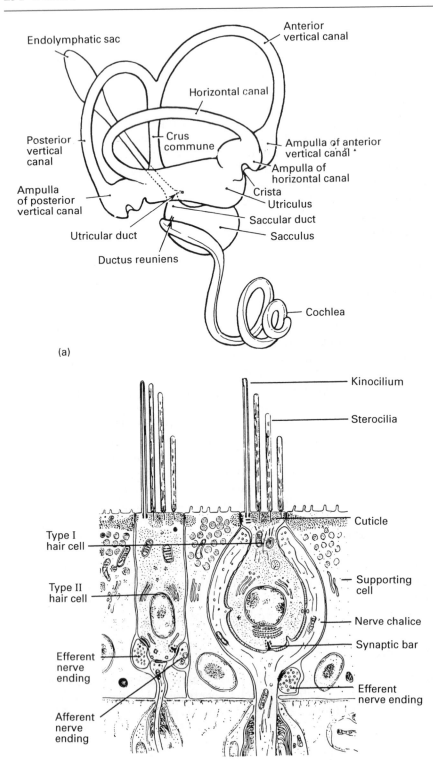

(a)

(b)

The type I cell is a globe-shaped cell with a narrow neck surrounded at its base by a large nerve terminal. Type II cells are rectangular with small synaptic terminals at their base. The functional difference between the two is not known. At their base, they synapse with both an afferent and efferent fibre from the vestibular nerve. In type I cells, there is no direct synapse between efferent fibres and the sensory cell. Efferent synapses are located on the afferent terminal. On their free surface, hair cells contain 40–70 **stereocilia** and a single motile **kinocilium** (which disappears soon after birth in mammals). The cilia are graded in height and are arranged with the longest nearest to the kinocilium, and the shortest furthest away. A line drawn from the kinocilium through the centre of the bundle runs parallel to the steepest gradient in stereocilia heights, and forms an axis around which the whole bundle is bilaterally symmetrical. This is termed the 'directional axis' of the cell.

There is continuous release of transmitter at rest from the hair cell to the afferent fibre, and hence continuous firing in the afferent nerve. Bending of the cilia causes modulation of this discharge (Figure 8.2). Stimuli which bend the hairs towards the kinocilium open ion channels and causes depolarization of the cell. This produces release of more transmitter, and increased firing in the afferent nerve. Bending the hairs away from the kinocilium has the opposite effect. It closes ion channels open at rest, hyperpolarizes the cell and reduces afferent discharge. Bending the hairs perpendicular to their morphological axis has little effect.

The ion channels which are opened are relatively non-selective to most positively-charged ions. However, since the endolymphatic fluid, in which the stereocilia are bathed has a high concentration of K^+ ions (which is quite unlike the normal composition of extracellular fluids), it is thought that it is K^+ which is responsible for much of the transducer current. From measurements of either the mechanotransducer noise or from knowledge of the total transducer current and the conductance of individual ion channels, it has been calculated that there are only 50–200 active transducer channels on each hair cell. This means that there are likely to be only 1 or 2 channels per stereocilium.

Figure 8.1 (a) The structures of the inner ear. The cavities within the petrous bone contain the membranous labyrinths, consisting of the cochlea, the semicircular ducts, the saccule and the utricle. Sensory cells of the vestibular system lie in the ampullae, and in the maculae of the utricle and saccule. (b) Hair cells of the mammalian vestibular system. The cells are separated from each other by supporting cells, to which they are attached by tight junctions. The hair cells are of two morphological types (I and II). Each hair cell receives afferent and efferent innervation from fibres in the vestibular nerve. On the apical surface is a single kinocilium and 40–70 stereocilia. (a, From Roberts, 1979, Figure 9.2; b, from Wersall and Bagger-Sjoback, 1974, Figure 30; with permission.)

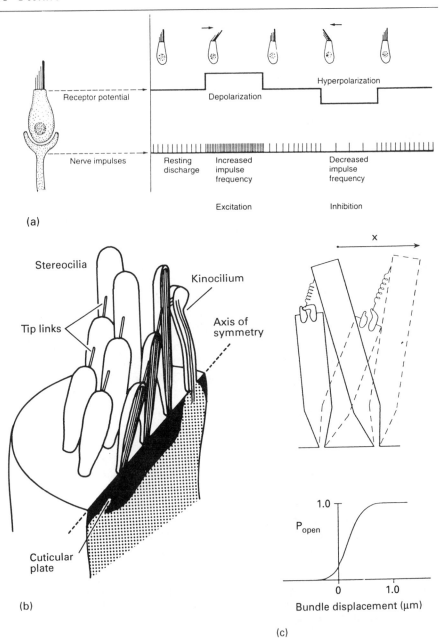

(a)

Stereocilia

Kinocilium

Tip links

Axis of symmetry

Cuticular plate

(b)

P_{open}

1.0

0   1.0

Bundle displacement (μm)

(c)

The current model of mechanoelectrical transduction is as follows. Rows of stereocilia parallel to the axis of the cell are connected together at their tips (Figure 8.2). Cilia in adjacent rows are not connected. It is thought that when the hairs are displaced along the axis of the cell, the links between cilia are put under stress, and that this mechanical event somehow opens channels in the membrane which change the receptor potential. Movement of the hairs in the opposite direction is thought to close some channels normally open at rest. Movement perpendicular to the major axis would not stress the links between cilia at all, and hence there would be no change in the number of open channels.

This is illustrated in Figure 8.2c where the connections between the tips of each stereocilium are envisaged as springs, directly opening transducer channels in the membrane of adjacent cilia. As with many ion channels, the gate is either open or closed; there is no intermediate, semiconducting state. The more the spring is stretched, the more likely the channel is to be found in the open state. Even at rest, there is thought to be some tension on the spring which ensures that the channel is open part of the time. This results in a constant relative depolarization of the hair cell and a constant release of transmitter to the afferent fibre. Confirmation of the model comes from the observation that destroying the tip links between stereocilia abolishes the transducer properties of hair cells.

In the ampullae, the hair cells are oriented along the axis of the semicircular duct and respond well to movement of the cupula when

Figure 8.2 (a) Bending the cilia of hair cells along the axis of the cell causes modulation in the afferent discharge rate. At rest, the afferent fibre discharges continuously. When the hairs are bent towards the kinocilium, firing increases due to depolarization of the cell and increased transmitter release. Bending the hairs in the opposite direction causes hyperpolarization, a reduction in transmitter release and a decrease in afferent firing. (After Flock, 1965, Figure 14.) (b) A cross-section of three rows of stereocilia on the surface of a hair cell. The broken line is the axis of symmetry of the cell. The single kinocilium is situated on the axis of symmetry, adjacent to the tallest row of stereocilia. Tip links attach adjacent cilia in each row, but not between rows. (c) A model of the mechanoelectrical transduction process. An ion channel is located in the tip of one stereocilium, and this is opened by increasing the tension in a 'gating spring' attached to the adjacent (taller) stereocilium. Displacing the cilia to the right increases the spring tension and this opens the ion channel. The diagram below shows how the probability of the ion channel being open varies with the displacement of the cilia. Note that there is a very narrow operating range: a displacement of only about one third of a micrometre is enough to ensure that the channel state changes from being always closed (P_{open} = 0) to always open (P_{open} = 1). This distance is the approximate diameter of a single stereocilium. At rest (displacement = 0), there is still a small probability that the channel will be open. Movement in the opposite direction (to the left) would reduce this to zero. The channels in this figure are shown to be at the lower end of each tip link, but equally well could be at the upper end, or both. (From Pickles and Corey, 1992, Figures 1 and 3; with permission.)

the head rotates. For example the cells in the horizontal canals are arranged with the kinocilium nearest the utricle so that forces which push the hairs in that direction produce increased afferent activity. For the right side of the head, this is caused by counterclockwise motion of the fluid in the semicircular ducts, whereas on the left, clockwise motion is needed. Thus, the two horizontal canals work as a pair: any angular acceleration in the horizontal plane which increases firing on one side will decrease firing on the other and vice versa. If the head is turned to the right, fluid in the right duct rotates in the opposite (counterclockwise) direction, resulting in increased firing from the right duct; on the left, the counterclockwise fluid motion decreases firing. The anterior duct on one side and the posterior duct on the other also form a functional pair in which increased firing on one side is usually accompanied by decreased firing on the other.

In the macula of the utricle, the hair cells are oriented in all directions relative to a curving landmark known as the **striola**. They can monitor forces exerted by the otoliths in all directions. Tilt of the head in any one plane will always cause some receptors to depolarize and others to hyperpolarize. How the nervous connections are organized to couple such pairs together is not known.

Summary of vestibular connections

The central connections of the vestibular organs are discussed in Chapter 6. Briefly, the lateral vestibular nucleus of Deiters, which is the origin of the lateral vestibulospinal tract, receives a major input from the maculae of the utricle and saccule. Some neurones in this nucleus even respond selectively to tilt of the head, increasing their discharge in one direction and decreasing it in the opposite. This pathway is likely to be important in mediating the positional reflexes onto limb muscles produced by changes in the orientation of the head. The semicircular ducts (and to an extent the maculae as well) project to the medial, inferior (descending) and superior nuclei, as well as giving a small input to parts of Deiters nucleus. The input to the medial nucleus, which is the origin of the medial vestibulospinal tract, may be important in vestibulocollic reflexes elicited by angular acceleration of the head. The medial vestibulospinal tract only projects as far as the upper thoracic cord. Other outputs of the vestibular nucleus are important in the control of eye movement.

In summary, the hair cells in the vestibular system are stimulated by movement of the cilia along the axis of orientation of the cell. The force applied is related to the acceleration of the head. The cells of the semicircular ducts respond to angular acceleration of the head; those of the utricular and saccular maculae to linear accelerations caused either by linear movement of the body, or, during head tilt, by the force of gravity. The latter allows the otolith organs to signal head position.

8.2 *Quiet stance*

When normal subjects stand upright, without moving other parts of their body, there is extraordinarily little EMG activity in postural muscles. The knee is locked in a hyperextended position, and the spinal column assumes its natural curve through the action of the long muscles and ligaments which tie together the dorsal spines on the vertebrae. Movement is possible at the ankle, hips and neck, but all three joints, once balanced are maintained by minimal activity in sets of antagonist muscles which span the joints like the springs of an anglepoise lamp. The criterion for stability in a stationary subject is that the vertical projection of the centre of gravity of the body (which usually lies around the middle of the hips) should fall within the base of support outlined by the outer edges of the feet. In normal standing, the centre of gravity projects just forwards of the ankle joint.

The question is, once the various segments of the body have been balanced one on top of the other, how much work does the central nervous system have to do in the way of reflex corrections, to make sure that they stay there? Even if we remain as still as possible, the beating of the heart, breathing, as well as any external forces all tend to disturb this delicate equilibrium. Indeed, when standing still, the projection of the centre of gravity moves continuously, by about 1 cm in the anterior–posterior direction and some 0.5 cm from side-to-side.

Role of muscle properties

In the first instance it seems as if equilibrium could almost be maintained by relying solely on the mechanical properties of the postural muscles, without any reflex interference. The way to demonstrate this is as follows. Most authors have modelled the quietly standing human as what is termed an 'inverted pendulum', pivoted around the ankles. This is, they have ignored possible movement at the hips and neck (which undoubtedly do occur, but are small). If the centre of gravity of the body is located exactly above the pivot of the ankles, then the system is perfectly balanced. If the centre of gravity rotates, say, θ degrees forwards from this position, then the body will tend to fall over unless a torque is applied at the ankle in the opposite direction. The amount of torque needed is given by the formula $T = m\,g\,h\,\theta$, where m is the mass of the of the body, h is the height of the centre of gravity above the axis

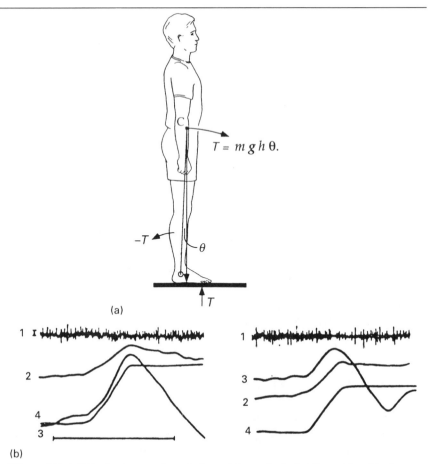

(a)

(b)

Figure 8.3 (a) The body as an inverted pendulum, pivoted at the ankles. The torque, *T*, needed to oppose a forwards sway angle of θ degrees is $T = m\,g\,h\,θ$. The stiffness of the ankle is given by $T/θ$. *T* can be measured from the forces acting on the support surface; θ can be estimated by measuring the forwards and backwards movement of the tibia. (b) Two records of the effect of platform rotation on the torques and ankle angle of a normal quietly standing subject. Traces are: (1) EMG from gastrocnemius/soleus, (2) angle of the ankle joint, (3) moment of force at the ankle, (4) movement of the platform. Time calibration is 1 s. When the platform is moved upwards, the ankle dorsiflexes and the force on the platform increases at the same time. There is no obvious change in the EMG over this time. (a, From Fitzpatrick *et al.*, 1992, Figure 2; b from Gurfinkel, Lipshits, and Popov, 1974, Figure 1; with permission.)

of rotation of the ankle, and *g* is the acceleration due to gravity. If the torque is distributed between both legs, then for each ankle it would be $0.5\,m\,g\,h\,θ$ (Figure 8.3).

To pursue the example further, when the body rotates θ degrees forwards, it will stretch the gastrocnemius/soleus muscles, and, because of their tension–length characteristics, produce an opposing torque at the ankle. The question of whether the central nervous system needs reflexes to stabilize posture then comes down to the question of whether the tension–length relationships of the muscles at the ankle are sufficient to compensate for the torques produced by the body sway.

The total torques produced by all muscles acting at the ankle can be measured by having subjects stand on a moveable platform which can be rotated about the ankle joint. Gurfinkel and colleagues (1974) found that if they rotated the platform at about the same velocity and through the same extent as the movements seen under quiet stance, then the torques generated at the ankle were the same or slightly more than those which were necessary to maintain balance. Furthermore, over the period that they measured these angle and torque changes, there was no appreciable change in the EMG activity of ankle muscles (Figure 8.3). The conclusion was that the intrinsic stiffness of the ankle muscles could be sufficient to stop the body falling over during normal quiet sway.

Role of afferent input

Although the length–tension relationships of muscles may help to stabilize the body, they cannot account for all aspects of postural stability. For example, in the experiments above, changes in EMG activity **were** seen (a decrease in soleus after dorsiflexion of the ankle, see also section on platform movements below), but only after completion of the applied disturbance, so that they had no effect on the estimated calculation of muscle stiffness. Similarly, it is easy to demonstrate that sway increases when subjects close their eyes. If sway were solely controlled by the mechanical characteristics of muscle, then this would not occur. Afferent input clearly does influence postural stability.

All classes of afferent input (visual, vestibular and somatosensory) contribute to stabilization of the body during quiet stance. Removal of all of these inputs makes it impossible to stand unassisted. This can be observed if patients with no vestibular function are asked to close their eyes and stand on foam rubber. Eye closure removes visual input, and the foam rubber support reduces the effectiveness of somatosensory input from the legs. The result is that the patients fall over. However, if they are given any one of three sources of input, then balance becomes possible even if not completely normal. For example, if the patients close their eyes and stand on a firm support, then their sway (at least initially) can be little different from that of normal subjects. Somatosensory input can be used on its own to maintain balance. Similarly, if the patients are put on a special platform which rotates about the ankles in an equal and opposite direction to the direction of sway (a 'sway-referenced support') so that the angle between the calf and the foot remains constant while the angle of the body changes relative to the force of gravity, then the patients can still stand. In this situation they must be relying mostly on visual input to balance (plus sensation from pressure receptors in the soles of the feet). Finally, to show that vestibular input alone can be used as the main source of afferent input to sustain balance, normal subjects can be asked to stand with their eyes closed on a sway-referenced platform. Under these conditions, visual and somato-sensory inputs are reduced or removed, but subjects do not fall over. The conclusion is that only one of the three main sources of afferent input is necessary for balance.

Within the somatosensory modality, the relative contribution of cutaneous/joint/muscle afferents can be investigated. One way of doing this is to use the technique of ischaemic anaesthesia. During ischaemia produced by inflation of a blood pressure cuff above arterial pressure, transmission in large sensory afferents is blocked before that in motor fibres. Thus, a stage is reached (after about 15 min) when motor power is relatively well preserved, yet sensation is much reduced. If the cuff is inflated around the ankles, input from the intrinsic foot muscles and cutaneous receptors in the soles of the feet is reduced. If the cuff is placed above the knee, the input from muscles in the calf also is reduced. The latter produces the most pronounced effects on sway. Inflation of a cuff above the knee leads, within 30–40 min to a pronounced an-teroposterior sway at about 1 Hz which is unaffected whether or not the subjects open or close their eyes. Ischaemia at the ankle has relatively little effect on postural sway when subjects have their eyes open. However, with the eyes closed, the body becomes unstable at low frequencies. It is thought that the input from the soles of the feet is important in stabilizing posture at low frequencies. This input can be replaced by input from the visual system, which is also sensitive to low frequencies (< 1 Hz) of sway. At high frequencies, input from the calf muscles (probably muscle spindles) may be more important. Visual input cannot substitute for this high frequency input.

Most studies have investigated anteroposterior sway. However, the body sways in the mediolateral direction as well. Abnormalities of lateral sway, as seen in some patients with cerebellar disease, cause a typical 'broad-based' gait. Spreading the feet apart increases the area of the postural base and minimizes the chances of overbalancing in the lateral direction. Spreading the feet apart, however, has another intri-guing effect: even in normals it decreases the amount of postural sway in the lateral direction. It is easy to see why increasing the lateral distance between the feet makes balance safer in the lateral direction, but why should it actually decrease the amount of sway that is there?

One reason is purely mechanical. Increasing the lateral distance between the feet allows the hip muscles to play a part in stabilizing lateral sway in addition to those at the ankles. For example, if it were possible to stand with both ankle joints coincident, the only muscles able to prevent lateral rotation about the ankles would be those that invert or evert the foot. As the feet are moved apart, a change in ankle angle is accompanied by a change in the hip angle, the amount of which increases with stance width. In other words the structure is made stiffer when the feet are moved apart simply because more muscles are involved in the movement. If the structure is stiffer, then the amount of movement will decrease.

A second possible reason for stabilization of lateral sway when the feet are moved apart is that with more muscles involved in the movement, there will be more afferent input on which the nervous system can base any reflex corrective forces which might reduce sway. (See Gurfinkel, Lipshits and Popov, 1974; Diener *et al.*, 1984a; Fitzpatrick, Taylor and McCloskey, 1992; Day *et al.*, 1993; for work covered in this section.)

8.3 *Postural reflexes*

Even if it were possible to stand quietly by relying on the mechanical stiffness of muscle alone, it is quite clear that, in bipedal man, balance is a precarious state of affairs. Any small disturbance to the body might cause the centre of gravity to fall outside the postural base and balance would be lost. Such disturbances to balance can be produced either by external forces acting on the body or by active movements on the part of the subject. The former are corrected by postural reflexes, the latter also involve anticipatory postural action which is covered in the next section.

In essence, the role of postural reflexes, like that of the reflexes described in previous chapters, is to maintain the status quo. That is, any movement of the centre of gravity, even if not immediately threatening to balance, is detected by afferent input and opposed by contraction of postural muscles. Simply by the fact that all the forces in a free-standing subject must be transmitted to the feet, postural reflexes often involve the contraction of many muscles all over the body, and hence are more complex than, for example, the simple stretch reflex. Two of the features which are rare in more simple reflex reactions are particularly important in postural reflexes. First, the responses can be initiated by input in somatosensory, visual or vestibular pathways. Often, the same response can be elicited (perhaps with different latencies) by any one of the inputs. Alternatively, inputs in one channel can modulate the responses evoked by stimulation of another. A second feature of postural reflexes is that the form of the responses, in terms of the pattern and amount of activity recorded in different muscles, varies depending upon the strategy adopted by the postural system to maintain balance.

Effects of pulls to the body

The polysensory nature of postural reflexes and the influence of

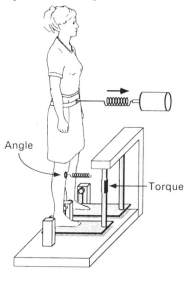

(a)

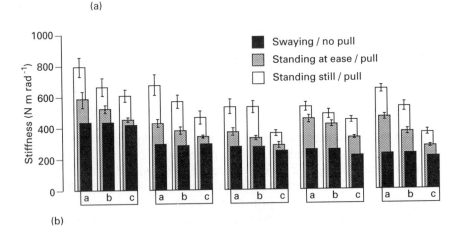

(b)

postural strategy is well illustrated by the responses which are produced by disturbances applied to the trunk. In these experiments, pulls are given to the body at waist level in order to cause sway at the ankles. The stiffness of the ankles is then measured with different sensory inputs available to the subject, or with the subject using different postural 'strategies'. If the pulls are very small, about the same size as the trunk movements seen during normal quiet stance, then the subject will not perceive them. In such circumstances, the sway is opposed only by the mechanical properties of the ankle muscles and by reflex activity, without the complicating effects of voluntary intervention. If the background level of EMG in ankle muscles remains constant, the muscle properties must remain the same, and any change in ankle stiffness represents a change in reflex resistance to body sway. When the subject closes his eyes, removing visual information about balance, ankle stiffness decreases. If vestibular input is removed as well (by a clever mechanical arrangement), then stiffness decreases further. Thus both forms of input can increase ankle stiffness, presumably by postural reflexes during quiet stance (Figure 8.4).

The effect of postural strategy can be demonstrated by asking subjects to 'stand as still as possible' rather than stand 'at ease'. In the former case the stiffness measured at the ankle is much greater than in the latter. The increase is not caused by voluntary co-contraction of ankle muscles since there are no changes in baseline levels of EMG activity. Instead, the increase in stiffness is thought to have been produced by increased reflex feedback caused by the change in voluntary intent of the subject.

The question in these experiments is which reflexes have changed? Measurements show that pulls to the trunk produce EMG responses in ankle muscles which are modulated by afferent input and voluntary intent. Since most of the movement in these experiments occurs around the ankle, it is usually assumed that local ankle stretch reflexes provide

Figure 8.4 (a) Method for perturbing standing subjects by pulling on the waist. The ankle angle is calculated from the movement of the tibia, whilst the torque at the ankle is calculated from the forces on the platform. The bars in (b) show the stiffness measured under different conditions in five different subjects. The dark bars indicate the stiffness measured in normal quiet sway (i.e., no pulls given to the body). The pale and open bars show the stiffness measured when the body was being pulled forwards by a motor with the subject either standing 'at ease' (pale bars), or 'standing still' (open bars). The three sets of bars (*a,b,c*) for each subject are measurements made with full visual, somatosensory, and vestibular feedback (*a*), with just vestibular and somatosensory feedback available (eyes closed) (*b*), and with just somatosensory feedback (*c*). The latter was achieved by asking fully supported subjects to balance a weight equivalent to that of the own body through a pulley system. Note that maximum stiffness occurs when all three types of feedback are available, and the subject is trying to stand as still as possible (open bars, column *a*). (From Fitzpatrick, Taylor and McCloskey, 1992, Figures 1 and 5; with permission.)

the major contribution to stiffness measurements. The conclusion is that when these muscles are being used in a 'postural' mode, the effectiveness of their local reflex pathways can be changed by other, remote, inputs. Such effects are not seen if similar ankle perturbations are applied in seated individuals.

The complexity of postural responses, in terms of the pattern of muscular activity produced, is illustrated by applying pulls to the arm rather than the trunk. A forwards pull to the arm tends to move the body's centre of gravity forwards as well. If the pull is brisk, then reflex responses can be recorded with very clear onsets and short latency in many muscles of the body. As expected these disturbances generate reflex responses in the stretched arm muscles, but their interest lies in what happens in the leg. In fact, a small pull forwards on the arm results in reflex EMG responses in the soleus muscle, which oppose the small forward sway of the body. Unlike the responses to perturbations of the trunk, these leg responses are not primarily local stretch reflexes. Their onset latency is only about 80 ms, which is shorter than the time taken for any measurable mechanical disturbance to be transmitted to the leg. In fact, they appear to be 'driven' by afferent input from the arm. The reflex system has **anticipated** the postural disturbance that will arrive with the leg, and has produced advance responses which are more effective in minimizing the consequences of the arm pull than local stretch reflexes set up by the leg movement. Such anticipatory action is a common occurrence in postural responses to self-generated body movements (see below), but rather less common in the reflex system.

As with pulls to the trunk, anticipatory reflex responses are modulated by postural strategies. If the shoulder is braced against forwards movement by a firm padded support, then the soleus responses to a forwards arm pull disappear. At the same time the reflexes in the arm become larger. It is argued that when the shoulder support is present, no body sway is produced by the arm pull, and hence the reflexes in soleus are no longer needed and are reduced in size. The arm, in contrast, is now working from a more stable base, and can therefore generate greater power to oppose the disturbance (Traub, Rothwell and Marsden, 1980; Nashner, 1982; Fitzpatrick, Taylor and McCloskey, 1992).

Platform studies

By far the largest number of investigations on reflexes of balance in man have used specially-built platforms on which the subject can stand, and which may be moved by powerful motors backwards or forwards (translation), or rotated about the ankle joint. The responses evoked provide excellent examples of the complexity of postural responses, their modulation by postural strategy, and the polysensory nature of the afferent input.

Platform translation

If a subject stands on a platform which is moved briskly backwards, the body sways forwards. This sway is opposed by postural reflexes in limb and trunk muscles. When the sway is forwards, muscles in the posterior leg and trunk contract, pulling the body backwards to restore the

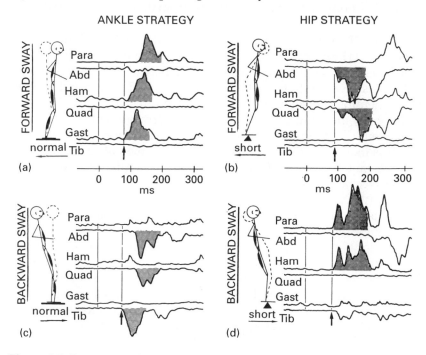

Figure 8.5 Responses of normal subjects to platform translations. In (a) and (b), the subject stands on a full support surface, which is moved backwards (a) or forwards (c), to induce anteroposterior sway of the body. When the body sways forwards, there is a sequence of distal to proximal activation of the muscles on the dorsal side of the body [gastrocnemius/soleus (Gast), hamstrings (Ham), and paraspinals (Para)]. When the induced sway is in the opposite direction, the muscles on the opposite side of the body contract in a similar distal to proximal order [tibialis anterior (Tib), quadriceps (Quad), and abdominal muscles (Abd)]. This distal to proximal pattern is reorganized when subjects have to stand on a narrow base (b and d). In this case, the ball of the foot is no longer in contact with the platform and ankle torques cannot be used to stabilize the body. Instead, a 'hip strategy' is used. In response to forwards sway, the hip is flexed by the abdominal and quadriceps muscles. This pushes the hip backwards, and restores equilibrium (see dotted stick figure). The opposite happens when the body is made to sway backwards. (From Horak and Nashner, 1986, Figure 2; with permission.)

position of the centre of gravity with respect to the base. When the sway is backwards, the muscles in the anterior leg and trunk contract and pull the body forward.

These responses occur with shortest latency in distal leg muscles, and are followed by activity in proximal and trunk muscles. Thus, backwards displacement of the platform (to produce forward sway) elicits responses in the gastrocnemius and soleus muscles at 80 or 90 ms, these are followed by responses in hamstrings some 10–15 ms later and finally by responses in the paraspinal muscles 25–30 ms after those in hamstrings. The body is stabilized, as it were, from the base upwards (Figure 8.5a).

Two mechanisms could explain the distal–proximal timing of the responses: (1) the responses in each muscle could be driven by local stretch reflexes. Hamstrings would be activated after gastrocnemius and soleus because stretch of hamstrings occurred later than that of the gastrocnemius and soleus muscles. (2) Responses in all muscles might be driven by the **same** afferent input, in which case a central delay must be evoked to explain the later onset of activity in more proximal muscles. It is likely that the latter is the case since the pattern of response depends on the position of the subject and the length of the supporting surface (see below).

As is typical of postural reflexes, the size of the responses can be modulated not only by changing the size of the disturbance, but also by many other factors. An intriguing example of this is the influence of body weight. If the subject's weight is decreased by immersion (of both apparatus and subject!) in water, the EMG responses to platform translation are reduced. If the weight of the subject is then increased by adding lead weights to the body, then the responses increase in size. Largest responses are seen at normal weight. If subjects stand in air and are loaded with extra weight, their responses do not get larger. The conclusion is that there is a 'gain control' mechanism for these responses which depends upon body weight. It may receive its input from receptors along the vertical axis of the body.

The responses are also influenced by the conscious expectation of the subject. For example, if subjects are expecting a large amplitude of displacement, the size of the early component of the response may be larger than if they are expecting a small amplitude displacement, even though the perturbation may not have reached its final amplitude before completion of the first part of the EMG response. The system 'sets' to some extent the gain of the early component of the response according to the amplitude of the disturbance it expects to receive. Repeated presentations of the same size of displacement lead to **habituation**. If subjects are trained to expect a disturbance of a given size (say by receiving twenty trials in a row of the same displacement), then inclusion of a novel displacement always results in a larger response, whether or not the displacement was greater or less than expected. Repeated presentation of the novel displacement eventually results in a normally-scaled response.

The most remarkable example of the modulation of reflexes to platform translation comes in situations when the whole postural strategy of the subject is changed. To appreciate this it is necessary to understand a little about mechanics of postural correction. The responses in anterior and posterior leg and trunk muscles which compensate for body sway produce their effect by exerting torques at the ankle. When the body sways forwards, the ankle plantar flexors (gastrocnemius and soleus) are activated and pull the body back towards the vertical. The activity in hamstrings and paraspinal muscles stiffens the rest of the body so that the trunk does not flex at the hips when the legs move backwards. However, there are some situations when it is impossible for the muscles to exert the necessary compensatory torques at the ankle. For example, if subjects try to balance on a see-saw pivoted at the feet, changes in torque at the ankle simply move the see-saw and have no effect on the position of the body. Balance in this condition is controlled by movements of the arm and trunk (rescue reactions). A complete change of postural strategy is produced by making the ankle muscles irrelevant to balance. It is a classic example of the postural system achieving the same end by different means.

This effect can be investigated in detail using platform studies. Rather than have subjects stand on a large platform, subjects stand on a base (9 cm) which is shorter than their feet. In this situation, balance can only be compensated over a small range of body movement by applying torques at the ankle. Large displacements necessitate a change of strategy. Under normal conditions, subjects would probably produce stepping reactions in response to large perturbations (see below). However, in most platform studies, subjects must keep their feet stationary. They then maintain balance by moving their arms and trunk.

Figure 8.5 illustrates the difference in the responses of standing subjects to translations of a platform when they stand on a normal support, and when they stand on a short surface. Movement of the normal surface evokes the usual distal–proximal pattern of activity in appropriate dorsal and ventral muscles. However, responses on the short platform are quite different, and tend to involve only the trunk, and not the ankles. Thus with forwards sway (produced by backwards translation of the platform), there is contraction of the abdominal and quadriceps muscles, which causes the hip to flex. The trunk is pushed forwards, and the legs are pushed backwards, stabilizing the centre of gravity. This results in quite different forces exerted by the feet on the platform than in the normal state. The ankle torque is much reduced, but the shear forces along the platform are much increased. Balance is maintained, but posture is not restored. This type of reaction is said to involve a 'hip strategy' to distinguish it from the 'ankle strategy' of the inverted pendulum model which normally occurs with platform translation. However, it should be noted that it is simply a special case of the rescue reactions discussed below, limited by requirements to keep both the feet and the arms still.

Another way of changing the order of muscle activation in response to platform translations is to ask subjects to stand on all fours. Backwards movement of the platform produces forwards movement of the trunk, which in this instance is compensated by flexing the hip. Instead of the gastrocnemius/soleus and hamstrings being activated in the usual distal–proximal sequence, it is the quadriceps and tibialis anterior which are activated in a proximal–distal sequence. Interestingly, even though the arms take half the weight of the body, they do not participate to any great extent in the postural corrections. They act as rigid supports which, during forwards displacements of the trunk (backwards platform translations), take more weight, and during backwards displacements take less weight. The greater the weight supported, the more the agonist and antagonist muscles on either side of the arm co-contract. Platform translations produce a similar asymmetry of leg and arm actions in the cat.

In all the platform studies referred to so far, the whole of the support surface has been translated in the same direction. However, if two separate platforms are used, different perturbations can be given to each leg. In this way, it is possible to investigate interlimb reflexes. In normal standing subjects, displacement of one leg forwards or backwards produces EMG responses in both legs: tibialis anterior is activated for displacements which produce forwards sway of the body, and gastrocnemius/soleus for those that produce backwards sway. The contralateral responses are quite large, being about half the size of those on the perturbed side in gastrocnemius/soleus, and 80–90% of their size in the tibialis anterior. The responses occur slightly earlier (about 85 ms) in the displaced than in the non-displaced leg (about 100 ms), but since movement of the non-displaced ankle joint is minimal, the crossed response is thought to be driven by input from the displaced limb. Presumably such crossed responses become even more important in locomotion. (Nashner, 1976, 1982, 1983; Dietz and Berger, 1982; Horak and Nashner, 1986; Diener, Horak and Nashner, 1988; Horak, Diener and Nashner, 1989; Macpherson *et al.*, 1989; Horstmann and Dietz, 1990; for work covered in this section.)

Platform rotation
Some of the most interesting platform studies have involved rotation of

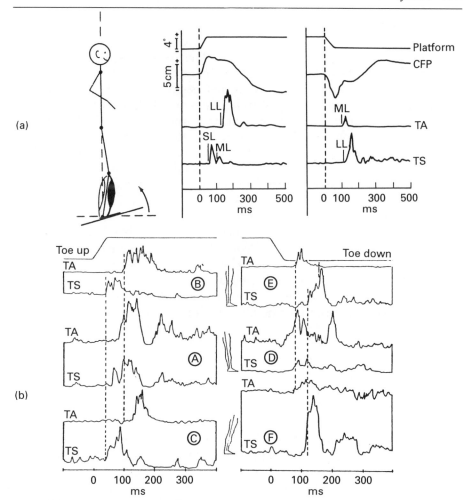

Figure 8.6 Responses of a normal subject to rotation of the support surface about the ankle joint. (a) The pattern of response which is seen after several trials of either toes-up (left) or toes-down (right) rotation. Motion of the platform (60°s⁻¹) and the position of the centre of foot pressure (CFP, forwards movement produces an upwards deviation of the trace) on the platform are shown together with the average surface rectified EMG records from the tibialis anterior (TA) and gastrocnemius/soleus (TS) muscles. Note the small stretch reflex in the stretched muscle (in the TS muscles during toes-down and TA during toes-up), which consists of a short latency (SL; in TS only) and medium latency (ML) component. This is a potentially destabilizing response, and is followed by a larger burst of activity in the antagonist muscle (LL). Panel (b) shows the effect of changing the initial stance position on the latency of responses to platform rotations. The upper traces (*B* and *E*) show the responses in normal upright stance (see stick figure in inset). The middle traces show the response when the subject is initially leaning backwards (*A* and *D*), and the bottom traces the response when the subject leans forward (*C* and *F*). The vertical dashed lines indicate the onset of EMG activity in TA and TS in the neutral position. (a, From Diener *et al.*, 1984b, Figure 1; with permission; b, from Diener *et al.*, 1983, Figure 2; with permission.)

the platform about the ankle joint rather than translation. Rotation of the support surface poses a particular problem for the control of posture. Consider a toes-up rotation. This produces a stretch reflex contraction after about 50 ms or so in the gastrocnemius/soleus muscles which tends to pull the body further backwards and destabilize the posture even more. The postural system only begins to correct for the disturbance after about 120 ms with a contraction of the tibialis anterior, a muscle shortened by the displacement. This contraction pulls the body forwards and restores the position of the centre of gravity (Figure 8.6).

If the same rotation is repeated a second and a third time, then the responses change. The stretch reflex in the gastrocnemius/soleus decreases in size so that the destabilizing action of local stretch reflexes is reduced. The later postural contraction of the tibialis anterior assumes greater importance and balance becomes more stable. Finally, the size of both the stretch reflex and corrective responses increases with the amplitude to the displacement. However, the slope of the relationship between the displacement amplitude and size of response (gain) is less steep for the stretch reflex component in gastrocnemius/soleus than the corrective response in the tibialis anterior. This indicates that the latter assumes greater importance in large displacements which are more likely to lead to overbalancing of the whole body.

There is some confusion over the precise form of these responses. The problem is that their latency depends on the initial posture of the subject. If subjects lean forwards before the platform is rotated, then the latency of the gastrocnemius/soleus responses to a toes-up displacement is increased, whilst the latency of the corrective response in the tibialis anterior is decreased. Indeed, in such a posture the corrective tibialis anterior response begins at approximately the same latency as the major part of the stretch reflex response. When the subject leans forwards, the responses in gastrocnemius/soleus retain the same latency, but those in tibialis anterior are now later than in the normal upright position. Figure 8.6 illustrates these effects, and also indicates a commonly used nomenclature for each component of the response.

A second point of debate about the responses to toes-up rotation is the extent to which the responses change on repeated presentation. Initially, it was thought that the stretch reflex component decreased whilst the corrective response increased in amplitude on the first five or so trials. This was termed functional adaptation, and was said to be absent in patients with cerebellar disease and postural instability. It was believed to be a very specific property of the postural control system designed to decrease unwanted reflex responses. More recently, it has been shown that the stretch reflex response in the gastrocnemius/soleus (to toes-up rotation) and the corrective response in the tibialis anterior both decrease on repeated presentation on the same displacement. A general decrease in the size of responses to repeated stimuli is usually termed habituation, and occurs to some extent in almost all polysynaptic reflexes (the R2 component of the blink reflex is a good example). Thus, the effect cannot be regarded as a specific adaptation of the part of the postural system. (Nashner, 1976, 1983; Diener *et al.*, 1983, 1984b; Keshner, Allum and Pfaltz, 1987; for work covered in this section.)

Receptors involved in responses to platform displacements
The leg and trunk muscle responses to platform displacements are produced mainly by somatosensory inputs which signal movement at the ankles. This input is then thought to trigger the EMG activity in many different muscles according to the particular type of displacement being studied. The reasons for excluding other channels of sensory input are:

1) The responses in leg muscles to most types of isolated visual and vestibular inputs occur at latencies longer than those seen after platform displacements (see below, but note that vestibular responses to galvanic stimulation can be surprisingly fast).
2) If visual and/or vestibular inputs are removed, the responses to platform displacement are still present. For example, closing the eyes in normal subjects does not decrease the size of EMG responses to platform perturbations. Similarly, patients with vestibular dysfunction also have normal responses to platform translations, although they have been reported by some groups to have reduced responses to platform rotation.
3) There are occasions when the motion of the head (which contains the visual and vestibular receptors) can be dissociated from that at the ankle. In these conditions, the EMG responses in the leg are those expected from ankle, rather than head movement. Thus, if a platform is moved backwards and then, 30 ms later, is reversed to move forwards, the body sway at the ankle will be initially forwards and then backwards. In contrast, the motion of the head is complex and undergoes several cycles of oscillation. With such rapid displacements, the body no longer behaves as an inverted pendulum. The EMG responses in the leg, however, are those predicted from the change in ankle angle: an initial response in gastrocnemius/soleus followed, at an appropriate latency, by a response in tibialis anterior.

Although they may not be the major source of input for the postural response after platform displacement, both visual and vestibular inputs can influence its final size and pattern.

For visual inputs the experiment is as follows. Subjects stand on a moveable platform with their head enclosed inside a large moveable

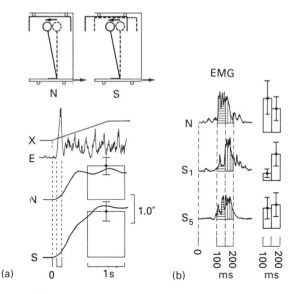

(a)

(b)

Figure 8.7 Effect of moving the visual field on the responses evoked in the gastrocnemius/soleus muscle by backwards translation of the support surface. (a) Experimental set-up. In the upper diagrams, the subject stands on a moving platform which tilts his body forwards when it starts to move (dotted and full stick figures). In the normal condition (N), the visual surround (box around the head) is fixed; in the stabilized condition (S), the box moves forwards at the same velocity as the head moves, thus stabilizing the visual field with respect to the head. X is the platform displacement; E, the EMG responses in the calf; N, the angular displacement of the body in the normal condition; and S, the angular displacement of the body in the stabilized condition. Note the larger sway in the stabilized condition. The vertical dotted lines indicate the onset of movement, and the initial EMG burst in the muscle which is shown in more detail in (b). (b) Typical EMG responses in gastrocnemius/soleus over the first 300 ms following platform displacement in the normal condition (N), and during stabilization of the visual field (S1, first occurrence of the stabilized condition; S5, fifth occurrence). Note the reduction in size of the EMG burst even over the first 100 ms of the response. The bars on the right show the average size of the first and second 100 ms of the response in the three trials. The biggest change is seen in the first 100 ms of the response on the first trial (S1) with stabilized vision. (From Nashner and Berthoz, 1978, Figures 1 and 2; with permission.)

box (Figure 8.7). Both box and platform can be moved independently. If the box remains stable while the platform moves, this is equivalent to a normal, stabilized visual field. However, if the box is moved at the same velocity as the head, then the visual field will appear stationary, even though the platform movement causes the body to sway. Under these conditions, the responses in the leg are smaller than usual, and hence visual input must have influenced the response at a latency much shorter than expected from the effect of vision alone (see below). The responses are also reduced if the box is moved in the opposite direction to the head in order to increase movement of the visual field during platform movement. It is thought that it is the *conflict* between visual signals indicating that no movement (or too much movement) has occurred, and somatosensory and vestibular input indicating the true amount of movement, which results in an attenuation of the reflex response.

For the vestibular system, as noted above, patients with loss of vestibular input appear to have normal responses to platform translation, but reduced responses to rotation. It is thought that during platform rotation vestibular input modulates the size of a primarily somatosensory evoked response. Patients with lack of vestibular input have also been reported to be unable to adopt the 'hip strategy' to deal with platform translations when they stand on a shortened support surface. Trunk and head movement are greatly increased in the 'hip strategy'. Perhaps vestibular input is important in controlling the amount of trunk flexion that occurs.

Finally, receptors in the feet have been shown to influence the type of postural response elicited by platform disturbances. If the feet are anaesthetized by ischaemia induced by a cuff around the ankle, then subjects tend to respond to translations with a 'hip strategy' even when they are standing on a normal support surface. The reason for this is thought to be uncertainty in defining the extent of the postural base when sensation from the feet is absent. (Nashner and Berthoz, 1978; Dietz *et al.*, 1988; Horak, Nashner and Diener, 1990.)

Postural reflexes elicited by visual inputs

Movement of the visual scene is well-known to produce illusions of movement. For example, if we are seated on a stationary train, and the train on the next track starts to move slowly away, then we experience the sensation that our train is moving in the opposite direction. In addition to this sensation of motion, visual inputs can also produce body sway, particularly in standing subjects. The direction of sway is opposite to the direction of the illusory movement.

If the visual scene is made to approach the body (either by physically moving the surround, or by using a special projector), then subjects experience a sensation that they are moving forwards and their body begins to sway backwards (i.e. in the direction of the visual stimulus). If the scene moves forwards, then the body sways forwards, and similarly for side-to-side or rotatory motion (Figure 8.8). The latency of the effect is of the order of 1 s. It is not certain why the body sways in the direction of the image motion. If we imagine an approaching visual scene, one possibility is that the subject tries to maintain a constant

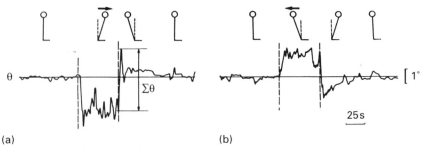

(a)   (b)

Figure 8.8 Postural adjustment to a moving visual scene. The two panels show the effect of moving the visual field forwards (a, see arrow over top stick figure), and backwards (b). The bottom trace is the angular rotation of the body about the ankles. Downwards deflection indicates dorsiflexion of the ankle. When the scene moves forwards (0.55 m s^{-1}) at the time of the first vertical dotted line, the subject sways forwards by about two degrees, after a latency of 1.2 s. This posture was maintained for the duration of the movement. When the movement was stopped, there was a transient sway in the opposite direction. (From Lestienne, Soechting and Berthoz, 1977, Figure 4; with permission.)

distance between the approaching object and his body. Maintenance of a constant spatial relationship would also explain the responses to other types of visual motion. Alternatively the sway may be a reaction which compensates for the lack of vestibular input during perceived forwards motion of the body. The effect of moving the visual field is usually quite small during normal standing. However, if subjects are placed in a moving vehicle and then the visual input is manipulated, the responses become much larger. Under such conditions, visual inputs are thought to be given more 'weight' than during normal stance.

The effect of visual input on sway depends upon the speed at which the visual field is moved. In most experiments, the motion has been relatively rapid, and under these conditions, the body sway is maintained for the duration of the movement. The responses are also still present after many repetitions of the same stimulus. The results are different if very slow displacements of the visual field are given at velocities approaching the normal velocity of body sway (about 0.03 m s^{-1}). With slow displacements, the subject sways initially in the direction of the visual displacement, but this is compensated after about 2–3 seconds by a compensatory sway in the opposite direction. If the visual stimulus is repeated, then both the initial and compensatory responses are very much smaller, and in normals are almost completely suppressed. Interestingly, patients with loss of proprioception from the knees down (due to tabes dorsalis, for example) sway much more in response to slow movement of the visual field than normals, and they do not habituate on subsequent trials. Conversely, patients with absent vestibular function and normal somatosensory input sway the same amount and habituate to the same extent as normals.

The results suggest that under the conditions of these experiments, the postural system weights the inputs it receives in the order somatosensory > visual > vestibular. Thus, in patients with no vestibular function, somatosensory input can still cause adaptation of responses to visual stimuli, because the somatosensory cues that the body is stationary override the visual impression that the body is moving. In contrast, in patients with no sensory input, vestibular signals alone cannot override visual cues, and the responses to visual field displacement fail to habituate.

Although postural responses to visual input are usually considered to have very long latencies (1 second or more) it may be possible for more rapid responses to occur under special conditions. There is a single report of a patient with absent vestibular function and severe sensory loss below the knees who relied solely on visual information to maintain balance: he fell over after about 1 s if he closed his eyes, and could walk only with great concentration. His sway was measured under conditions of flicker illumination. This showed that he needed a visual input rate of at least 17 Hz to stand upright. Lower frequencies of illumination caused him to fall over. A 17 Hz input theoretically provides a new head position value every 59 ms, which would be consistent with the idea that under special circumstances there may be a rapid visual contribution to sway. A possible candidate pathway for this rapid effect is the tectospinal pathway, which in the cat has been shown to be able to mediate very rapid visual corrections to aimed arm movements. (See papers by Lestienne, Soechting and Berthoz, 1977; Berthoz *et al.*, 1979; Bronstein, 1986; Paulus, Starube and Brandt, 1987).

Postural reflexes elicited by vestibular inputs

Several studies have used H-reflex monitoring of monosynaptic excitability in leg and arm muscles to show that small, subthreshold changes in spinal excitability are produced by vestibular inputs caused by changes in head position. The effects are small and variable. Three other ways of investigating vestibular reflex effects in normal balance have been used. In order to exclude possible complicating effects of voluntary intervention, they have concentrated on the early phasic, components of the responses.

1) As noted above, platform studies can be adapted to allow both horizontal translation and ankle rotation. If the platform is rotated in exactly the same direction as body sway produced by platform translation, then the angle at the ankle will be stabilized even though the body has been allowed to sway. Under these conditions

('sway-referenced' platform motion), the somatosensory input from the ankle is reduced, and if the subjects close their eyes, the major source of input about body position is the vestibular system. The reflexes elicited are very similar to those evoked by platform translation on its own. That is, activation of gastrocnemius/soleus and hamstrings during perturbations which induce forwards sway, and activation of tibialis anterior and quadriceps during backwards sway. The order of muscle activation is also the same as after platform translation, progressing from distal to proximal muscles. This suggests that the pattern of activation is a package of preset commands in the postural system which can be elicited by both somatosensory and vestibular input. However, the latency of the response is much longer after vestibular input (> 200 ms) than after pure platform translation (80–90 ms).

2)   Reflexes from the vestibular system can be elicited by dropping subjects from a moderate height (Figure 8.9). The drops produce activity in leg muscles after about 80 ms, and are absent in patients who have no labyrinthine function. The size of the response is proportional to the acceleration experienced, but saturates at about 1 g. Like many of the responses described above, it is influenced by visual input. If the visual scene is moved upwards at the same velocity as the subject falls, the responses are increased in size. The initial part of the response is believed to be due to stimulation of otolith organs since it is still present in cats in which the semicircular canals have been plugged. Although the response is robust, it is not yet clear to what extent this is a startle reaction elicited by otolith input, or a separate, purely vestibular response. The essence of the matter is that the startle reaction is thought to be a reticulospinal phenomenon, whereas the response to dropping may be either a direct vestibulospinal effect, or a reticulospinal response elicited by vestibular input.

3)   As explained in Chapter 7, DC electrical stimulation across the mastoid processes is thought to activate vestibular afferents, and can cause the body to sway. Sway is greatest when the eyes are closed, since this removes one source of conflicting information about the state of the body. For reasons not fully understood, the sway is always in the direction of the stimulating anode, and this is true even when the body or trunk is twisted relative to the feet. If the anode is behind the left ear, and the subject looks straight ahead, then the body will sway to the left. If he turns to look over his left shoulder, the body will sway backwards (the left ear is now at the back of the body). If he then twists his trunk to the right, whilst keeping the head turned to the left, the body will once again sway to the left (the left ear is now on the left again). The outcome

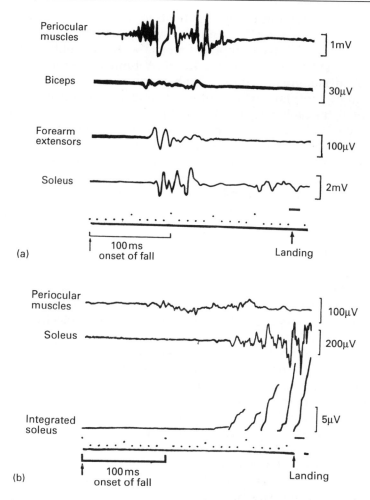

Figure 8.9 (a) EMG activity in the periocular muscles, biceps, forearm extensors and soleus in a normal subject during an unexpected free fall from 20 cm. (b) EMG activity in the periocular and soleus muscles in a subject with absent labyrinthine function during an unexpected free fall from 25 cm. Note the absence of early EMG activity [note that the EMG gain is much greater here than in the traces in (a)]. (From Greenwood and Hopkins, 1976, Figures 3 and 7; with permission.)

is that the same stimulus to vestibular afferents produces different actions depending on the relative position of the body parts. This indicates that the system receives somatosensory inputs signalling the position of the body segments relative to each other, and can incorporate them into the final output command. It is a clear example of how the action of one set of balance-related inputs can be modulated by the action of information in other sensory channels.

Finally, there is one group of patients in whom there is a mechanical method of stimulating the vestibular system directly, in the absence of visual or somatosensory inputs. Patients with the otolith–Tullio phenomenon suffer from sound-induced vestibular phenomena such as vertigo, nystagmus and postural imbalance. It is caused by a hypermobile stapes footplate which allows sound of normal intensities to stimulate the otolith organs. In standing patients, a brief sound causes activity in tibialis anterior and gastrocnemius/soleus muscles with a latency of about 50 ms. This is presumed to be a direct vestibulospinal effect caused by sound-induced movement of the otoliths. Its latency is similar to the earliest EMG responses to galvanic vestibular stimulation in normals (Greenwood and Hopkins, 1976; Watt, 1976; Nashner, 1983, Dieterich, Brandt and Fries, 1989; Britton *et al.*, 1993).

Rescue reactions

The reactions which help to restore balance once the centre of gravity has fallen outside the postural base are perhaps the most familiar. These **rescue** reactions involve obvious, gross movements of the whole body. They may be classified as stepping, sweeping and protective reactions (Figure 8.10).

If a standing subject is given a large unexpected push from behind, this will cause him to take a step forwards. This increases the area of his base of support, and captures the forwards-moving centre of gravity. In normal circumstances the postural system does not wait until the centre of gravity has actually fallen outside the postural base before initiating the step; it predicts whether or not an initial displacement is likely to cause such a movement to occur, and if so it initiates the step before the point of no return is passed. This prediction can be demonstrated by giving the same push several times to the same subject. Unless the push is so strong as to overbalance him on every trial, it is likely that stepping will be produced on the first and maybe the second trial, but that afterwards, the subject will just allow his body to sway, knowing from experience that the disturbance is short-lasting and hence not sufficient to move his centre of gravity to an unstable position.

It is possible that many different categories of sensory input could contribute to stepping reactions, even those which are not normally thought to be involved in control of posture. In some experiments, stepping has been elicited by cutting a rope which was stabilizing forwards-leaning subjects around the hips. Under these conditions, the initial response is activity in the gastrocnemius/soleus muscle at about 60 ms after release. This is followed by a burst in tibialis anterior after about 200 ms in the leg which takes the first step. The responses are present in labyrinthine defective patients, and are smaller but still

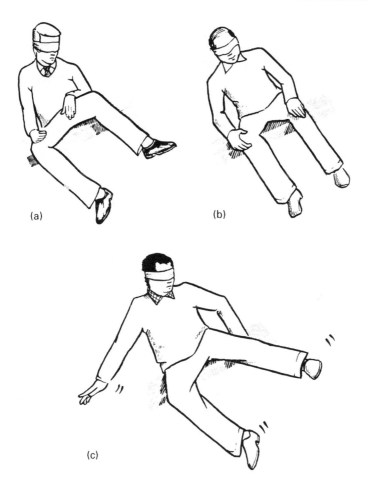

(a) (b)

(c)

Figure 8.10 Behaviour of a normal subject (a) and labyrinthine-defective subjects (b and c) when seated blindfold on a tilt-table. The normal subject shows stabilization of the head and trunk relative to gravity. The freely-dangling legs show sweeping reactions consisting of rotation about the thighs. Panel (b) shows the behaviour of a labyrinthine-defective patient. There is no stabilization of the trunk and head, and he is just about to overbalance. Note that the right hand is about to move off the knee to break the impending fall. Panel c shows complete fall-breaking reactions in another labyrinthine-defective patient. Both legs show strong sweeping reactions, the right arm is thrust out to break the fall, and the left hand clutches onto the support. This patient also shows slight stabilization of the head and trunk with respect to gravity. (From Roberts, 1979, Figures 10.24, 10.25, 10.26; with permission.)

present if the foot sole is anaesthetized or the calves made ischaemic with a cuff around the thigh. It is therefore possible that in these conditions, stepping is driven by receptors in the abdomen which sense the sudden release of force when the rope is cut.

Under some conditions, stepping reactions are inappropriate and can be suppressed. For example, if a swimmer begins to overbalance when standing on the edge of a pool, he will rapidly swing his arms forwards in an attempt to use the reaction forces that the arm movements make on the trunk to push the body back into equilibrium. In some circumstances, these sweeping reactions may also locate stable objects in the environment, onto which the subject can hold to maintain balance.

Finally, if all else fails, and balance is lost, powerful protective reactions occur, which are designed to protect the head and body during falls. The arms are thrown out and the trunk rotated to break the fall. The reactions are not easily suppressed. The arms will be thrown out through panes of glass or into fire in order to fulfil their protective role. These reactions are not dependent on vestibular function, since they are present in labyrinthine defective individuals (Figure 8.10). However, all the rescue reactions are depressed in certain diseases of the basal ganglia (see Chapter 11). (Roberts, 1979; Do, Breniè and Bouisset, 1988; for work covered in this section.)

8.4 *Postural reflexes described in animals*

A large number of postural reflexes have been described in animals, but since they are difficult, if not impossible to demonstrate in man, they will be discussed only briefly here. All these reactions are profoundly depressed in decorticate animals.

A series of reactions (the placing, magnet and supportive reactions) are available to bring a non weight-bearing limb onto a stable surface and help it support the weight of the animal. Thus, if a cat is held with all four limbs off the ground and then brought towards the edge of a table, the slightest touch of the table against any aspect of one of the feet will cause an immediate placing of the foot squarely down on the table. This is an example of a **placing reaction**: correct orientation of the limbs ready to take the weight of the animal.

If the support surface has a certain 'give', and moves down when the foot is placed upon it, then the leg will extend, and follow the support downwards. This is the **magnet reaction**. It can be seen in normal cats if they are lifted slowly from a lying position into the air. The legs will remain in contact with the ground as long as possible, gradually extending as the body is raised.

If the support surface is stable, and the weight of the animal is taken by the leg, this will cause the toes to splay out and stretch the intrinsic foot muscles. This input causes the **positive supporting reaction**, a stiffening of the whole leg into a rigid pillar to fix the limb for standing. This can be demonstrated in new-born babies if the weight of the child is taken around the chest, and the baby lowered until the feet touch the table. Its legs stiffen up because of the supporting reaction, although in this instance the force developed is not sufficient to take the body's weight.

If the body of an animal showing a positive supporting reaction in one leg is moved forwards or backwards or side-to-side, the leg hops in the direction of the displacement, in order to take the weight directly through the shoulders or hips. This is achieved by a gradual reduction in the supporting reaction as the body is pushed over the leg and then its sudden disappearance, followed by a leg flexion and subsequent extension before reappearance of the full supporting reaction in the new position (**hopping reaction**).

Finally, if the animal is lying with its body turned over to one side on the ground, a number of **body righting reflexes** are elicited which (usually in cooperation with neck and vestibular reflexes) place the body so that the weight is distributed symmetrically with respect to the support. The head will be lifted up approximately to the vertical (even in labyrinthectomized animals), and the trunk twisted so as to equalize the forces acting on either side. The responses are due to asymmetrical stimulation of the body surface. They do not occur if a weighted board is placed on the uppermost side of the body.

Postural reflexes in animals elicited by inputs from the vestibular system and from the neck

Vestibular reflexes are produced by activity in the receptors of the semicircular canals and the otolith organs. The reflexes from the semi-circular canals are provoked by angular acceleration of the head. These reflexes mainly affect the extraocular and neck muscles and have very little effect on the limbs. Briefly, any angular acceleration detected by movement of the skull will excite the vestibulo-ocular reflex to maintain the direction of gaze constant with respect to the external world. The same stimulus also engages the vestibulocollic reflex to cause contraction of neck muscles to oppose the movement of the skull. In animals like the frog, with limited neck movement, the reflex works on the limbs rather than the neck, again with the effect of opposing the angular acceleration applied to the head.

Vestibular reflexes which act on the limbs arise principally from the otolith organs. These receptors can monitor the position of the head with respect to gravity and they elicit tonic reflex responses to sustained changes in head position. They are supplemented by positional reflexes from the neck which arise from the large number of muscle spindles which innervate the small paraspinal muscles. Both sets of reflexes produce movements of the eyes in the orbit as well as movements of the body: only the latter are dealt with here.

Under normal conditions, movement of the head is usually accompanied by movement of the neck, so that the vestibular and neck reflexes interact. To investigate the vestibular or neck reflexes in isolation is difficult. For example, if we try to keep the head position constant in space, and rotate the neck by rotating the body with the head fixed, the contact forces along the body surface will alter. These contact forces can themselves give rise to reflex reactions in the limbs, which will interfere with our assessment of the reflexes from the neck. In order to separate neck and vestibular reflexes a special animal preparation is needed.

Cats are decerebrated, and the trunk, the axis vertebra, and the head are supported separately. The joints between the axis and atlas vertebrae and the atlas and the skull are denervated. In this state, when the axis vertebra is held firm, the head can be moved without rotation of the neck (apart from the denervated atlantoaxial and atlanto-occipital joints), and the neck (i.e. the remainder of the cervical vertebrae) can be rotated by holding the head still and moving the axis vertebra. This latter produces a twist of the cervical spine, but the reflexes evoked are not as brisk as expected in an intact animal since most of the displacement in normal movements of the head on the neck occurs at the atlanto-occipital and atlantoaxial joints (which are denervated here).

These and other experiments show that the reflexes from the neck and the vestibular system onto the limbs are of opposite sign. The effect of vestibular reflexes (that is those produced by changing the position of the head) can be summarized by saying that they tend to restore the head to its standard orientation. For example, if the head is tilted down, the forelimbs extend, and the hindlimbs flex; if the head is rotated so that the left ear is nearest the ground (left side down), then the left limbs extend and the right limbs flex. Vestibular effects are also seen on neck muscles, again fulfilling the same stabilizing function on the head. If the head is tilted down, the neck is extended, etc.

Reflexes from the neck on the limbs are the inverse of those just described. Rotation of the neck equivalent to left side down produces extension of the right limbs and flexion of the left. Flexion of the neck elicits flexion of the forelimbs and extension of the hindlimbs. Movement of the neck also evokes local stretch reflexes in the neck muscles. Unlike the neck reflexes onto the limbs, the stretch reflexes in the neck work in the same way as the vestibular reflexes, to restore the neck to a standard position. Interestingly, despite the large number of muscle spindles in the neck, there are very few monosynaptic connections from large afferents onto neck motoneurones. Thus the monosynaptic stretch reflex is small or non-existent. The stretch reflexes are mediated through

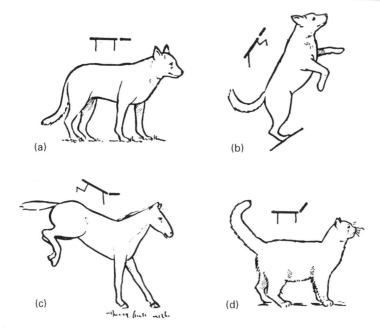

Figure 8.11 Characteristic poses to ilustrate the relationship between neck and vestibular reflexes. (a) head normal, neck straight: neck and vestibular reflexes minimal, weight evenly distributed; (b) head tilted back, neck straight: posture assisted by vestibular reflex, hindlimbs extended, forelimbs flexed ('prepare to jump'); (c) head normal, neck bent back: posture assisted by neck reflexes, hindlimbs flexed, forelimbs extended ('prepare to land after jumping'); (d) head tilted back, neck bent back: vestibular and neck reflexes cancel, weight evenly distributed. (From Roberts, 1979, Figure 10.14; with permission.)

pathways with several synapses, the details of which have yet to be fully investigated.

Normally the opposing reflexes from the vestibular system and neck work together (see Figure 8.11). Consider a dog on a platform tilted nose down. In the initial state, the head is tilted down, but the neck is still straight. The vestibular reflexes act to extend the neck and to extend the forelimbs and flex the hindlimbs. Once the neck extension is in progress then neck reflexes can assist the vestibular reflexes in extending the forelimbs and flexing the hindlimbs. The reflexes cooperate in a two-stage stabilization of the skull. The tonic reflexes from neck and vestibular systems are much more difficult to demonstrate in man than in the decerebrate animal because it is so easy to override them by voluntary intervention. Some characteristic poses are consistent with input from these reflexes, but these inputs cannot be the only ones which sustain that particular posture. For example, the neck reflexes may be involved in producing the characteristic posture assumed in fencing (neck to right, right arm extended and left arm flexed), and may contribute to driving errors when we move our heads (turn the head and neck to the right, extend the right arm on the steering wheel and turn the car towards the left). However, these are readily overridden voluntarily.

8.5 *Postural adjustments produced by voluntary movement*

Voluntary movement of the body displaces the position of the centre of gravity. For example, abduction of the right arm to the shoulder level would, if no action were taken, cause the centre of gravity to shift to the right. Under normal conditions this shift is minimized by postural mechanisms: the body sways to the left as the arm is raised. If we bend forwards to pick up a chair, the ankles plantarflex, and the hip is displaced backwards to keep the centre of gravity over the feet. As demonstrated in the party trick, if the hips are prevented from moving back, as they are if the subject is made to stand with both heels against a vertical wall, then the chair can no longer be picked up because the centre of gravity falls outside the postural base of the feet when the subject bends down.

The important feature which distinguishes the postural reactions accompanying voluntary movement from those which occur in response to external disturbances from the environment, is that the former occur at the same time, or before, the prime mover is activated. They anticipate the postural displacement that the prime movement will generate, and hence reduce the movement of the centre of gravity to a

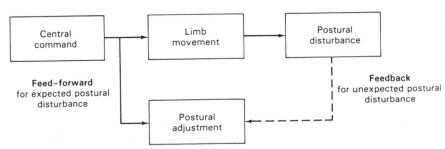

Figure 8.12 Illustration of the idea that postural and prime mover activity are coordinated by a central command. Thus, the command to move a limb is linked to ('feed-forward') commands to postural muscles designed to anticipate the effect that the movement will have on the position of the body. If the postural activity is incorrect, or if the limb encounters some unforeseen disturbance, then reflex ('feedback') corrections of posture also occur. (Adapted from Gahéry and Massion, 1981.)

minimum. It has been proposed that the central command to move produces instructions for both the prime movement and the postural adjustment, as shown in Figure 8.12. If the postural adjustment is incorrect, or if unexpected external forces are encountered, then feedback mechanisms can come into operation to stabilize body posture further. The postural command adapts very quickly to new conditions. If the body is supported, rather than free-standing, postural muscle activity in the legs is no longer needed when the subject makes, say, an arm movement. It takes only 2–3 trials for the postural activity to decline from that seen in the unsupported condition.

Figure 8.13 illustrates the typical pattern of activity in postural muscles during a brisk, self-paced, forwards movement of the arm made against an opposing load. The prime mover muscle is the anterior deltoid. However, its activity is preceded by changes in many muscles of the leg and trunk which may begin 50–100 ms beforehand. Like the EMG responses to external platform translations, this anticipatory postural activity begins earliest in distal muscles, and then spreads proximally. The activity not only stiffens the joints, it also produces movement of the body which is in such a direction as to reduce the postural disturbance produced by the anterior deltoid action. Thus, in this case, the hamstrings and erector spinae muscles contract, extending the hips and trunk, and pulling the body backwards to balance the outwards movement of the arm. The activity in the soleus (decrease) and tibialis anterior (increase) muscle in this example is difficult to interpret. One would predict that this EMG activity would lead to ankle dorsiflexion, and forwards sway of the body. However, this need not be the case. Ankle position also depends on the position of the knees, hip, etc. and in such circumstances it is often found that the actual movement of body segments does not correspond to what is expected from activity in local muscles. Movement was not measured in this case so that the precise behaviour of the ankle joint is not known.

The size and timing of the anticipatory EMG responses depend, to a first approximation, on the initial force applied by the prime mover. If this is small, as it is for slow movements against no load, the anticipatory activity is small and occurs nearer in time to that in the prime mover. Presumably the postural disturbance produced by the prime movement is relatively small under these conditions. When the initial force in the prime mover is large, as in very fast movements against a large load, then anticipatory activity is large and begins earlier in relation to the prime mover. The exact relationship between the force generated by the prime mover and the latency of postural activity is still debated at present.

Anticipatory activity is best seen in self-paced movements. In reaction time conditions, the onsets of prime mover and postural activity are much closer together, although in most cases, some postural activity still precedes that in the prime mover (Figure 8.13b). In situations where balance is not threatened, there can be said to be a trade off in the amount of postural displacement which is allowed, and the final reaction time

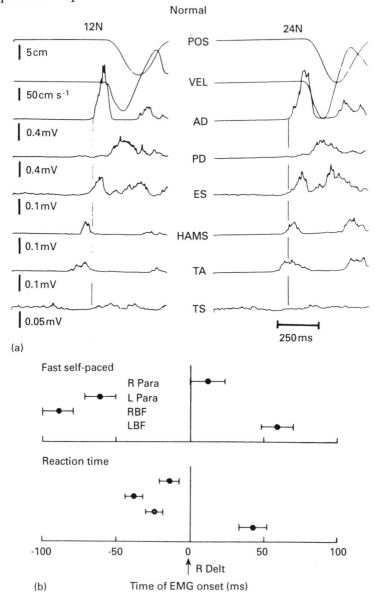

(a)

(b)   Time of EMG onset (ms)

of the prime mover. The implication of this is that the commands for postural and prime mover activity are not inextricably linked. They can vary independently and hence must be computed by two separate processes.

The pattern of postural muscle activity can be modulated by training. An example is the technique that ballet dancers use to lift one leg sideways into the air. To do this, the centre of gravity has to be transferred to the support leg. Normal subjects achieve this by tilting the whole body to the supporting side, whilst at the same time keeping their head approximately horizontal by lateral flexion of the neck. Ballet dancers achieve the same weight transfer by tilting the support leg sideways and at the same time counter-rotating the trunk at the hips so that the whole trunk and head remain upright over an externally rotated support leg. The advantage of the ballet dancers' method is that the head does not move with respect to the trunk and the torso remains vertical with respect to the ground. This is thought to simplify the calculation of the forces acting on different body segments during any subsequent movements.

Postural adjustment has also been investigated in two simpler situations. In the first, subjects are asked to raise themselves onto tip-toe. The prime mover in this task is the gastrocnemius/soleus, but its activity is preceded by activity in tibialis anterior. The latter brings the body forwards and thus minimizes the backwards displacement of the centre of gravity which gastrocnemius/soleus activation would produce if operating in isolation. A second example of anticipatory postural adjustment is in a bimanual task in which subjects support a heavy weight in one hand by flexion at the elbow. If the weight is lifted off the hand by the experimenter, then the arm flexes upwards. If the subject uses his other hand to remove the weight, then the weight-bearing arm remains stationary. The reason for this lack of movement is that at the same time as the opposite arm takes the weight of the object, the activity

---

Figure 8.13 (a) Averaged arm position (POS), velocity (VEL) and EMG activity from a number of muscles during a brisk, self-paced forwards flexion of the arm against a force of 12 N (left) or 24 N (right). The prime mover muscle is the anterior deltoid (AD), and its antagonist the posterior deltoid (PD). Activity in many postural muscles on the same side of the body precedes the onset of activity in the prime mover (vertical line). ES, erector spinae; HAMS, hamstrings; TA, tibialis anterior; TS, gastrocnemius/soleus. (b) Average EMG onset times relative to activation of the right anterior deltoid muscle (t = 0 ms) during rapid forwards flexion of the arm. In the fast, self-paced mode similar to that illustrated in (a), the right hamstrings (RBF) muscle precedes anterior deltoid by almost 100 ms; the left paraspinals (L Para) precede it by some 60 ms. In a reaction time version of the same task, in which the subject had to lift the right or left arms depending on the position of an imperative signal, the difference in latencies was much reduced. (a) From Dick *et al.*, 1986, Figure 1; (b) from Horak *et al.*, 1984, Figure 3; with permission.)

in the weight-bearing arm is decreased. The nervous system controls both prime mover and postural activity together. Interestingly, this cooperation between the two actions is preserved in totally deafferented individuals, indicating that the instructions for this movement are organized centrally, and can be delivered independently of afferent feedback. (Gahéry and Massion, 1981; Lee, Buchanan and Rogers, 1987; Zattara and Bouisset, 1988; Forget and Lamarre, 1990 give further details of the work covered in this section.)

Bibliography

Review articles and books

Gahéry, Y. and Massion, J. (1981) Coordination between posture and movement, *Trends Neurosci.*, **4,** 199–202.

Nashner, L. M. (1983) Analysis of movement control in man using the moveable platform, *Adv. Neurol.*, **39,** 607–619.

Pickles, J. O. and Corey, D. P. (1992) Mechanoelectrical transduction by hair cells. *Trends Neurosci.*, **15,** 254–259.

Roberts, T. D. M. (1979) *Neurophysiology of postural mechanisms.* Butterworths, London.

Original papers

Berthoz, A., Lacour, M., Soechting, J. F. and Vidal, P. P. (1979) The role of vision in the control of posture during linear motion, *Prog. Brain Res.*, **50,** 197–209.

Britton, T. C., Day, B. L., Rothwell, J. C., Thompson, P. D. and Marsden, C. D. (1993) Postural electromyographic responses in the arm and leg following galvanic vestibular stimulation in man, *Exp. Brain Res.* (in press).

Bronstein, A. M. (1986) Suppression of visually-evoked postural responses, *Exp. Brain Res.*, **63,** 655–658.

Day, B. L., Steiger, M. J., Thompson, P. D. and Marsden, C. D. (1993) Influence of vision and stance width on human body movements when standing: implications for afferent control of lateral sway, *J. Physiol.* **469,** 479–99.

Dick, J. P. R., Rothwell, J. C., Berardelli, A. *et al.* (1986) Associated postural adjustments in Parkinson's disease, *J. Neurol. Neurosurg. Psychiatr.*, **49,** 1378–1385.

Diener, H. C., Bootz, F., Dichgans, J. and Bruzek, W. (1983) Variability of postural 'reflexes' in humans, *Exp. Brain Res.*, **52,** 423–428.

Diener, H. C., Dichgans, J., Bootz, F. and Bacher, M. (1984a) Early stabilization of human posture after a sudden disturbance: influence of rate and amplitude of displacement, *Exp. Brain Res.*, **56,** 126–134.

Diener, H. C., Dichgans, J., Guschlbauer, B. and Mau, H. (1984b) The significance of proprioception on postural stabilisation as assessed by ischaemia, *Brain Res.*, **296,** 103–109.

Diener, H. C., Horak, F. B. and Nashner, L. M. (1988) Influence of stimulus parameters on human postural responses, *J. Neurophysiol.*, **59,** 1888–1904.

Dieterich, M., Brandt, T. and Fries, W. (1989) Otolith function in man. Results from a case of otolith-tullio phenomenon, *Brain*, **112,** 1377–1392.

Dietz, V. and Berger, W. (1982) Spinal coordination of bilateral leg muscle activity during balancing, *Exp. Brain Res.*, **47,** 172–176.

Dietz, V., Horstmann, G. and Berger, W. (1988) Involvement of different receptors in the regulation of human posture, *Neurosci. Lett.*, **94,** 82–87.

Do, M. C., Brenière, Y. and Bouisset, S. (1988) Compensatory reactions in forward fall: are they initiated by stretch receptors? *Electroencephal. Clin. Neurophysiol.*, **69,** 448–452.

Fitzpatrick, R. C., Taylor, J. L. and McCloskey, D. I. (1992) Ankle stiffness of standing humans in response to inperceptable perturbation: reflex and task dependent components, *J. Physiol.*, **454,** 533–547.

Flock, A. (1965) Transducing mechanisms in the lateral line canal organ receptors, *Cold Spring Harbor Symp. Quant. Biol.*, **30,** 133–145.

Forget, R. and Lamarre, Y. (1990) Anticipatory postural adjustment in the absence of normal peripheral feedback, *Brain Res.*, **508,** 176–179.

Friedli, W. G., Hallett, M. and Simon, S. R. (1984) Postural adjustments associated with rapid voluntary movements 1. Electromyographic data, *J. Neurol. Neurosurg. Psychiatr.*, **47,** 611–622.

Greenwood, R. and Hopkins, A. (1976) Muscle responses during sudden falls in man, *J. Physiol.*, **254,** 507–518.

Gurfinkel, V. S., Lipshits, N. I. and Popov, K. Y. (1974) Is the stretch reflex a main mechanism in the system of regulation of the vertical posture of man? *Biofizika*, **19,** 744–748.

Horak, F. B. and Nashner, L. M. (1986) Central programming of postural movements: adaptation to altered support-surface configuration, *J. Neurophysiol.*, **55,** 1369–1381.

Horak, F. B., Esselman, P., Anderson, M. E. and Lynch, M. K. (1984) The effects of movement velocity, mass displaced, and task certainty on associated postural adjustments made by normal and hemiplegic individuals, *J. Neurol. Neurosurg. Psychiatr.*, **47,** 1020–1028.

Horak, F. B., Diener, H. C. and Nashner, L. M. (1989) Influence of central set on human postural responses, *J. Neurophysiol.*, **62,** 841–853.

Horak, F. B., Nashner, L. M. and Diener, H. C. (1990) Postural strategies associated with somatosensory and vestibular loss, *Exp. Brain Res.*, **82,** 167–177.

Horstmann, G. A. and Dietz, V. (1990) A basic posture control mechanism: the stabilisation of the centre of gravity, *Electroencephal. Clin. Neurophysiol.*, **76,** 165–176.

Keshner, E. A., Allum, J. H. J., Pfaltz, C. R. (1987) Postural coactivation and adaptation in the sway stabilising responses of normals and patients with bilateral vestibular deficits, *Exp. Brain Res.*, **69**, 77–92.

Lee, W. A., Buchanan, T. S. and Rogers, M. W. (1987) Effects of arm acceleration and behavioural conditions on the organisation of postural adjustments during arm flexion, *Exp. Brain Res.*, **66**, 257–270.

Lestienne, F., Soechting, J. and Berthoz, A. (1977) Postural readjustments induced by linear motion of visual scenes, *Exp. Brain Res.*, **28**, 363–384.

Macpherson, J. M., Horak, F. B., Dunbar, D. C. and Dow, R. S. (1989) Stance dependence of automatic postural adjustments in humans, *Exp. Brain Res.*, **78**, 557–566.

Nashner, L. M. (1976) Adapting reflexes controlling the human posture, *Exp. Brain Res.*, **26**, 59–72.

Nashner, L. M. (1982) Adaptation of human movement to altered environments, *Trends Neurosci.*, **5**, 358–361.

Nashner, L. M. and Berthoz, A. (1978) Visual contribution to rapid motor responses during postural control, *Brain Res.*, **150**, 403–407.

Paulus, W., Starube, A. and Brandt, T. H. (1987) Visual postural performance after loss of somatosensory and vestibular function, *J. Neurol. Neurosurg. Psychiatr.*, **50**, 1542–1545.

Traub, M. M., Rothwell, J. C. and Marsden, C. D. (1980) Anticipatory postural reflexes in Parkinson's disease and other akinetic-rigid syndromes and in cerebellar ataxia, *Brain*, **103**, 393–412.

Watt, D. G. D. (1976) Responses of cats to sudden falls: an otolith-originating reflex assisting landing, *J. Neurophysiol.*, **39**, 257–265.

Wersall, J. and Bagger-Sjoback, D. (1974) Morphology of the vestibular system, in H. H. Kornhuber (eds.) *Handbook of Sensory Physiology* (VI/I), Springer Verlag, Berlin.

Zattara, M. and Bousset, S. (1988) Posturo-kinetic organisation during the early phase of voluntary upper limb movement. 1. Normal subjects, *J. Neurol. Neurosurg. Psychiatr.*, **51**, 956–965.

Cerebral cortex 9

9.1 *Structure of cerebral cortex*

The cerebral cortex is the most distinctive feature of the human brain. It is composed almost entirely of neocortex, the most recent part of the cortex to develop in the evolutionary timescale. Archicortex (hippocampus) and paleocortex (olfactory cortex) form only a small fraction of cerebral cortex in man.

The internal structure of cortex is covered in detail in many textbooks. Only a brief summary is given here. There are two main types of cell in the cerebral cortex: pyramidal and stellate (or granule) cells. The relative position and preponderance of the cell types gives the cortex a laminar appearance when seen in vertical sections (Figure 9.1). Six layers are defined:

1. The molecular or superficial plexiform layer. This lies immediately below the pia and contains very few cell bodies. Its main components are the axons and apical dendrites of neurones in deeper layers.
2. The external granular layer. This is a layer of densely packed small cells including small pyramidal and stellate cells.
3. The external pyramidal layer. A layer of medium sized pyramidal cells.
4. The internal granule layer, composed of densely packed stellate and pyramidal cells.
5. The ganglionic layer. A layer of large to medium-sized pyramidal cells.
6. The multiform layer. Irregularly shaped cells, many with axons leaving the cortex.

The pyramidal cells have axons which leave the cortex, whereas the stellate cells (of which there are several different types) are the true cerebral interneurones. The pyramidal cells in all layers have vertical dendrites which may extend into layer I. Their axons leave the cortex

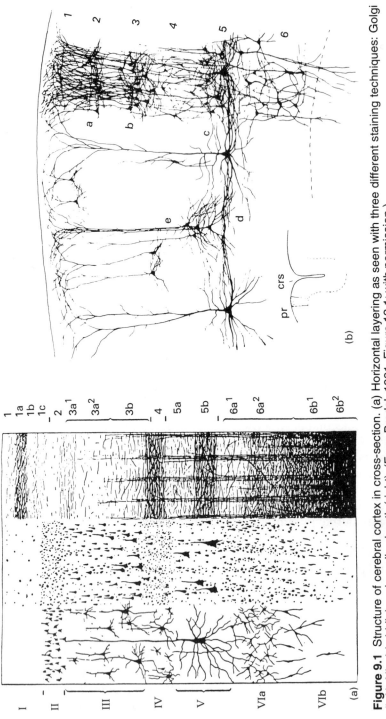

Figure 9.1 Structure of cerebral cortex in cross-section. (a) Horizontal layering as seen with three different staining techniques: Golgi (left), Nissl (middle) and myelin sheath (right). (From Brodal, 1981, Figure 12.1; with permission.) (b) Golgi preparation from an adult cat showing vertical orientation of apical dendrites of pyramidal cells and the horizontal layering of their basal dendrites in layer 5. (From Schiebel *et al.*, 1974, Figure 1; with permission.)

through a layer of basal dendrites which may extend horizontally for several millimetres. These dendrites form a conspicuous band in layers II, III and IV. Afferents to cortex come from either the thalamus or from other areas of cortex, via corticocortical connections. Afferents from the specific sensory nuclei of the thalamus terminate in discrete zones in layer IV, whereas those from the non-specific nuclei branch in the subcortical white matter and end widely in layer I with horizontally directed axons. Corticocortical afferents are found in several layers.

The dendrites of pyramidal cells give the cortex strong perpendicular connections. Indeed, electrophysiological studies have shown that groups of cells work together in vertical units called cortical columns. Within the columns, the cells share common input and output connections, forming the basic unit of cortical processing.

Because thalamic afferents from specific sensory nuclei terminate on the dendrites of stellate cells in layer IV, there are few direct connections between specific inputs and the pyramidal output cells of the cortex. Instead, many of the stellate cells of layer IV have vertically oriented axons which make contact with the apical dendrites of pyramidal (and other) cells. These vertical connections may form the basis of a strong input–output coupling within a cortical column. Corticocortical connections, particularly those from the opposite hemisphere (transcallosal connections) also respect the columnar organization and form discrete terminal patches in the cortex. Afferents from non-specific thalamic nuclei branch widely in the subcortical white matter and end in layer I. Their terminations, unlike those of the specific afferents, do not respect the columnar pattern of organization.

In addition to the vertical connectivity, there are also strong horizontal connections between the columns. Pyramidal cells form horizontal plexuses of dendrites in layers II, III and IV of cortex, which may extend for several millimetres. Presumably, these dendrites are capable of receiving input from surrounding cortical columns. In addition, the pyramidal axons send off recurrent collaterals before they descend into the white matter. These travel horizontally for several millimetres and may be another route for intercolumn communication. Finally, some stellate cells have processes which extend laterally for several millimetres.

The description above of cortical layering only gives a 'typical' picture of cortical organization. There are considerable variations in the number and thickness of the layers in various regions of cortex. It was on this basis that Brodmann distinguished more than 50 separate regions in the human brain (a smaller number are present in the monkey). The numbering of Brodmann's areas seems confusing at first sight. This is because the sections of brain that he examined were cut horizontally,

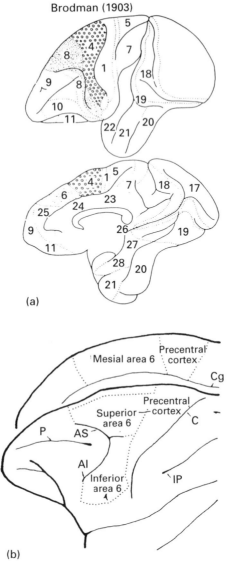

Figure 9.2 (a) Brodmann's subdivisions of the monkey cerebral cortex. Top, lateral aspect; bottom, medial aspect of hemisphere. Shaded areas 4 and 6 are the classical motor areas of cortex. (b) General subdivision of the agranular frontal cortex according to Matelli and Luppino. Dorsolateral and mesial views of the brain are shown. Dotted lines show the borders between various cortical areas. P, principal sulcus; AS, superior limb of the arcuate sulcus; AI, inferior limb of the arcuate sulcus; C, central sulcus; IP intraparietal sulcus; Cg, cingulate sulcus. The superior precentral sulcus is not labelled but can be seen as a horizontal line between the labelling for superior area 6 and the precentral cortex. (From Matelli and Luppino (1991), Figure 1a; with permission.)

beginning at the vertex. Precentral and postcentral regions appeared in the first few sections and were given low numbers, jumping from front to back as new cytoarchitectonic features were encountered. This began on the crown of the postcentral gyrus with areas 1 and 2, then area 3 in the depths of the sulcus, followed by area 4 in the precentral region, and so on (Figure 9.2).

9.2 *Frontal motor areas of cortex*

Cytoarchitectural subdivisions

Much of the work on the subdivisions of the frontal cortical areas has been performed on monkeys. Areas 4 and 6 of Brodmann contain the main motor areas of the frontal cortex. They are distinguished from surrounding cortical areas by their small or non-existent internal granular layer (IV). They are bounded anteriorly by area 8, the frontal eye field, and posteriorly by area 3a which is part of the primary sensory cortex within the central sulcus. Area 6 is said to differ from area 4 in the absence of giant Betz cells. This is relatively clear in the medial part of area 4, where the Betz cells which project to the leg are particularly large. However, Betz cells in lateral area 4 (arm and face representations) are smaller than those in the leg area, so that the distinction is less clear.

Area 6 itself was not subdivided by Brodmann, but it is now usually recognized to contain at least three separate regions, an inferior and superior section on the lateral aspect of the hemisphere, and a mesial area. The superior and inferior areas are separated by the spur of the arcuate sulcus; the mesial and superior regions by their locations on different surfaces of the cortex. Further subdivisions of these areas have also been proposed (see Figure 9.2) on the basis of both classical cytoarchitectonics, and by newer methods of staining for cytochrome oxidase.

Mesial area 6 borders ventrally on the agranular cingulate cortex (areas 23 and 24). This area was virtually ignored in previous studies of the motor system, but is now regarded as an important motor region.

An antatomical definition of frontal motor areas

Area 4, the precentral cortex is known as the primary motor area. Area 6 and regions of the cingulate cortex are known as the non-primary motor areas. These non-primary motor areas are now thought to contain several separate motor regions. Strick and colleagues have suggested that a non-primary motor area should be defined by two criteria: (i) presence of pyramidal neurones with axons terminating in the spinal cord (i.e. corticospinal axons), and (ii) corticocortical connections with the region of primary motor cortex which sends axons to the same spinal segments. On this basis, there are 6 subregions of the non-primary motor cortex devoted to control of arm movement. Thus, injections of retrograde tracers into cervical segments of spinal cord grey matter label groups of neurones in:

(1) the inferior limb of the caudal bank of the arcuate sulcus, a region sometimes termed the arcuate premotor area (APA);
(2) the lateral edge of the superior precentral sulcus (SPcS), just anterior to the primary motor cortex proper;
(3) mesial area 6, in the region corresponding to the supplementary motor area (SMA), at about the same anterior extent as the genu of the arcuate sulcus on the lateral surface of the hemisphere;
(4) three small areas of cingulate cortex:
(a) a rostral cingulate motor area (CMAr) on the dorsal bank of the cingulate sulcus, anteror to the genu of the arcuate sulcus

Nomenclature of frontal motor areas

(a)

Figure 9.3 (a) A guide to the bewildering variety of names used by different authors to describe the motor areas of cortex. The nomenclature used in this book is that of Dum and Strick.

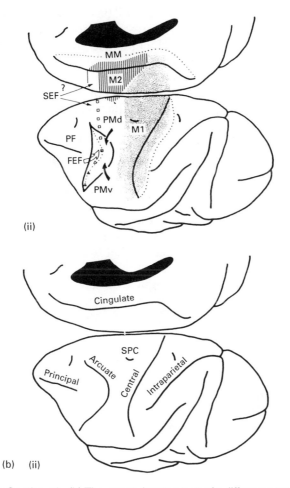

Figure 9.3 Continued. (b) The approximate areas of a different nomenclature used by Wise *et al.* Missing from both (a) and (b) is the ventral cingulate motor area of Dum and Strick. Abbreviations: APA, arcuate premotor areas; CMA, cingulate motor areas; FEF, frontal eye field; MC and MI, primary motor cortex; MII, M2 and SMA, supplementary motor area; MM, medial motor areas; MsI, primary motorsensory area; PMa, anterior premotor cortex; PMd, dorsal premotor cortex; PMv, ventral premotor cortex; SEF, supplementary eye field; SPC superior precentral sulcus. (From Wise *et al.*, 1991 Figure 1; with permission.)

(b)   a caudal area on the ventral bank of the sulcus (CMAv), and

(c)   a second caudal area on the dorsal bank of the sulcus (CMAd).

The corticospinal output from these six areas in the monkey is very substantial: it represents about 60% of the total frontal lobe projection to the spinal cord. Each of these regions also has connections with the arm area of the primary motor cortex.

This subdivision of the motor areas will be followed throughout this book. However, there are a bewildering variety of other nomenclatures that have been, and are being used. Figure 9.3 gives an idea of some of these and shows a 'rival' parcellation of the cortex made by Wise *et al.* (1991). Correspondances between the diagram and the nomenclature used here are given in the two bottom rows of the table.

A particularly thorny problem, which has yet to be addressed in any detail is the relationship between the motor areas described in the monkey and those seen in humans. In the monkey, the arcuate sulcus represents the forward boundary of area 6 and the major motor fields. Anteriorly is area 8, a region concerned with eye movements. In the human brain, the precentral sulcus is thought to be the homologue of the arcuate sulcus, but area 6 is much larger than it is in the monkey. The forward boundary of area 6 in the human brain stretches anterior to the precentral sulcus particularly in the dorsal regions of the cortex. Indeed, this anterior region is usually what has been referred to as the premotor cortex in humans. Much of this area is relatively unexplored in the monkey. The nearest homologue would be the dorsal bank of the arcuate sulcus (PM_d) distinguished by Wise and colleagues. (Reviewed in Humphrey and Freund, 1991; Dum and Strick, 1991.)

Thalamic inputs to motor and sensory cortex

Subcortical inputs from spinal cord, basal ganglia and cerebellum terminate within the ventrobasal nuclei of the thalamus in separate regions, with little or no overlap. Neither are there any interconnections between cell groups in nearby nuclei. Each region, however, spreads over several architectonic subdivisions, making the anatomical organization quite complex.

Sensory afferents from the body and face terminate in VPLc (ventroposterior lateral pars caudalis) and VPM (ventroposterior medial) respectively. Within their fields of termination, a 'core' of afferents from cutaneous receptors is surrounded by input from deep receptors in muscle and joint. These thalamic nuclei send input to postcentral cortex only: the cutaneous 'core' projects to sensory area 3b and 1, whereas the deep surround projects to areas 3a and 2 (Figure 9.4). Some spinothalamic afferents end in VPLo (ventroposterior lateral pars oralis), and are in a position to send direct sensory input to primary motor cortex.

Fibres from the basal ganglia and cerebellum end in thalamic nuclei which send their output to frontal motor areas. Basal ganglia afferents from the internal pallidum and substantia nigra pars reticulata end in VLm (ventrolateral pars medialis), VLo (ventrolateral pars oralis) and VA (ventroanterior), with VLo and VLm receiving a major input from

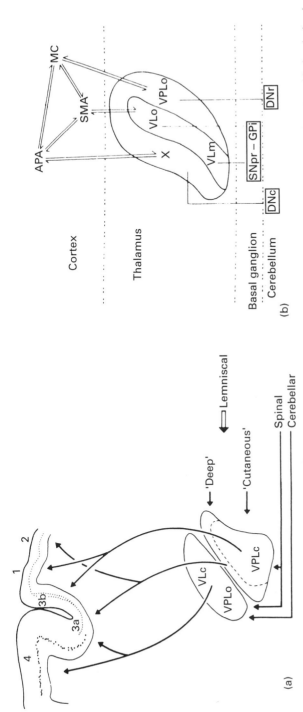

Figure 9.4 Thalamic inputs to pre- and postcentral areas of cerebral cortex in the monkey. (a) Inputs to areas 1 to 4, with the thalamic nuclei shown in saggital section. In the VPL$_c$, lemniscal inputs separate into a core of afferents from cutaneous receptors, surrounded by a shell of afferents from deep receptors (including muscle afferents). All of these fibres then project to postcentral areas of cortex. Spinothalamic afferents also terminate in VPL$_c$ but also project to VPL$_o$ and VL$_c$, whence they may travel to precentral cortical areas. These nuclei also are the relay station for cerebellar afferents to motor cortex. (From Friedman and Jones, 1981, Figure 20; with permission.) (b) Summary of thalamic projections to precentral motor areas, including MI (MC), SMA and the premotor (here called the arcuate premotor area, APA) cortex. As in the diagram on this page, cerebellar input is shown travelling via the VPL$_o$ nucleus to MI. In this case, extra detail is included to make the point that this input comes mainly from the rostral portions of the deep cerebellar nuclei (DNr). Caudal portions of the deep cerebellar nuclei (DN$_c$) project via nucleus X to the premotor (APA) area. Input from basal ganglia travels via VL$_m$ and VL$_o$ to the SMA. However, it should be noted that basal ganglia output is now thought to reach both MI and APA from its thalamic target nuclei. (From Schell and Strick, 1984, Figure 16; with permission.)

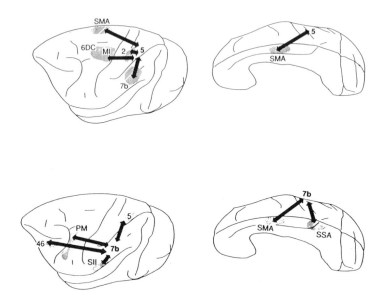

Figure 9.5 (a) Main corticocortical connections of the proximal arm representation of area 5. (b) Main corticicortical connections of area 7b. The view on the left is the lateral surface and that on the right the medial surface of the hemisphere of a monkey's brain. Abbreviations: MI, primary motor cortex; SMA, supplementary motor area; SSA, supplementary somatosensory area; PM, ventral premotor area; SII, second somatosensory area; 6DC, dorsal area 6 (dorsal premotor area). (From Johnson, 1990, Figures 7 and 8; with permission.)

regions processing sensorimotor information. VA receives input mainly from the basal ganglia-frontal cortex 'loop'. Cerebellar input from the rostral portion of the cerebellar nuclei ends in VPLo and VLc (ventro-lateral pars caudalis); that from the caudal regions of the nuclei in thalamic nucleus X.

The cortical projection areas of these various nuclei were described before the recognition of all 6 of the subareas of the non-primary motor cortex described above. The description that follows refers mainly to the arm areas of cortex, and only distinguishes between APA, SMA and primary motor cortex. Thalamic nucleus VA projects to prefrontal cortex; VLo and VLm (basal ganglia output) project mainly to SMA; VPLo and VLc (cerebellar output) mainly to primary motor cortex, nucleus X to APA. These projections are summarized in Figure 9.4. However, it should be pointed out that this scheme is simplified, and that there is considerable overlap between the final terminations of the various subcortical structures. Indeed the precise pattern of thalamic terminations in the cortex is a matter for some debate.

From this description it can be seen that there are very few direct inputs to primary motor cortex from peripheral afferents. Only a proportion of the spinothalamic tract ends in a region of thalamus (VPLo) with projections to motor cortex. The majority of sensory input is indirect via either the dorsal column-medial lemniscal innervation of postcentral cortex, or through spinocerebellar pathways and the cerebellar nuclei.

Corticocortical inputs to motor areas

The motor areas of cortex are intimately interconnected and receive a large proportion of their corticocortical input from parietal areas. The inputs from the postcentral gyrus and areas 5 and 7 of the parietal lobe have been investigated in some detail and are summarized in Figure 9.5. As can be seen, primary motor cortex receives primary somatosensory input from area 2 of the postcentral gyrus (which itself is interconnected with areas 3a, 3b and 1 of the sensory cortex) and from lateral area 5. Areas 5 and 7b project to SMA, whilst the superior part of lateral area 6 receives from area 5, and the inferior part from area 7b. Generally speaking, cells in areas 5 and 7b have complex somatosensory responses which are modulated by the task a monkey performs. Their input to motor areas may be important in the somatosensory guidance of movement (see below). Motor cortical areas also receive input from frontal areas including cingulate cortex and prefrontal areas 8 and 46 (Wiesendanger, 1981; Johnson, 1991).

Outputs of cortical motor areas

The major descending output of the cortical motor areas is the corticospinal tract (see Chapter 7). However, area 4 also has corticocortical connections to parietal areas 2 and 5, and to the premotor and supplementary motor areas reciprocating the input from these regions. There are important outputs to basal ganglia, pontine nuclei (and thence to the cerebellum), red nucleus and reticular formation. The latter two pathways allow for the possibility of indirect cortical actions from area 4 on the spinal cord. Non-primary areas of motor cortex (except MII, the lateral motor area defined by Woolsey and colleagues), contribute

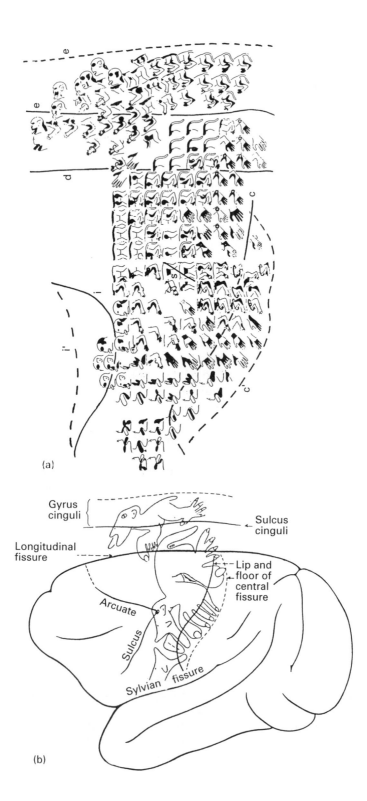

(a)

Gyrus cinguli

Sulcus cinguli

Longitudinal fissure

Lip and floor of central fissure

Arcuate

Sulcus

Sylvian fissure

(b)

the remaining 60% of fibres in the corticospinal tract which arise in frontal motor areas. These areas also have other indirect pathways to the spinal cord. The SMA sends efferents to the red nucleus and areas of the reticular formation, and the premotor area projects to the medial tegmental region of the brainstem. In addition, there are fibres going to the pons (and thence to cerebellum) and the basal ganglia.

9.3 *Electrical stimulation of the motor areas*

The motor areas of cortex were defined originally as regions in which electrical stimulation produced movement of a part of the body. It is not a particularly useful definition, since stimulation of almost any area of cortex with sufficient intensity and under the appropriate conditions can evoke movement. Such effects depend critically on the depth and type of anaesthesia used and, if strong stimuli are used, there is always the possibility that the stimulus will spread electrically or via corticocortical connections to other areas distant from the stimulating electrode.

The areas with the lowest threshold for electrical stimulation are Brodmann's areas 4, 6 and 8. At higher intensities, movements also can be elicited from areas 1, 2 and 3. These latter are conventional sensory areas of cortex and it is believed that much of the movement provoked by electrical stimulation is mediated by corticocortical connections to the precentral cortex. There is a large pyramidal projection from these sensory areas to spinal cord, but this is probably used in modulating

Figure 9.6 (a) Figurine map of the movements elicited from left primary motor cortex and SMA in a single monkey; (b) composite simunculus derived from experiments on several monkeys. Stimulation of the cortical surface was performed with a 0.5 mm diameter stainless steel wire using 2 s trains of stimuli (up to 2 mA) at 60 Hz. Except for responses from points in an ipsilateral face area (extreme bottom), the responses are all from the right side of the body, but to maintain orientation in the total pattern, left-sided figurines are used. In the figurine map, shading indicates the area of body in which movement occurred. For example, shading over the knee indicates extension of the knee joint. The depth of shading indicates the relative strength of movement: solid black signifies the strongest and earliest movement, crosshatching intermediate, and dots the weakest effect. Symbols with a cross on the ankles indicate eversion of the foot; those with open centres on the hip and ankle signify adduction and inversion. The solid lines indicate the lips of the central sulcus (c), the arcuate sulcus (i), and the cingulate sulcus (e); the dotted lines indicate the floor of the fissure, which has been opened out for the purposes of the diagram. Line d indicates the division between the medial and lateral aspects of the hemisphere. The figurine diagram has the same orientation as the simunculus in (b). (From Woolsey *et al.*, 1952, Figures 131 and 132; with permission.)

transmission of sensory input to brain rather than in production of movement.

The electrical excitability and localization of motor cortex was discovered independently in the 1870s by Ferrier and by Fritsch and Hitzig in experiments on the brains of monkeys and dogs. Detailed maps of the monkey motor cortex were made later by Woolsey *et al.* (1952) and in man by the neurosurgeon Penfield and his colleagues (Penfield and Jasper, 1954). In humans the investigations were performed during surgery for epilepsy and were necessarily limited in time by ethical considerations. Final maps of human cortex were made by combining the results obtained from many different patients. In these pioneering investigations, the authors noted the visible movements that were evoked when the cortex was stimulated electrically with a 60 Hz alternating current applied to the surface with a small ball electrode. Woolsey and colleagues (1952) illustrated the results by drawing small figurines on a map of the cortex to show the movements observed at each site (Figure 9.6). They then collapsed these maps into the more familiar simunculus and homunculus representations. An important point, though, is that these final maps are only **approximate** representations of the data. A look at the figurine data shows that the true representation for movement is far more complex than the impression given by the final drawings.

Two major representations of the body were found in areas 4 and 6 of both humans and monkeys. The lowest threshold area was area 4 lying along the surface of the precentral gyrus and extending into the anterior wall of the sulcus. This is now termed the primary motor cortex. The movements evoked from this area were simple, flicklike twitches of discrete parts of the body. Leg movements occurred after stimulation of the most medial parts, arms and face movement after lateral stimulation. As well as this mediolateral organization, Woolsey and colleagues described a rostrocaudal gradient: distal extremity muscles were more easily activated by stimulation in the central sulcus, whereas proximal body parts were commonly represented rostrally, on the surface of the gyrus. Although most movements were contralateral, there were exceptions: bilateral activation of the soft palate, larynx, masticatory muscles and the upper part of the face, and, quite frequently, bilateral trunk movements.

A second complete map of the body was found on the medial surface of the hemisphere in area 6, anterior to the leg representation of the primary motor cortex. This area was termed the supplementary motor area (SMA). It is quite different to the primary motor region. The threshold for movement was higher than for primary motor cortex, and the somatotopic representation of the body parts was poor. The movements that were evoked were complex, involving activation of muscles at more than one joint. Stimulation often produced tonic sustained contractions involving the assumption and maintenance of limb postures. Complex acts like vocalization or yawning occurred, and bilateral movements were common. The SMA is discussed in more detail later in the chapter.

A third area was also described in some early work, located in the most lateral part of the precentral gyrus, extending into the Sylvian fissure. It is sometimes referred to as MII. The two body representations in MII and the primary motor cortex (MI) lie lip to lip, with that of MII being reversed compared to MI. MII had been little investigated in recent times. There is a strong bilateral representation of the face, which may be a factor in the relative immunity of the facial muscles to hemiplegia.

With the recognition of the multiple non-primary motor areas, the detail of these early maps has had to be revised. The main points of difference are that on Woolsey's map, the forward boundary of the primary motor area is placed too far rostral. Microstimulation in the bank of the arcuate sulcus is now known to produce movements of the hand and arm not shown on the classical map. Stimulation around the superior precentral sulcus also is now thought to be activating a separate motor representation, with a slightly higher threshold and with different cell discharge properties to those observed in the primary motor area. Another main difference is that Woolsey's map of the SMA representation also includes portions of the cingulate gyrus. These are now believed to contain their own separate motor representations (see sections on non-primary motor areas below and discussion of Wise *et al.*, 1991).

Detailed motor cortex mapping

There are now many maps of the organization of the primary motor cortex. However, the interpretation of such maps is not as straightforward as it may seem. There are two complicating factors. First, a true output map should describe the projection targets of the output (pyramidal) cells in each small area of cortex. Unfortunately, there are no methods of electrical stimulation which activate **only** pyramidal neurones. All excite several cortical elements in addition, such as interneurones or afferent fibres. These elements can then excite further pyramidal cells at a distance from the stimulating site through synaptic connections. The result is that cortical maps represent, to some extent, the organization of inputs to pyramidal cells as well as the pattern of projection to their targets.

A second complicating factor is the index used to monitor cortical output. Some authors use visible movement of a part of the body, others record EMG activity, and others insert microelectrodes into spinal motoneurones to detect synaptic activity. The latter method is obviously the most sensitive since it can detect even subthreshold inputs, yet at the same time, it is the most limited in the number of output targets that it can monitor at one time. Visible movement is the least sensitive indicator of cortical output, and has the added disadvantage that it can fail to detect obvious muscle activity if this produces co-contraction about a joint. Nevertheless, it is simple to use and can monitor a very wide variety of output targets. Indeed, it can be argued that movement is the most important result of cortical output.

Four different varieties of electrical stimulation have been used to map the motor cortex: single pulses versus repetitive trains of stimuli, and surface stimulation versus intracortical microstimulation (ICMS). There are advantages and disadvantages of each.

(1) Single stimuli activate the smallest area of cortex, but never produce visible movement in anaesthetized animals. The descending activity set up by a single stimulus does not produce enough excitatory input to spinal motoneurones to raise them above their firing threshold.

(2) Repetitive stimuli can evoke movement since they set up more descending activity. But because of temporal summation and facilitation at cortical synapses, repetitive stimulation activates a large proportion of the pyramidal output neurones indirectly, at some distance from the actual site of stimulation.

(3) Surface stimulation produces relatively widespread current flow, but if anodal stimuli are used, the majority of pyramidal neurones can be activated directly, rather than trans-synaptically. This is because an anode on the surface of the cortex produces electrical currents which can flow into the vertically-oriented dendrites of pyramidal cells, or into their cell bodies, and then exit at the initial segment region, or the first node of the axon. Outward current at the latter sites depolarizes the neurones and causes direct activation of the pyramidal cell. Interneurones in the cortex, which are oriented perpendicular to the pyramidal cells, are not so well stimulated by a surface anode, and trans-synaptic pyramidal activation is rare.

(4) ICMS is the most focal form of stimulation. It uses a thin microelectrode inserted into the cortical grey matter. Movements can be obtained with stimuli as small as 5–10 μA. However, although the physical spread of current is limited, the physiological actions are more widespread. Repetitive ICMS is a particularly good method of activating cortical circuitry, and can result in synaptic activation of pyramidal cells up to 0.5 mm from the microelectrode itself.

The maps of motor cortical representation produced by combinations of stimulation methods and indices of motor output differ in detail, but in general they all confirm the topographic representation of the major body parts (head, arms trunk and legs). There are slight differences in the detail of the representation of the head and arm regions. For example, Kwan and colleagues (1978), in an influential paper, proposed a 'nested ring' organization of the hand and arm, with hand movements being obtained after stimulation of a central area, surrounded by rings from which movements at progressively more proximal joints could be obtained. In contrast, Strick and Preston (1982) thought that there were two parallel maps in the motor area, so that each part of the body was represented twice, in mediolateral strips.

Such schemes are inevitably simplifications of the total data that was collected in the experiments. In fact, despite the efforts to impose some form of organization on the output, one thing which stands out is the extent to which there is intermingling of sites within motor cortex which project to widely varying portions of the arm or hand (see Figure 9.7). This is true even more for the gross surface maps of Woolsey and Penfield.

Three points emerge from such mapping studies of motor cortex. (1) A single muscle can be activated from a large surface area of cortex. (2) Because of the large number of muscles represented, there is considerable overlap of the representation for each muscle. (3) Within the total projection area of a muscle, there are 'hot spots' of high excitability, from which activation can be produced at very low stimulus intensities. (For recent reviews of motor cortex mapping: Asanuma, 1981; Lemon, 1988, 1990.)

Corticomotoneuronal mapping
The organization of motor cortical output is well-illustrated by the corticomotoneuronal (CM) projection. CM neurones project monosynaptically to spinal motoneurones, hence their activation can be identified by the presence of monosynaptic EPSPs recorded using intracellular microelectrodes.

Phillips and colleagues (reviewed in Phillips and Porter, 1977) mapped the distribution of CM cells projecting to motoneurones innervating different muscles of the baboon's arm. In order to minimize indirect activation, they used surface anodal stimulation. As pointed out above, this method of stimulation preferentially stimulates pyramidal output cells in layer 5 of the cortex with little effect on cortical interneurones. To determine how far the current from a single pulse spreads in the cortex, single corticospinal tract axons were recorded in the lateral funiculus of the spinal cord. The intensity of stimulus needed to activate a given axon then was charted from different points on the

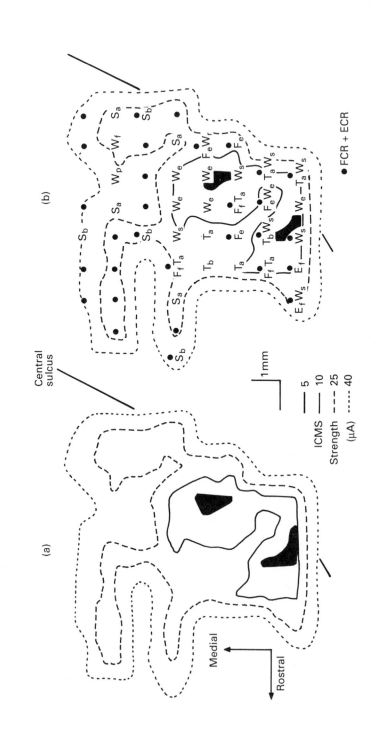

(a)

(b)

Central
sulcus

Medial

Rostral

1mm

ICMS
Strength
(μA)

——— 5
——— 10
– – – 25
······· 40

● FCR + ECR

surface. There was a single point of minimum threshold, with current intensity increasing beyond that point with the square of distance moved by the electrode (see Figure 9.8). From such curves, it can be seen that, for example, a 1 mA pulse might activate any corticospinal neurones within a circle 6 mm in diameter centred on the stimulating electrode.

After this, a single α-motoneurone was impaled in the spinal cord, and the size of the monosynaptic EPSP was measured following corticospinal stimulation. First, the best point for that cell was found – that is, the point of minimum threshold. Then, leaving the electrode on that point, the intensity of stimulation was increased until a maximum EPSP was recorded in the motoneurone. At this stage, all pyramidal tract neurones projecting to that cell must have been stimulated, and the spatial spread of the stimulus could be read off the previous graph. For example, in the motoneurone illustrated in Figure 9.8 the maximum EPSP in a distal motoneurone occurred with an intensity of 2 mA, indicating that pyramidal tract cells within a circle of diameter 10 mm had been activated. This area represents the total cortical field for the motoneurone under study. The neurones within this area which project to the motoneurone are known as a cortical colony. Within the colony, there are 'hot spots' of localization. Stimulation at these points at very low intensity produces activity in the motoneurones of only one muscle. However, the majority of the area occupied by the cortical colony consists of a mixture of cells. Some project to the motoneurones of the muscle under study, while others project to motoneurones of other muscles.

During these mapping experiments it was found that larger maximum EPSPs could be elicited in motoneurones of distal rather than proximal muscles. Thus, all other factors being equal (for instance, motoneurone size, synaptic density and location, synaptic efficiency),

Figure 9.7 Multiple representation of muscles (a) and movements (b) evoked from primary motor cortex by ICMS in a conscious macaque monkey. Parts of the map overlapping the central sulcus indicate areas buried in its rostral bank. (a) shows a surface projection of the representation of a single wrist muscle. Repetitive ICMS was delivered within lamina V; the contours enclose areas from which activity in extensor carpi radialis (ECR) was evoked by stimulus trains of 5, 10, 25, and 40 µA. Note the discontinuity of the lowest threshold area, and the large areas occupied by the 10 and 25 µA maps. Movements observed from the same sites are shown in (b). Wrist extension (We) occurs only on stimulation within or immediately adjacent to the two low threshold sites for ECR. At all other locations, ECR is activated as one of a set of stabilizing muscles during another primary movement. Ef, elbow flexion; Fe, finger extension; Ff, finger flexion; Ta, thumb adduction; Sa, Sb, shoulder adduction and shoulder abduction; Wf, wrist flexion; Wp, wrist pronation; Ws, wrist supination. The filled circles show all locations from which coactivation of ECR and FCR (flexor carpi radialis) was obtained. (From Humphrey, 1986, Figure 2; with permission.)

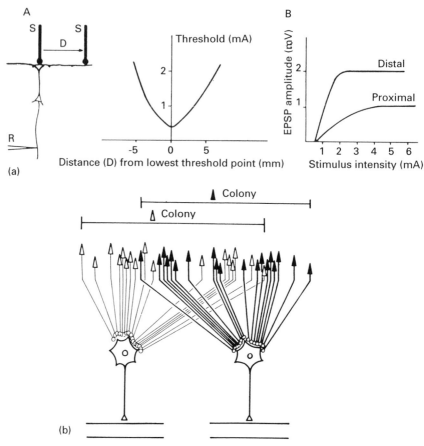

Figure 9.8 (a) A, Experimental arrangement for determining how far stimulating current spreads from a focal cortical anode. A pyramidal tract axon is recorded and single stimuli (S) are applied at various points on the cortical surface. The size of the threshold stimulus necessary to produce activity in the cell is then plotted as a function of distance from the lowest threshold point. In B, two spinal motoneurones were impaled, one innervating a distal muscle and the other innervating a proximal muscle. A ball anode was then used to explore the cortex and find the best point for activating the motoneurones. With the electrode at this point, a graph was constructed of the size of the recorded EPSP in the motoneurone versus the stimulus intensity applied to the cortex. Maximal EPSPs are produced when all the corticomotoneuronal cells of a cortical colony are active. Thus, the spread of the stimulus intensity necessary to produce a maximal EPSP can be read off the graph in A, to give an estimate of the spatial extent of the cortical colony projecting to the impaled spinal motoneurone. (From Ruch and Fetz, 1979, Figure 3.21; with permission.) (b) Diagram summarizing the structure of the monosynaptic corticomotoneurone connection to the baboon's hand. The colonies of cells projecting to motoneurones of two muscles are shown. Note the overlap between the colonies and the variation in the distribution of cells belonging to each colony. (From Andersen *et al.*, 1975, Figure 12; with permission.)

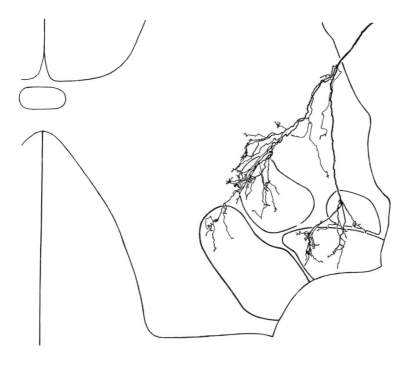

Figure 9.9 Transverse reconstruction of a corticospinal axon at C7 originating from the hand area of the monkey motor cortex. The figure was reconstructed from 12 serial transverse sections. The solid lines show the areas of four different motor nuclei innervating ulnar motoneurones (upper two nuclei), and radial motoneurones (bottom two nuclei). The branches of this axon projected to at least four different motoneurone groups and close contact of terminal boutons with proximal dendrites of some motoneurones (not shown) could be identified in three of them. (From Shinoda *et al.*, 1981, Figure 2; with permission.)

more cortical cells project to spinal motoneurones which innervate distal muscles. Despite the larger number of corticomotoneurone cells, the **area** of the cortical colony projecting to each distal motoneurone was found, in contrast, to be smaller than that projecting to motoneurones innervating proximal muscles. Thus, for distal muscles, there are a large number of corticomotoneuronal cells concentrated in a small area of cortex. For proximal muscles, there are fewer corticomotoneuronal cells within a wider area.

Spike-triggered averaging
The studies above document the area of cortex from which

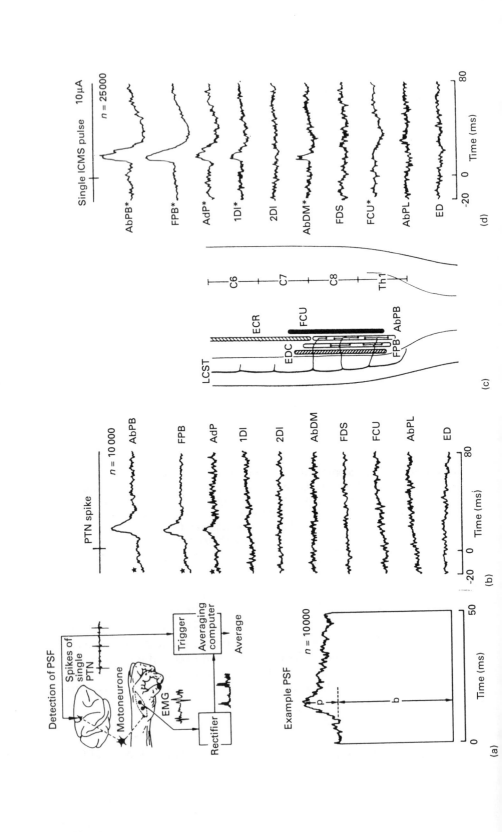

(a) Detection of PSF

Spikes of single PTN

Motoneurone

EMG

Rectifier → Averaging computer → Trigger

Average

Example PSF

n = 10000

p b

Time (ms)

(b) PTN spike

n = 10000

AbPB
FPB
AdP
1DI
2DI
AbDM
FDS
FCU
AbPL
ED

Time (ms)

(c)

LCST

ECR
EDC
FCU
FPB AbPB

C6 C7 C8 Th1

(d) Single ICMS pulse 10μA

n = 25000

AbPB*
FPB*
AdP*
1DI*
2DI
AbDM*
FDS
FCU*
AbPL
ED

Time (ms)

corticomotoneuronal cells project to a single motoneurone. Techniques are also available to answer the inverse question: how many muscles are innervated by a single CM cell?

Anatomical studies show that pyramidal tract axons terminate very widely within the spinal cord (see Chapter 7). An example of the ramifications of a branch of a single pyramidal tract axon from a monkey is shown in Figure 9.9, which illustrates a section of the spinal cord at the low cervical level. Even at this one level, the axon sends terminals to widespread locations in the intermediate and ventral grey matter, covering the area occupied by motoneurones projecting to four different muscles. More branches could be seen in other spinal segments suggesting that this one axon could contact the motoneurones of many different muscles. The implication is that the CM projection may be very diverse.

Even though the anatomical divergence of pyramidal tract axons is wide, the important question is how diverse are the **physiological** actions? That is, how many synaptic contacts does the axon actually make, and how effective are they? The extent of effective CM projections can be investigated by the spike-triggered averaging (STA) method. The discharge of a single identified pyramidal tract neurone (for technique see below) is recorded in the motor cortex of an awake animal at the same time as EMG activity from a large number of different muscles in the hand and arm as the animal performs various tasks. The EMG

Figure 9.10 (a) Method of detecting pyramidal neurone influences on muscle activity in the monkey. A single pyramidal tract neurone is recorded in the primary motor cortex at the same time as EMG activity is recorded from muscles in the periphery. The spikes of the pyramidal tract neurone are used to trigger the sweep of an averager which samples the EMG activity. An example of the averaged EMG in one muscle after 10 000 spikes in the pyramidal neurone is shown below. There is a peak of facilitation with an onset latency of about 8–10 ms. (b) The distribution of post-spike facilitation from a single CM cell in different muscles of the hand and forearm. The average EMG after 10 000 spikes has been recorded from 10 different muscles. There is a peak of facilitation in three of them, the abductor pollicis brevis (AbPB), the flexor pollicis brevis (FPB), and the adductor pollicis (AdP). (c) A schematic horizontal section through the spinal cord at the low cervical level, with the relatively restricted output from the axon of a single CM cell running in the lateral corticospinal tract (LCST), and with collaterals contacting motoneurone cell columns innervating two thumb muscles (FPB and AbPB), but no wrist muscles (EDC, extensor digitorum communis; FCU, flexor carpi ulnaris; ECR, extensor carpi ulnaris). (d) The distribution of facilitation in the same group of muscles after single pulse microstimulation through the same electrode as used to record the activity in the pyramidal tract neurone shown in (b). Additional muscles are facilitated (1DI, first dorsal interosseous, and AbDM, abductor digiti minimi), and there is inhibition in FCU. Other abbreviations are: 2DI, second dorsal interosseous; FDS, flexor digitorum superficialis; AbPL, abductor pollicis longus. (From Muir and Lemon, 1983, Figure 1; Lemon, 1988, Figure 3; with permission.)

activity is then averaged with respect to the firing of the cell over a very large number of discharges (typically 10 000). If the pyramidal cell has a monosynaptic input to the motoneurones of any muscle, then on average, that muscle will show an increase in EMG activity (post-spike facilitation, PSF) shortly after the cell discharge. The number of muscles showing such activity and the size of the EMG changes gives an idea of the distribution and effectiveness of the pyramidal neurone terminations. An example is shown in Figure 9.10.

It is usually assumed that any post-spike facilitation which shows up in the surface EMG after STA is due to activation of a monosynaptic connection between the pyramidal cell and the spinal motoneurones of the muscle under study (i.e. a CM connection). This is because if there were more than one synapse in the pathway, the extra temporal dispersion at each relay, would smooth out the final input to motoneurones so much that it would be difficult to detect in the surface EMG. This is one reason why inhibitory effects during STA occur only infrequently: inhibitory inputs from pyramidal tract cells to motoneurones are disynaptic. A second factor also contributes to the difficulty in identifying inhibitory connections from pyramidal tract cells. If a pyramidal tract cell has an inhibitory connection to a muscle, then under normal circumstances, the muscle will be silent when the cell fires, and, conversely, when the muscle is active, the cell will be silent. This can be overcome in the following way. In order to make pyramidal tract cells fire during contraction of a muscle that they inhibit, local injection of glutamate can be made into the cortex. This excites the pyramidal neurones and hence reveals their effect on active muscles.

It should also be noted that the technique of spike-triggered averaging depends on the assumption that the firing of individual CM neurones is uncorrelated. For example, if two CM neurones always fired together, but actually projected to different muscles, then the effects could wrongly be attributed to activity in a single CM neurone rather than two. Luckily, the firing of single CM neurones is usually uncorrelated.

With this method, it seems that the actions of CM cells to hand and arm are limited to a set of synergists and/or antagonists acting about a small number of joints. Many of the CM cells that project to forearm muscles are organized in a reciprocal fashion, facilitating extensors and inhibiting flexors or vice-versa (Figure 9.11). CM cells never seem to facilitate functional antagonists, even though co-contraction of antagonists often occurs during voluntary movements requiring stabilization at the wrist. The CM projection to hand muscles is more restrictive than that to the forearm: Buys and colleagues (1986) found that a single CM cell usually produces post-spike facilitation in 2–3 intrinsic hand muscles, whereas Cheney and Fetz (1985) found that 42% of the CM cells

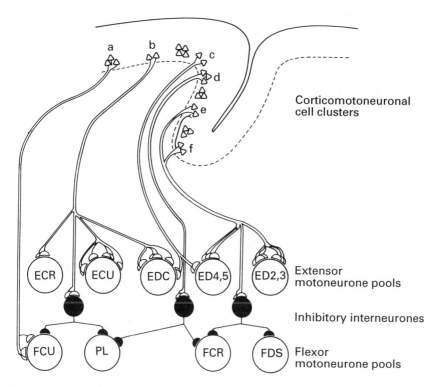

Figure 9.11 Simplified diagram showing the organization of the corticomotoneurone projection to the hand and forearm muscles in the monkey. Some CM cells facilitate agonist muscles with no effect on antagonists (a,c). Others facilitate agonist muscles (extensor motoneurone pools) and simultaneously suppress antagonist muscles (flexor pools) via a reciprocal inhibitory pathway (b,e,f), or produce suppression alone (d). The clustering of cells with common 'muscle fields' is also indicated. (From Cheney *et al.*, 1985, Figure 12; with permission.)

facilitated more than half of the tested group of flexor or extensor muscles at the wrist.

Maps have been made of the spatial distribution of CM cells in the cortex which produce post-spike facilitation in different target muscles. These maps, like those of the stimulation studies above, show that: (i) the cells are distributed over a wide area of cortex, and (ii) that within any one area neurones projecting to different muscles are intermixed.

The microelectrode used to record CM activity in the cortex can also be used for intracortical microstimulation (ICMS). If single pulse ICMS is used, the average effects on EMG activity in target muscles are usually much larger than the results of spike-triggered averaging from a single CM cell. In the example of Figure 9.10, spike-triggered averaging produced a peak of increased EMG activity in the APB muscle which was

40% greater than the background level, whereas a single pulse ICMS produced a peak of 162% greater than the background. This indicates that there must be more than one CM cell near the electrode (in addition to the one used for STA) which has a facilitatory effect on the APB muscle. In other words, the CM cells projecting to a given muscle are grouped together in clusters.

One other point is evident in the figure: ICMS produces effects on more muscles than spike-triggered averaging from a single cell. Thus, as expected from the overlapping territories in the cortical map, CM cells projecting to different muscles must be intermingled even within the small area of cortex activated by single pulse ICMS. (Reviews by Cheney and Fetz, 1985; Kasser and Cheney 1985; Lemon, 1988, 1990; Schieber, 1990; for further details.)

Plasticity of the motor cortical map

It has been known for many years that electrical stimulation at the same site in motor cortex does not always produce exactly the same movement on every occasion over the course of a period of several hours. However, it is only within the past ten years that this capacity of the cortex to change its organization, even in the adult, has been investigated in detail. The majority of experiments have analysed the somatosensory system and the post-central cortex; relatively few have been conducted on the motor system.

Sanes and Donoghue (1991) examined changes in the motor cortical map of adult rats before and after transection of the facial (VII) nerve. Animals were anaesthetized and their cortical organization mapped very carefully using short trains of ICMS. Movements of the forelimb, the vibrissae (in the rat, the whisker representation is prominent in the primary motor cortex), and the periocular muscles were noted. In all cases, the map showed a relatively clear separation between the cortical areas devoted to these movements. After mapping was complete, the branches of the contralateral facial nerve, which contain the motor supply to the vibrissae were cut and the animals allowed to recover. One week to 5 months later, the cortex was remapped using the same stimulation sites. Stimulation of the vibrissae area, of course, no longer produced any whisker movement. Instead, the same low intensity stimuli which had previously evoked vibrissae movement now produced movement of either the forelimb or the periocular muscles (Figure 9.12). There appeared to have been an 'expansion' of adjacent motor representations into the disconnected region.

This reorganization occurs very rapidly. In a separate series of experiments, the cortex was mapped, the VIIth nerve sectioned and the cortex remapped all within the space of a few hours. Figure 9.12 shows that

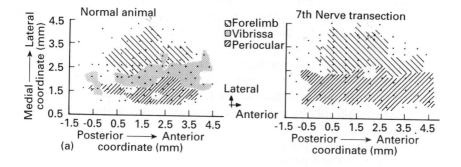

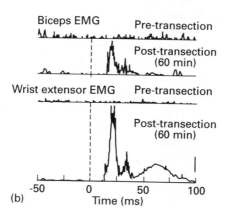

Figure 9.12 Reorganization of the motor map of primary motor cortex in the rat after transection of the buccal and mandibular branches of the facial nerve. (a) Surface views of the cortex illustrating the major functional regions in a normal animal and an experimental animal 123 days after nerve transection. Each dot indicates an electrode site at which movement was evoked at stimulation currents of less than 60 μA. Dashes indicate sites at which no movement could be evoked. Shading shows the three main regions of the cortex (see key). Note the apparent expansion of the forelimb and periocular sites into the vibrissa zone in the experimental animal. (b) EMG responses in biceps and wrist extensor muscles to microstimulation in part of the vibrissa representation before and 60 minutes after nerve transection in the same animal. Prior to nerve transection, no responses were seen in these muscles. Afterwards, responses could be obtained from both of them, showing the speed at which the reorganization occurred. (From Sanes and Donoghue, 1991, Figure 2; with permission.)

stimulation at a site which before nerve section produced only movement of the vibrissae, could evoke forelimb activity some 60 minutes after section.

Such rapid changes in the motor map are thought to be caused by changes in afferent input after nerve section. Although a pure motor nerve, section of branches of the facial nerve, by paralysing the vibrissae, will dramatically alter the pattern of afferent input from the sensitive nerve endings which surround the follicles. It is thought that this could disturb the normal balance between afferent input and motor output, and somehow cause reorganization in the pattern of movements produced by cortical stimulation. Indeed, changes in the forelimb map can be produced simply by continuous passive movement of the limb for an hour or so.

Although it is tempting to suppose that the changes in the motor map occur because of reorganization of the motor cortex, it should be remembered that changes at any point between cortex and muscle can affect the motor map. In the rat, there are very few (if any) CM projections: corticospinal axons first synapse on interneurones in the spinal cord or brainstem which then activate motoneurones. Alterations in the efficiency of these synapses could result in reorganization of the map obtained at the motor cortex.

There is one experiment which suggests that at least some of the reorganization occurs in the cortex itself. The forelimb and vibrissae areas of motor cortex were mapped in rats (Figure 9.13). After the mapping was complete, a local iontophoretic injection of the GABA$_A$ antagonist bicuculline was placed at a site in the forelimb area. When the vibrissae area was remapped, there were several sites at which low threshold stimulation produced effects on the forelimb, rather than evoking movement of the vibrissae as it had before the application of bicuculline. Such changes are unlikely to have been due to subcortical effects. Instead, they are consistent with a model in which local GABAergic inhibitory connections maintain the motor cortex output map (Figure 9.13).

The hypothesis is that when trains of ICMS are used, the motor cortical output map mainly reflects the organization of synaptic inputs to pyramidal cells projecting to the target muscle. Changes in the map therefore reflect changes in the synaptic weighting of connections within the cortex. GABAergic interneurones may be a critical element in shaping the pattern of these connections.

Summary of motor cortex maps

1. The population of pyramidal cells which produces effects on motoneurones of any one muscle (either monosynaptically, or via spinal interneurones) is spread over a wide area of cortex. Within this area, there may be clusters of cells which yield a low threshold point for a particular muscle in the motor map.

2. Cells which project to different muscles are intermixed to a large

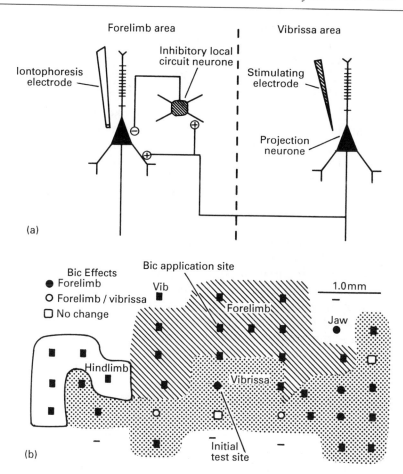

Figure 9.13 Reorganization of primary motor cortex in the rat after blockade of local inhibition with bicuculline. (a) A hypothetical circuit to explain the reorganization. Ordinarily, stimulation in the vibrissae area results only in vibrissae movement because lateral spread of excitation (see branching axon from vibrissa output neurone to forelimb area) is limited by simultaneous activation of local circuit inhibitory interneurones. During bicuculline application, this inhibition is removed. Stimulation in the vibrissa area then activates output neurones in the forelimb area via the intracortical axon collateral and evokes forelimb movement. (b) The results of one experiment in which bicuculline was placed in the forelimb area. The cortex was mapped first, and then selected sites were remapped after application of the drug. At each filled circle, the movement elicited at threshold switched from vibrissa to forelimb. At open circles, both vibrissa and forelimb movements were evoked after bicuculline. Squares mark sites at which there was no change in the movement evoked. (From Sanes and Donoghue, 1991, Figure 4; with permission.)

extent, although not sufficiently to obliterate the gross somatotopic organization of the face, arm and leg.
3.   Each pyramidal cell exerts effects on more than one muscle. Those projecting to the hand innervate the smallest number of motoneurone pools.

Lesions of primary motor cortex leave monkeys and humans with a permanent deficit in the control of fine, fractionated, finger movements. How can such a distributed system, as evident in the motor maps, produce individual finger movement? The results above show that one muscle is not controlled by one small area of cortex. Thus, individual movements cannot be made by activating discrete cortical zones. Instead, it is the pattern of activity within a large population of cells which produces fractionated movements. Some cells will be involved in activating the primary agonist muscle, others in inhibiting antagonists, and other in stabilizing proximal joints, etc. Some cells may have connections which participate in all the appropriate actions, in others some outputs may be appropriate, but others may result in unwanted activity at different joints, which has to be cancelled by the action of other cells.

Although this system is complex, it has great advantage over a simple point-to-point pattern of muscle (or movement) representation on the cortex. It can cope very well with damage caused by small lesions. In a point-to-point organization, a lesion to the area controlling, say, the APB muscle, will result in permanent paralysis of that muscle. In the distributed organization, the function of the lesioned cells can probably be taken over, almost as well, by another group of surviving cells.

9.4 *Electrophysiology of the corticospinal projection*

All corticomotoneuronal synapses are excitatory. IPSPs may be produced by pyramidal tract or cortical stimulation, but their latency is about 1 ms longer than the shortest latency EPSP. Early IPSPs are disynaptic and employ the Ia inhibitory interneurones of the spinal intermediate zone.

As noted above the size of the maximal EPSP in spinal motoneurones (i.e. the EPSP produced by stimulation of all CM cells projecting to a muscle) is greatest in motoneurones innervating distal muscles of the forearm, which correlates with the density of terminals in spinal motor nuclei. However, they are still quite small in absolute terms, being about 1–2 mV, and are not capable of raising a resting motoneurone above its firing level (which usually takes about 10 mV). This is compensated by 'frequency potentiation' at the corticomotoneuronal synapse. Repetitive discharges in the pyramidal tract result in progressively increasing sizes of EPSP in the motoneurone. Such an effect is not seen at the synapses of Ia muscle afferents. It is a presynaptic action produced by augmenting transmitter release in the second and subsequent volleys of a train. The degree of potentiation depends on the frequency of activation, and is maximum at frequencies of about 500 Hz. This is more than the maximum steady firing rate of pyramidal cells, which is around 100 Hz. However, higher frequencies are observed at the onset of pyramidal cell firing, before the rate declines slowly to more tonic levels. Such a phasically decreasing discharge rate shows better frequency potentiation at corticomotoneurone synapses than steady high frequency firing. It is probably used most effectively in the initiation of rapid movements.

As well as their connections with α-motoneurones, there are also direct corticomotoneuronal connections with γ-motoneurones. The fibres originate in the same area of cortex as those which project to α-motoneurones innervating the corresponding extrafusal muscle. This organization may indicate a degree of α-γ coactivation in the pyramidal tract projection. Indeed, this may be the anatomical basis for the results from microneurographic recordings of muscle spindle firing during voluntary contractions in man (see Chapter 6).

Although both large and small diameter pyramidal tract neurones can form monosynaptic connections with spinal motoneurones, the corticomotoneuronal connection represents only a small proportion of the total corticospinal projection. Most of the spinal terminations of the motor cortex component of the pyramidal tract are in the intermediate zone. There they may synapse onto spinal interneurones, such as the Ia inhibitory interneurones.

Finally, the pyramidal tract cells show extensive branching via axon collaterals within the motor cortex itself. These collaterals may spread within a radius of 30 μm and extend vertically to layer 2 of the cortex. The action of these collaterals is complex. When a single motor cortical pyramidal tract cell is recorded and the pyramidal tract stimulated in the medullary pyramids with an intensity subthreshold for direct antidromic activation of that cell, two later phases of membrane potential change can be recorded: an initial depolarization followed by a hyperpolarization. The initial phase is thought to be caused by monosynaptic excitation from collaterals of pyramidal cells which were excited antidromically by the stimulus. The hyperpolarization is probably due to inhibition transmitted via cortical interneurones (e.g. Figure 9.12). Repetitive stimulation greatly increases the size and duration of the hyperpolarization, and it may be that during normal activation, a group of discharging pyramidal tract cells can cause 'surround inhibition' of neighbouring areas of cortex through this action of axon collaterals on cortical inhibitory interneurones (Phillips and Porter, 1977; Asanuma, 1981).

Motor cortex cell activity during voluntary movements

One of the most important advances in recent years has been the ability to record single cell activity from the brains of conscious animals while they perform particular learned movements. The technique was pioneered by Evarts in the 1960s (1966, 1968). A small chamber enclosing an hydraulic microdrive is fixed above a hole in the skull during an operation in an anaesthetized animal. The drive is then capable of inserting a tungsten microelectrode into parts of the underlying cerebral cortex at various depths in the behaving animal to record extracellular unit activity. This technique can be used to record activity in any part of the brain. When used to study motor cortex activity, another (non-movable) set of electrodes is usually implanted within the medullary pyramids. These electrodes are used to stimulate the axons of pyramidal tract (PT) cells so that PT neurones in the cortex can be identified, and their conduction velocity measured.

The procedure to identify a motor cortex output cell is as follows: a unit is encountered in the motor cortex while moving the microelectrode. A stimulus is then given to the pyramid. This usually produces an impulse in the cell which may be an antidromic potential conducted along the axon of the cell, or a synaptic potential produced via activity in other cells. To distinguish these modes of activation, the motor cortical cell is itself stimulated through the recording electrode shortly before the arrival of the antidromic volley. If the latter is an antidromic impulse, it will collide with the descending (orthodromic) potential and will be unable to activate the cortical cell. In contrast, a synaptically conducted potential will still activate the cell. This test is known as the collision test. If it is positive, then the motor cortical cell which is being recorded from is a pyramidal tract (PT) neurone. If not, it is likely to be an intrinsic interneurone, or an output neurone which projects elsewhere than the pyramidal tract.

Using such techniques, many workers have investigated how neuronal discharge varies in different types of movement. Unfortunately, it is sometimes difficult to compare the results since some studies were made on identified pyramidal tract neurones, others on the specific class of corticomotoneurones, and yet others have included data from any neurone in the cortex (including interneurones). In addition, some tasks involve hand movements, others shoulder movement and others involve both. Given that the corticospinal projection to proximal muscles may well be more indirect than that to distal muscles, then it may not be surprising to find a difference in the relation of neural activity to movement in these tasks.

Pyramidal tract neurone activity

Many of the studies on identified pyramidal tract neurones have examined how their activity relates to relatively simple (e.g. flexion/extension) of the wrist and fingers. In such tasks, pyramidal tract neurone activity usually (a) precedes the onset of EMG activity, and (b) is related in intensity to the force of muscle contraction.

An example of the timing of pyramidal tract discharge is shown in Figure 9.14. Monkeys were trained to perform reaction time movements of the hand to release a key as soon as possible after a visual stimulus. The time between the signal and the EMG response in wrist extensor muscles was of the order of 150 ms. However, many motor cortex cells including PT neurones began to change their firing rate up to 100 ms before the EMG activity (Figure 9.14). Because the timing of this activity was more closely correlated with EMG onset than with the time of the visual stimulus, it was suggested that it played a direct role in the production of movement.

Some of the extra delay may have been caused by the time taken for a volley of pyramidal tract discharge to bring resting motoneurones to threshold. However, this should only account for 20 ms or so, at most, of the total 90 ms. The reason for the remaining shortfall is unknown. One possibility is that the initial part of the pyramidal tract volley 'presets' excitability of spinal circuits before the onset of movement. For example, it has recently been demonstrated that presynaptic inhibition of Ia afferent terminals on the agonist motoneurones is decreased before movement (see Chapter 6). This produces increased excitability of the monosynaptic reflex which may be useful in assisting the muscle contraction.

The idea that pyramidal tract neurone activity may sometimes be important in preparing for a movement is consistent with the high proportion of cells that show **instruction-dependent** changes in activity. In these tasks, the monkey has either to pull or push a lever by moving the whole arm at the shoulder and elbow. The trigger to move is a short perturbation applied to the handle by a torque motor; the direction in which to move is indicated by a red or green instruction light which comes on 2–5 s before the 'go' signal. In the period between the instruction and the response, over half of pyramidal tract neurones in the primary motor cortex show changes in activity which are related to the direction of the forthcoming movement (Figure 9.15). This occurs without any obvious change in EMG activity. The activity may be involved in presetting the activity of reflex circuits prior to movement.

The timing of motor cortical cell discharge can be contrasted with that of postcentral cells. In general, postcentral cells only fire after the onset of movement, whereas most precentral cells fire before (Figure 9.14 Evarts, 1966, 1981; Phillips and Porter, 1977).

The second question to be tackled with this technique was whether

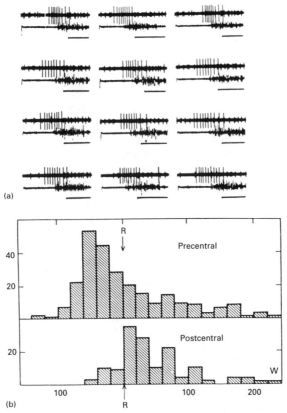

(a)

(b)

Figure 9.14 (a) Timing of discharge in pyramidal tract neurones (PTN) before a voluntary movement in a monkey. Upper traces are a series of 12 trials in which a monkey had been trained to react as rapidly as possible to a visual stimulus (presented at the start of each sweep) by extending its wrist. Recordings are from a single PTN neurone (upper traces) which was silent during flexion movements, EMG from wrist extensor muscles (middle traces), and onset of movement (bottom traces). The timing of the PTN discharge is more closely related to the timing of EMG activity than it is to onset of the visual stimulus. This is seen most clearly in the second and third trials of the first row. In the second, the PTN onset latency was short, and the EMG reaction time was also short; in the third, the onset of PTN discharge was long, as was that of the EMG activity. The duration of the sweeps was 500 ms. (From Evarts, 1966, Figure 2; with permission). (b) Distribution of onset latencies of a sample of precentral (upper histogram) and postcentral (lower histogram) neurones in relation to abrupt flexion and extension movements of the wrist. Latencies have been measured in relation to the onset of movement (R). The reason for using movement, rather than EMG onset as a reference, was that responses to several different types of movement were combined in this graph, and it was not possible to discover which muscle started its activity first in each case. The y-axis plots the number of cells responding at a particular latency. Note that precentral neurones commonly changed their firing 60 ms before movement, whereas postcentral neurones only began to discharge after movement. The cells which were sampled were both PTN and non-PTN neurones. (From Evarts, 1972, Figure 5; with permission.)

the firing of pyramidal tract cells was more closely related to the **force** of muscle contraction or to the **position** of the joint during the contraction. In the first experiments to tackle this question, monkeys were trained to move their wrist through various positions of flexion and extension against a series of different loads. Pyramidal tract cells showed a closer relationship of their firing rate to the force of contraction than to the position of the wrist (Figure 9.16).

The relationship between pyramidal tract neurone activity and force is linear only over a small range of forces. The majority of large pyramidal tract neurones have a sigmoid relationship of static firing frequency to force applied. They are usually silent at zero force, and saturate at high force levels. The phasic discharge of these cells, during small adjustments of muscle force depends on the level of static force. For a given small change in force, the largest phasic bursts of activity occur at low static forces (and firing rates). At high force levels, the same changes in force are accompanied by little or no phasic activity. Thus, pyramidal tract neurones show their greatest frequency changes for small adjustments of load near zero force, at a time when relatively few motor units in muscle are active. For these fine adjustments of force, activity of a large portion of pyramidal tract neurones appears to control a relatively small number of motoneurones.

Finally, although most pyramidal tract neurones show reciprocal activity for flexion and extension tasks, a separate subpopulation of cells with discharge related to co-contraction of antagonist muscles can also be observed in more rapid wrist movements in which the joint is stabilized by combined action of flexors and extensors. (This topic reviewed by Humphrey and Tanji, 1991.)

Role of small pyramidal tract cells
Microelectrodes inserted into the cortex record preferentially from large cells. Thus most studies have analysed behaviour of the fastest conducting pyramidal tract neurones. Recently, attention has focused on small pyramidal tract neurones which contribute the majority of axons to the pyramidal tract. Analysis of firing behaviour of small and large cells has revealed differences which, in some ways, are reminiscent of the size principle of spinal motoneurone recruitment. Small pyramidal tract cells are more likely to maintain a steady firing rate in the absence of load than large cells. Their firing rate also tends to be related to muscle force over a wider range. Large cells may only be recruited above zero force levels, and have an 'S' shaped relation between firing rate and muscle force. Therefore, during very small movements, or small adjustments of force, small cells are more likely to be firing than large cells, and hence they may be of particular importance in the control of fine movements at low force levels (Evarts *et al.*, 1983).

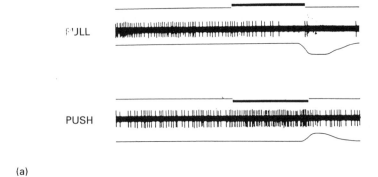

(a)

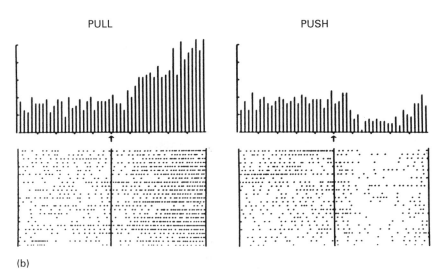

(b)

Figure 9.15 Preparatory activity in a pyramidal tract neurone of the monkey motor cortex in the delay period between a warning and a 'go' signal. Monkeys had to pull or push a handle with their arm according to the direction given by an instruction signal. (a) Two sets of traces from single trials. Corresponding to each trial are three traces: the solid black line in the top trace indicates the onset of the instruction; the middle trace shows the discharge of the pyramidal tract neurone; the bottom trace indicates the movement of the handle (upwards = push). When the monkey receives the pull instruction, the cell discharge increases in the delay period, without any visible movement of the handle. When the instruction is to push, the activity is reduced. (b) Histograms and rasters showing the discharge of another neurone for 1 s before and after the appearance of the instruction signal. The rasters show the discharge of the unit in each single trial; the histograms are the average of the activity in successive 40 ms periods. (From Tanji and Evarts, 1976, Figures 2 and 5; with permission.)

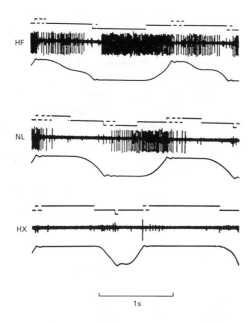

Figure 9.16 Relationship of discharges in a pyramidal tract neurone (PTN) during wrist flexion/extension movements made against different loads. Three experiments are shown: movements made against a load of 400 *g* opposing flexion (and assisting extension) (HF), movements with no load (NL), and movements with a load of 400 *g* assisting flexion (opposing extension) (HX). In each section, the traces are wrist position (lower trace: upwards deviation is flexion), PTN cell activity (middle traces) and a marker showing when the monkey had reached the end stops at the limit of the flexion and extension movements. In the 'No-load condition', the PTN can be seen to discharge before and during movement of the wrist into flexion. In the presence of an opposing load (HF), the discharge is greatly increased, and even continued during the extension phase of movement. In this instance, it may have been that extension was made by using the flexor muscles as a brake on the assisting extensor load. When a load assisting flexion was applied (HX), the cell was almost totally silent. Thus, even though the same wrist movements were made in the three experiments, cell discharge varied greatly and was related to the active force of movement rather than the position of the wrist. (From Evarts, 1968, Figure 7; with permission.)

Corticomotoneurone activity

Because corticomotoneurone (CM) cells have a direct input to spinal motoneurones, their firing pattern is usually closely related to the EMG activity of the muscles to which they project. Cheney and Fetz (1980) examined flexion/extension movements about the wrist joint, and classified CM discharge into six different groups depending on their firing pattern during a ramp and hold contraction. The majority of cells either had a phasic peak of discharge at the onset of movement, followed by a sustained, tonic activity, or showed just a tonic increase in firing throughout the movement. The discharge rate in the tonic phase was linearly related to the force of muscle contraction. Interestingly, this relation was steeper for CM cells projecting to extensor muscles than it was for CM neurones with projections to flexor muscles. That is, extensor-coupled CM cells increased their firing rate more for a given increment of force than flexor-coupled neurones. Since the EMG-force relationship for the flexors and extensors was the same in this task, then this implies a difference in the CM control of extensors and flexors.

Discharge of a CM cell must, by definition, produce synaptic input to motoneurones. If we take the average discharge rate at a given force level, and assume that this applies to all CM cells projecting to flexors or extensors, then we can obtain a very rough estimate of the net CM input to the motoneurone pool. Thus, for wrist muscles, the maximum EPSP produced by all CM cells projecting to forearm muscles is about 2 mV, and the mean firing rate for a moderate contraction is about 45 Hz. Since the mean excursion of the membrane potential of a regularly firing motoneurone is of the order of 10 mV, a randomly occurring EPSP of 2 mV will cause an extra discharge on one in every five trials. If EPSPs arrive at 45 Hz, then they will on average cause a 9 Hz increment in motoneuronal firing rate. This is about half of the 20 Hz rate of most motoneurones during a moderate contraction at the wrist, so that at a rough guess, CM input in this task represents about 50% of the total drive to the muscle.

This calculation does not hold for the onset of muscle activity at the start of the ramp contraction. The phasic burst of CM discharge begins 100 ms or so before EMG onset, and does not produce as much EMG activity as expected. As with the early onset of pyramidal tract neurone activity discussed above, the precise reasons for this discrepancy are not known. Three factors may contribute: the amount of depolarization need to bring a resting motoneurone to threshold, the activity of spinal inhibitory interneurones such as those mediating reciprocal inhibition, and activity in other descending pathways to the motoneurone. All may delay the final discharge of motoneurones after the onset of phasic CM activity.

Despite their direct excitatory connection to motoneurones, CM cells are not equally active in all tasks requiring contraction of their target muscle. For the CM cells projecting to the intrinsic hand muscles, activity is greatest during independent finger movement, and least in gross movements involving the whole hand. Figure 9.17 shows that the discharge of a single CM cell projecting to the first dorsal interosseous (FDI) muscle was much greater during the precision grip task than in the power grip, despite the fact that FDI activity was higher in the latter task. The preferential role of CM cells in such fractionated finger

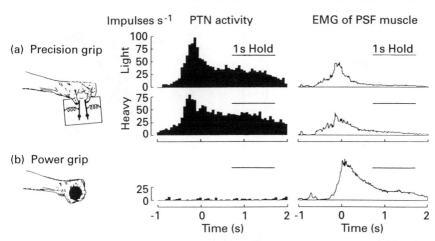

Figure 9.17 Difference in the activity of a CM cell in primary motor cortex during performance of a precision and power grip. The cell showed post-spike facilitation of the 1st dorsal interosseous (FDI) muscle. Histograms show the discharge for 16 repetitions of the tasks. The rectified EMG on the right is from the FDI muscle. (a) shows activity in the precision grip task, against either a light (top) or heavy (bottom) load; (b) shows activity in a power grip. Note that although the activity of the muscle is greater in the power grip task, there is virtually no discharge of the CM cell. Time 0 indicates the time at which the first detectable change in pressure began. (From Muir and Lemon, 1983, Figure 2; with permission.)

movements may be related to the small number of hand muscles supplied by any one CM cell. Interestingly, non-CM cells in motor cortex do not seem to show such a preference for fine finger tasks.

Why are CM cells not used to the same extent in the power grip task as they are in the precision grip? One possibility is that the very specific pattern of facilitation and inhibition that the cells produce is inappropriate for the power grip. In other words, some of the muscles inhibited by CM cells may need to be activated in the power grip whilst others, facilitated by some CM cells, may need to be relaxed. Thus if CM activity were used fully it would interfere with performance of the task.

A similar task-dependent firing pattern has been shown for motor cortex cells (not necessarily CM cells) related to activity of jaw closing muscles. During isometric biting, many cortical cells fire in relation to the force exerted, whereas during natural chewing, many of the same cells fail to show any modulation at all. Presumably, natural rhythmic movements are controlled by other centres (Fetz *et al.*, 1989; Lemon, 1990).

Population studies

The activity of CM cells projecting to the hand and forearm muscles can be easily related to the pattern of muscle activity. However, when the

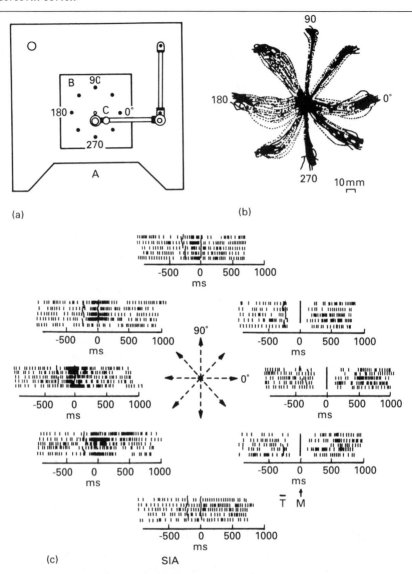

(a)

(b)

(c)   SIA

Figure 9.18 Population studies of motor cortex discharge in the primary motor cortex of monkeys. (a) The apparatus. The monkey had to hold the manipulandum in the centre of the working area. LEDs were placed at eight different positions on a concentric circle of about 20 cm diameter. When an LED was illuminated, the monkey had to move towards the light. (b) The trajectories that the animal made in 30 different movements to each target. (c) The discharge of a typical motor cortex neurone in the shoulder region of the cortex is plotted as rasters for each different direction of movement. The rasters are aligned on the onset of movement (M): five repetitions of each movement are shown. The time of the 'go' signal is indicated by the bar labelled 'T'. Movements to the left are associated with an increase in firing of this cell; movements to the right are associated with a decrease in firing.

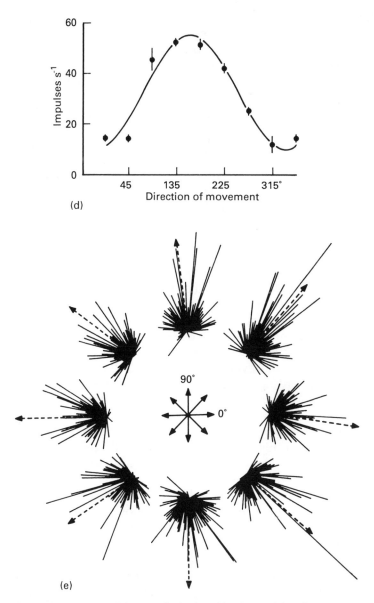

(d)

(e)

(d) The relationship between firing rate (calculated for the total time from appearance of the target to the completion of the movement) and the direction of movement. (e) A vector summation of the activity of 241 cells. Cell vectorial contributions (thin lines) and population vectorial sums (thick interrupted lines with arrows) for each of the eight directions of movement are shown. Note how the population vector approximates the true direction of the movement that was made. (From Georgopoulos *et al.*, 1982, Figures 1 and 4; with permission.)

activity of non-CM cells is examined, or if the movements involve activity of proximal, rather than distal muscles, it becomes much more difficult to relate the firing of any one motor cortex cell to the particular parameters of movement. A recent approach has been to examine the behaviour of the population of motor cortical cells (i.e. CM cells, pyramidal tract neurones, and interneurones) related to the movement.

Georgopoulos and colleagues trained monkeys to make free arm movements, involving the shoulder and elbow, from a central target to one of eight different target positions on a concentric surround. The monkey held the handle steady in the central position and then moved immediately to one of the targets when it was illuminated. The activity of cells in the shoulder area of motor cortex was recorded.

Three-quarters of cells with movement-related activity modulated their discharge according to the direction of movement (Figure 9.18). For any one cell, the frequency of discharge was highest with movement in a particular direction (the **preferred direction**), and decreased progressively with movements made in directions further and further away from the preferred one. Different cells had different preferred directions, and these were equally distributed over the eight directions of movement. In order to show that the discharge depended on the direction of movement, rather than the final end position, in some experiments monkeys were also required to move from the outside points on the target circle back into the centre. In this case the final end position was the same every time; nevertheless, neuronal firing was still modulated according to the direction of movement.

For any one direction of movement, the behaviour of all the directionally-selective cells that had been sampled (sometimes several hundred different neurones), was combined using a neuronal **population vector sum**. The idea of this analysis is that each cell 'codes' for its preferred direction. This direction is represented by a vector (a line pointing out from a central spot in the preferred direction of the cell), the length of which is proportional to the discharge rate of the cell. Thus, if the cell's preferred direction is in the same direction as the movement being made, then the vector will be long; if it is in the opposite direction then the vector will be very short. This has been done in Figure 9.18 for 241 cells. For each of the eight directions of movement, the vectors from all cells have been added together (vector summation), and the result plotted as a thick arrow. In every case, the vector sum of neuronal activity points in the actual direction of movement made by the monkey. The population discharge (at least for the directionally-selective, movement related cells) can therefore be thought of as coding some feature related to the direction of arm movement. (The non-directionally selective, and non-movement related cells are, by definition equally active in all directions of movement and therefore cannot contribute to this analysis.)

As with the single joint movements described above, it is possible that the population discharge is related to the direction of movement *per se* or to the muscle activity needed to achieve it (or both). In order to examine the latter, monkeys were trained to perform the task against different external loads. In the experiment of Figure 9.19, constant loads were applied in separate blocks of trials which pulled the arm in one of the eight different directions of movement. The plot in the centre illustrates the cell's discharge in the normal, unloaded task. The dashed

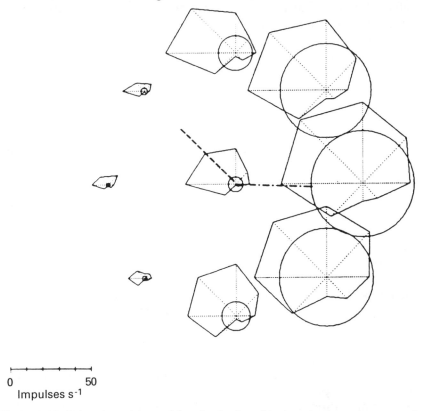

0   50
Impulses s^{-1}

Figure 9.19 Polar plots of the activity of a single cell in the primary motor cortex of a monkey during loaded movements of the arm (see text for details). The central polar plot shows the activity of the cell during unloaded movements. The discharge rate in each of the eight different directions of movement is indicated by the length of the faint dotted lines radiating from the centre of the circle. The tips of these lines have been joined together by a solid outline. The cell discharges maximally in movements to the upper left (thick dashed line; the preferred direction of movement). The position of the other polar plots corresponds to the direction in which the load pulled the arm away from the central position. Thus, when the load pulled the arm towards the right, there was a general increase in the firing rate of the cell for all directions of movement. This is the direction of the load axis of the cell (thick dot-dashed line). (From Kalaska *et al.*, 1991, Figure 1C; with permission.)

line indicates the preferred direction of movement. The eight plots surrounding show the behaviour when movements were made with constant loads pulling in different directions. When the loads pulled the arm towards the left, the discharge of this cell, for all directions of movement, was much less than in the unloaded condition. When the loads pulled the arm to the right, then the discharge was much higher. In fact, the **load axis** of this cell was in approximately the opposite direction to the axis of preferred movement. Most other cortical cells behaved in a similar fashion, consistent with a role in specifying the forces involved in moving the arm in different directions.

Such experiments have been extended in three ways. First, it has been shown that the cell firing is much more tightly coupled to the **change** in force needed to make the movement than to the static force needed to hold the arm in the initial or final position. It is as if the motor cortex was providing the phasic input for the movement whilst another system(s) was controlling the posture of the arm. Second, in some experiments, monkeys have been trained to point to the same target using different hand paths. For example, on some trials, the monkey would point straight to the target, whilst on others, it would move in a concave or convex trajectory. Neuronal firing depends on the trajectory used to approach the target, as well as the final end position of the movement. Finally, it has been possible to show that the same population coding can apply to movements made in three, rather than two dimensions.

How this pattern of cell firing is related to the complicated pattern of muscle commands around the shoulder is not clear from these experiments. Some of the sampled cells may be pyramidal tract neurones which project to particular groups of shoulder muscles. Their activity would be directionally-dependent because of the changing pattern of muscle activity for different directions of movement. Other cells may be interneurones, or may project to networks of interneurones in the spinal cord or brainstem. Their activity might be directionally selective because of the particular pattern of connectivity in the network.

The population vector can be calculated at any point during or before the onset of movement, by weighting the vectors for active cells according to their firing rate over the time period examined. In a simple reaction time task, as above, the population vector gradually builds up in the reaction period to point more and more clearly in the direction of the forthcoming movement (see Figure 9.20). In a warned reaction time task, a warning signal indicates the direction of the next movement, but the movement is withheld until presentation of a 'go' signal. As with pyramidal tract neurone activity discussed above, many of the cells (about 50% in this task) show instruction-dependent activity in the delay period. When a population vector sum is calculated in this period,

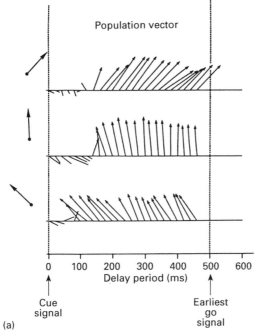

Population vector

0 100 200 300 400 500 600
Delay period (ms)

Cue
signal

Earliest
go
signal

(a)

Figure 9.20 (a) Time evolution of the population vector during the delay period of an instructed delay task. The cue signal giving the direction of the next movement was given at the start of each row. The three rows show the response to three different cue directions. The 'go' signal occurred at the end of the traces. The direction of the population vector is shown at different times during the delay period. After an interval of about 100 ms, the population activity gradually begins to code for the direction of the upcoming movement.

it points, as expected, in the direction of the forthcoming movement, even though there is no detectable muscle activity. Presumably, this preparatory activity is 'presetting' in some way, circuits (motor or sensory) involved in the next movement.

The population vector is a useful tool because it provides an indicator of cortical activity which seems valid even in the absence of movement. The most elegant way in which it has been used has been to show that the preparatory activity in motor cortex can reflect surprisingly complicated 'cognitive' processes relating to an upcoming movement. In the experiments above, the monkey always pointed **at** the target light. In a different set of experiments, a monkey was trained to point not directly at the target, but to a point which was, say 90° counterclockwise to the light. It has been shown in man that the reaction time is longer for rotated movements than for movements made directly to a target. The increase is proportional to the amount of rotation required, and it has been suggested that this indicates 'mental rotation' of the direction of

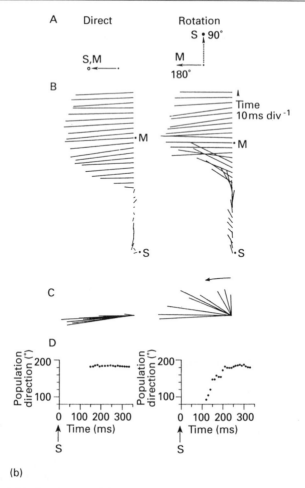

(b)

Figure 9.20 Continued. (b) The population activity in primary motor cortex preceding arm movements in direct and rotated task conditions. In the first column, the monkey had to point directly at the target light (there was no delay period in this task). The movement required was to the left, and the evolution of the population vector is shown below from the time of the 'go' signal (S), to shortly after the beginning of arm movement (M). Time increases upwards in this display. As expected, the vector begins to point towards the direction of movement some 150 ms after the target is turned on. In the column on the left, the task involves the monkey pointing 90° counterclockwise to the target. The evolution of the population vector shows that it begins to point in the direction of the stimulus (i.e. straight ahead), but then rotates in the reaction period towards the final direction of movement (i.e. to the left). The population vectors at different times have been collapsed together in the panels below, and a graph is plotted at the bottom showing how the direction of the vector changes over time in the reaction period. (a, From Georgopoulos *et al.*, 1989a, Figure 11; b, from Georgopoulos *et al.*, 1989b, Figure 1; with permission.)

movement. If it is supposed that the rotation proceeds smoothly, then this explains why the reaction time increases more, the larger the degree of rotation.

In the monkey experiments, it was possible to confirm that this idea of 'mental rotation' was true by calculating the motor cortical population vector for a large number of cells at different times in the reaction period. The population vector initially pointed in the direction of the target, and then rotated counterclockwise (i.e. by the shortest route) towards the final direction of the movement that was made (Figure 9.20) (Georgopoulos *et al.*, 1984, 1992; Georgopoulos, 1989).

9.5 *Sensory input to motor cortex*

The activity of pyramidal tract cells is influenced not only by central commands fed into motor cortex from other brain areas, but also by afferent input from the peripheral moving parts. Very short latency activation of precentral cells can be produced by electrical stimulation of cutaneous, muscle or mixed nerve trunks, or by natural stimulation of cutaneous or muscle receptors. Cutaneous input is directed mainly to rostral area 4, and muscle input to the caudal part, where it may lie buried within the central sulcus. There is a considerable overlap between these projection areas.

There is no major direct sensory input to motor cortex, except a small contribution from spinothalamic terminations in the VPLo nucleus of the thalamus. The majority of input comes via two routes. (1) A lemniscal projection to the VPLc nucleus of the thalamus. This then projects to sensory cortex and thence, via corticocortical connections to the motor areas. (2) Spinocerebellar pathways to interposed and dentate nuclei, which then send fibres to the 'cerebellar' thalamus (VPLo and nucleus X). This projects directly to motor cortex. The relative importance of the cerebellar versus sensory cortex route to the motor areas is not known. However, since at rest, the threshold and latency of somatosensory responses in cerebellar thalamus is higher and longer than that in the VPLc, it is thought that sensory cortex may provide the larger contribution (Wiesendanger and Miles, 1982; Butler, Horne and Rawson, 1992).

Relation between sensory input and motor output

Afferent input to area 4 is carefully related to the pyramidal tract output of that region. Intracortical microstimulation has shown how tight this coupling can be. The precentral cortex is explored with a stimulating microelectrode and a 'hot spot' may be found which produces, say,

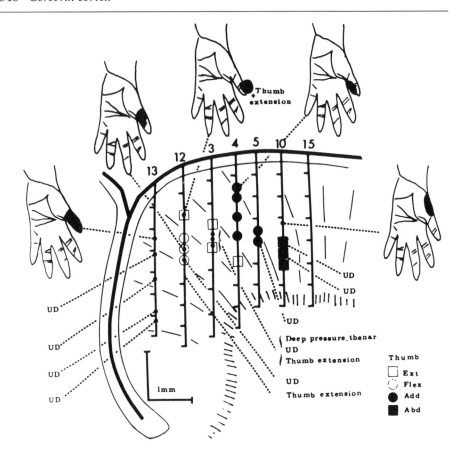

Figure 9.21 Relationships between receptive field of afferent input and movement produced by intracortical microstimulation (<5 μA) at various sites in the thumb area of monkey motor cortex. The diagram shows a reconstruction of electrode tracks (solid vertical line) and cell locations (small filled dots on the lines). Cortical spots stimulated at 5 μA without producing motor effects are shown by small solid lines perpendicular to the electrode tracks. The symbols and diagrams connected to various points show the thumb movements produced and receptive fields recorded at each location. For example, in electrode track 12, the first cell encountered a cutaneous receptive field on the distal tip of the thumb. Microstimulation at this point produced a thumb extension movement. Further down the same track, the next cell had a cutaneous receptive field in the middle of the thumb about the interphalangeal joint. Stimulation here produced thumb flexion. UD, undefined input. (From Rosen and Asanuma, 1972, Figure 2; with permission.)

flexion at a particular joint. Cortical stimulation is then stopped, and the same electrode is used to record the neuronal response of the nearby cells to peripheral sensory stimulation. Almost inevitably, the input is from receptors at or nearby the joint that moved during stimulation.

Cutaneous input

The first experiments by Rosen and Asanuma (1972) suggested that the cutaneous receptive field was likely to be on a region of skin which would encounter obstacles during the movement produced by microstimulation. (Figure 9.21). For example, in one experiment (not illustrated), they found a series of neighbouring cells, stimulation of which evoked, separately, extension, adduction and abduction of the thumb. The cutaneous receptive fields for each of these sites came, respectively, from the distal tip, medial surface, and lateral surface of the thumb – precisely the regions likely to be stimulated during the movements.

They speculated that such a tight input–output coupling could underlie the phenomenon of 'forced grasping' (see below) which sometimes is seen following lesions of prefrontal areas which project onto area 4. It was suggested that lesions might release the reflex circuit set up by these connections so that cutaneous input could drive selected pyramidal cells to maintain a grasp on an object.

Other authors have found the coupling between input and output to be less tight, particularly for proximal muscles. Cutaneous receptive fields are sometimes found on the side of a limb opposite to that expected from the results of Asanuma and Rosen.

Deep inputs

Muscle and joint afferent input is equally tightly coupled to motor cortical output. The predominant arrangement is that if microstimulation of cortex produces movement in one direction, then **passive** movement of the same joint in the **opposite** direction is likely to excite that area of cortex. The receptors responsible are the joint receptors, tendon organs and muscle spindles. For the spindles, this arrangement means that if a muscle is passively stretched, their afferent input goes to that area of cortex where it can excite cells which produce contraction of the same muscle.

The response of motor cortex cells to afferent input is enhanced during the performance of finely graded movements. The heightened sensitivity to afferent input, coupled with the increased discharge during small movements, suggests that motor cortex may be preferentially involved in fine muscular control at small force levels. It is not known what mechanism is used to change motor cortex responsiveness to afferent inputs. However, it is possible that motor cortex output itself can change the effectiveness of transmission through sensory pathways via its projections to the dorsal column nuclei and dorsal horn (Evarts, 1981, gives further references).

Transcortical reflexes

In previous chapters it has been pointed out that some portion of the long-latency cutaneous and stretch reflexes may operate through transcortical reflex pathways. The demonstration that there is a short-latency afferent input to motor cortex which can affect the firing rate of pyramidal tract neurones is consistent with this. Indeed, experiments using the technique of single unit recording in conscious monkeys have shown that spindle input to motor cortex is powerful enough to change the firing pattern of pyramidal tract cells during the course of a movement, and that this activity can influence spinal motoneurone discharge. For example, a monkey can be trained to hold a constant position of the wrist by contracting the flexor muscles against a constant load. If the wrist suddenly is extended by increasing the load, the active (flexor) muscles are stretched. The spindle discharge produces, at cortical level, a short latency (20 ms or so) increase in firing rate of the pyramidal cells which project to the flexor muscles.

Alternatively, if the load is suddenly removed, thereby unloading the active (flexor) muscles, the activity of the pyramidal cells is followed by corresponding changes in EMG activity of the flexor muscles with a latency appropriate for the known conduction times from motor cortex to periphery (Figure 9.22). This activity is the long latency reflex described in Chapter 6. Pyramidal cell discharges are believed to be responsible for part, if not all, of the long latency reflex EMG responses in some muscles. In such cases the muscle spindle projection to motor cortex is strong enough to function as part of an active 'long-loop' reflex pathway which operates in parallel with the conventional spinal monosynaptic reflex (Evarts, 1973; Cheney and Fetz, 1984).

9.6 *Non-primary motor areas*

The combined motor regions in area 6 are approximately the same size in the monkey as primary motor cortex, area 4. In humans, area 6 is **six** times larger than area 4 and may therefore prove to be particularly important in control of human movement.

Electrical stimulation

The original studies using electrical stimulation with trains of surface stimuli only distinguished between the mesial portion of area 6 and the primary motor cortex. Parts of the lateral portion of area 6 were regarded as part of the primary motor area. The newer subdivisions of these regions (APA, SPcS, SMA, CMAr, CMAv and CMAd) have been investigated only recently using microstimulation techniques.

The existence of an excitable region of cortex in the caudal bank of the arcuate sulcus (APA) is now well-recognized. Microstimulation here produces movement of the hand and arm. Stimulation immediately caudal to this produces no movement. If stimuli are applied in the most posterior portions of the inferior part of lateral area 6, then proximal trunk movements are seen. Whether this is part of the body map in primary motor cortex (as suggested by the simunculi of Woolsey) or whether it is part of a somatotopic representation in inferior area 6 is debated.

The recognition of a separate arm area representation around the superior precentral sulcus (SPcS) is quite recent. In the past, this area was thought to be a portion of the primary motor area. However, three main features distinguish it: (i) the stimulation threshold for evoking movement is slightly higher than in primary motor cortex, (ii) there are a larger proportion of cells with 'set-related' activity (see below), and (iii) the pyramidal neurones project to the reticular formation in addition to the spinal cord, whereas those in primary motor cortex do not.

Electrical stimulation of the supplementary motor area (SMA) has long been known to produce movement. However, there has been some debate over the precise pattern of the motor representation. Penfield and Welch, who originally described the SMA in man thought there was no topographic organization in the SMA, whereas Woolsey's map showed both a rostrocaudal (face, arm, leg) and dorsoventral (distal to proximal) organization. The most recent studies using microstimulation all confirm that arm movements can be obtained from stimulation of the mesial surface of the hemisphere, and that in most cases, stimulation of more anterior parts evokes face and eye movement. Leg movements are evoked from posterior sites, overlapping with the arm representation anteriorly and the leg area of primary motor cortex posteriorly. However, the distal–proximal organization is not clear, and it is possible that in previous studies, some stimulation sites within the cingulate sulcus may have activated the newly-discovered cingulate motor areas (Mitz and Wise, 1987, Gentilucci *et al.*, 1989).

Electrical recording and lesion studies

Much of the work in this section was performed before the distinction was made between the six different areas of the non-primary motor cortex. Most studies distinguish only between SMA and the superior portion of lateral area six, usually including both the APA and the SPcS. The latter are termed the premotor area.

Internally and externally guided movements

An important concept that has directed research in recent years is the distinction between movements that are guided by sensory inputs from the environment, and those that are made on the basis of internally-

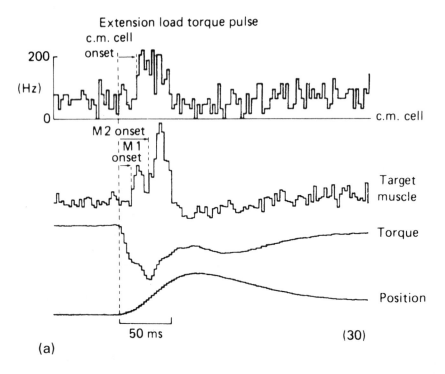

Extension load torque pulse

(a)

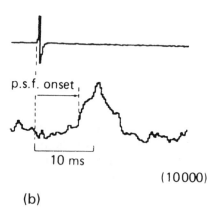

Spike-triggered average

(b)

generated motor commands (Passingham, 1987, 1988, 1989). The hypothesis is that the premotor area is concerned preferentially with control of the former, whereas the SMA is involved in the latter. This distinction may be related to the differences in the predominant input to the premotor and SMA cortex. The former receives mainly from cerebellar structures, whereas the latter has an important input from basal ganglia.

Good evidence to support these concepts comes from lesion studies. Passingham trained monkeys to make one of two different movements in response to a visual cue: they had to pull a handle if the cue was a blue light, and they had to turn it if the light was red. After the animals had learned the task, the APA was removed on both sides. Postoperatively, the animals could still make both movements, but they could not learn to associate the correct movement with the appropriate cue. In contrast, monkeys with bilateral removal of the SMA could readily relearn a visually-cued task (Figure 9.23).

The difficulty of the monkeys with premotor area lesions in relearning the task was not due to a problem in remembering a particular movement from trial to trial. In a separate set of experiments, the monkeys reached through a hole to grasp a handle which could be either squeezed or turned, depending on how the experimenter had locked

Figure 9.22 Demonstration that peripheral input generated by muscle stretch can change the firing frequency of a single corticomotoneuronal cell in the monkey motor cortex. The same cell could be shown to project monosynaptically to the motoneurones of the stretched muscle. (a) Average responses of the corticomotoneuronal cell (top) and extensor digitorum muscle (second trace) to imposed flexion disturbances applied to the wrist during maintenance of a constant extensor contraction. The firing of the cortical cell is plotted as a histogram in which the number of impulses in each 2 ms period have been counted for each of the 50 wrist disturbances. The EMG response to stretch consists of the usual short (M1) and long (M2) latency peaks of activity. In this animal, they had a latency of 10 and 24 ms, respectively. Cortical cell activity began to change some 14 ms after stretch. (b) A method for proving that this cell had a monosynaptic connection to motoneurones of the extensor muscle under study. The technique involves recording the discharge rate of the cell and the small fluctuations in EMG activity in the muscle for long periods of time. An average picture of the muscle activity is then constructed, with the average triggered by each of the spikes in the train of cortical activity. The idea is that any muscle activity time-locked to the cortical spikes will emerge as a peak, and that randomly occurring changes in EMG activity will average out to a straight line. The technique is known as spike-triggered averaging. In this case, the cell produced a 'post-spike facilitation' of the extensor EMG with a latency of 10 ms. Adding together 10 ms with the 14 ms delay between stretch and the change in cell firing shown on the left, gives 24 ms. This is the precise timing of the M2 component of the stretch reflex. (Note that spike-triggered averaging only detects relatively strong monosynaptic connections. With disynaptic pathways, the extratemporal dispersion produced by an additional synapse makes any post-spike facilitation very difficult to detect.) (From Cheney and Fetz, 1984, Figure 3; with permission.)

the mechanism beforehand. The monkeys had to perform whichever movement was possible and then remove their hand. The experimenter then unlocked the handle so that both movements were possible, and 5 s later, the animal had to reach out and perform the same movement as before. Lesions of the APA and surrounding tissue had very little effect on the performance of this task, indicating that the monkeys could hold

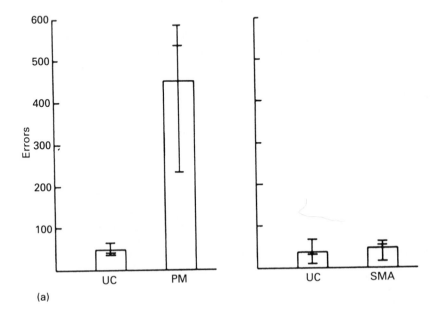

(a)

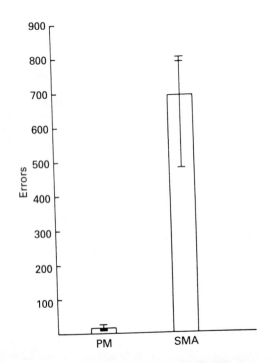

a movement in memory. Their problem lies in 'finding' that particular movement when they are presented with a particular cue.

A different motor task involved learning a sequence of three movements: monkeys had to operate a catch by first pushing it up, then twisting it and finally lifting it to uncover a food well. The appearance of the apparatus gave no cue as to which movement to make first, or even the order of the movements. Monkeys with APA lesions had no difficulty in relearning the task, whereas those with bilateral SMA lesions failed to learn in over 1000 trials. The monkeys could not retrieve the next movement in the sequence from their memory on the basis of their knowledge of the movement just performed (Figure 9.23).

The implication of these lesion studies is that the premotor cortex is concerned with retrieval of movements made on the basis of information provided by vision, whereas the SMA is concerned with retrieval on the basis of information within the animal's motor memory.

There have been several electrophysiological studies which have attempted to find a difference between the neuronal discharge in SMA and premotor cortex in these two types of movement (Tanji, 1987, for example). Most of these studies used relatively simple single joint movements made either in response to a visual signal, or self-paced by the animal itself. Few differences were found in the neuronal activity in premotor and supplementary motor areas. However, differences became quite clear when more complex tasks, similar to those used in the lesion studies were examined. The implication is that the more complex the task (i.e. if movements are made at more than one joint, or sequential movements at a single joint), the more likely the activity in non-primary motor areas is to be different from that of area 4.

Mushiake, Inase and Tanji (1991) trained monkeys to perform a sequential movement involving reaching out to touch three pads in a particular order. In one condition (VT; visually triggered), they followed lights illuminated one after another behind each pad. In the other (IT; internally triggered), they had to remember the sequence and touch the pads in order without any visual cues. The movements were started by a visual 'go' signal, and activity was analysed in the reaction period and the movement period. Most neurones in the primary motor cortex

Figure 9.23 (a) Errors in learning a visual conditional motor task (operating a handle in response to a colour cue). The histograms give the means and the bars the scores for individual monkeys (there were three in each group). UC, unoperated control animals; PM, animals with premotor lesions; SMA, animals with supplementary motor area lesions. Note the inability of animals with premotor lesions to learn the task. (b) Comparison of the ability of monkeys with premotor or SMA lesions to relearn a motor sequence task (pushing, twisting and lifting). The histograms give the mean scores, and the bars show the individual data (there were three animals in each group). Animals with SMA lesions were unable to relearn this task. (From Passingham, 1987, Figures 1 and 2; with permission.)

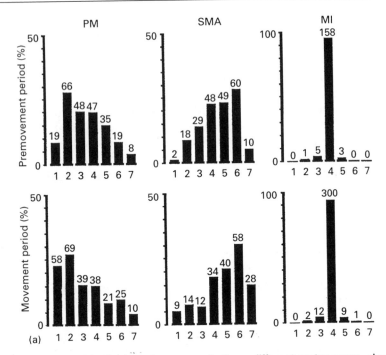

Figure 9.24 (a) Distribution of neurones in three different motor areas, classified according to their firing pattern relative to visually-guided (VT) and internally guided (IT) movement sequences (see text). The histograms show data from cells in the primary motor cortex (MI), premotor cortex (PM), and supplementary motor area (SMA). The bars indicate the percentage of cells with each of seven different classes of firing pattern. 1, exclusive relation to VT; 2, VT >> IT (P < 0.001); 3, VT > IT (0.001 < P < 0.05); 4, VT = IT; 5, VT < IT; 6, VT << IT; 7, exclusive relation to the IT task. The number of neurones studied is shown at the top of each bar. The two rows of histograms show activity in the delay period before the 'go' signal, and the activity in the reaction period between the 'go' signal and the onset of movement.

behaved similarly in both tasks. In contrast, premotor neurones were more active in the VT task, whereas SMA neurones were more active in the IT task. Thus, as with the lesion studies, it appeared as though premotor areas were more involved in visually-guided movement whereas SMA was preferentially active in internally guided movements (Figure 9.24).

Consistent with this interpretation, two special types of neurone were sometimes seen. In the SMA, some neurones were **sequence specific**: that is, they only discharged in relation to movements made in a particular order. For example, the cell might begin to fire in the reaction period, and throughout the movement period if the sequence was 1, 2, 3. If the sequence was 1, 3, 2 then the cell would not discharge, even though the first movement was the same as previously (Figure

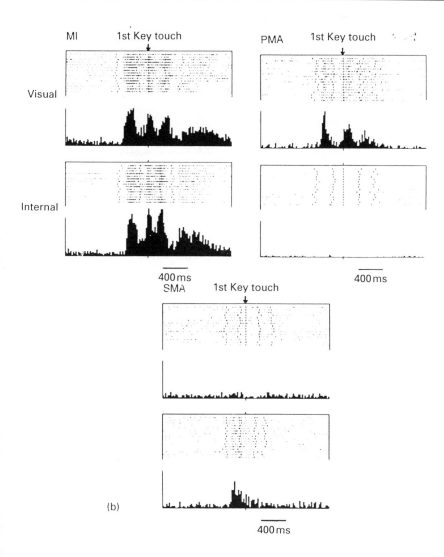

Figure 9.24 Continued. (b) Examples of cell activity in each of the three motor areas in the visually guided and internally guided tasks. The rasters indicate cell discharge in each movement, and the histogram shows the summed activity in all 16 trials. The trials are aligned to the onset of the first movement. The cell in the motor cortex (MI) fired equally well before either type of movement, whereas the premotor (PMA) cell only fired in the visually guided task, and the SMA cell fired in the internally guided task. Note that these are extreme examples of the behaviour of the cells in these areas. The histograms above give a clearer picture of the diversity of responses, although the tendency is the same. (From Mushiake, Inase and Tanji, 1991, Figures 4, 5, 7 and 8; with permission.)

9.25). In the premotor area, such cells were less common. Instead, there was a preponderance of another type of cell which would fire preferentially in the short training period when changing over form IT to VT (or vice versa) movements. They were termed **transition-specific** neurones.

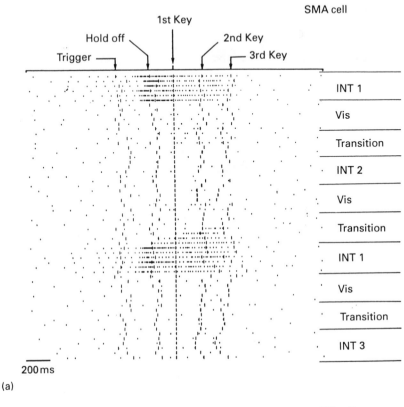

(a)

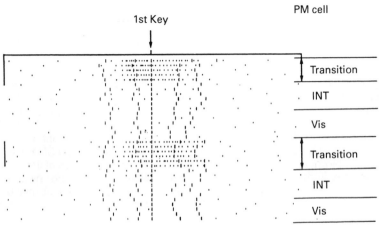

(b)

Finally, the authors found that some cells changed their firing rate in the waiting period before the 'go' signal. In the SMA such changes were seen more frequently before IT movements. Since the monkey knew what movement was to be made in the IT task, this activity may have been used in preparing for the forthcoming movement. Such preparation was not possible in the VT task since the sequences occurred randomly.

Set-related activity
Set-related or instruction-dependent activity has been referred to briefly in the section above, and also in relation to pyramidal tract neurone discharge in motor cortex. In a typical experiment, an instruction stimulus which indicates the direction of a forthcoming movement is given shortly before the 'go' signal. Three general classes of neuronal discharge can be seen in these tasks: movement-related, set-related and stimulus-related. Movement-related activity is linked to the onset of EMG activity; stimulus-related activity is linked to the appearance of the visual instruction stimulus. Set-related activity occurs throughout the delay period. All three types of activity can be seen in both primary and non-primary motor areas, although as a general rule, movement-related activity is predominant in primary motor cortex. Set-related, and movement-related activity are about equally common in the premotor cortex and SMA, whilst stimulus-related activity is predominant in prefrontal areas of cortex anterior to area 6. At present, it is thought that although there is a considerable mixture of cell types in all areas (usually regarded as indicating a substantial amount of parallel processing in these motor areas), there is a trend for stimulus-related processing to be concentrated in anterior aspects of frontal cortex, whilst movement-related activity is found posteriorly.

About half of cells in the superior portion of the lateral part of area 6, and the APA show set-related activity. A typical example is shown in Figure 9.26. Set-related activity begins 130–140 ms after the instruction

Figure 9.25 Examples of the activity of a sequence-specific neurone in the SMA (a), and a transition-specific neurone in the premotor cortex (b). The rasters show the time of cell discharge in individual trials aligned to the time of the first key press (middle of the traces). In (a) the monkey first performed six trials of an internally-guided sequence consisting of movements to top, bottom then right (INT 1). The neurone was very active in these trials. Then the monkey performed six visually-guided movements (Vis), and the cell was inactive. After a transition period the monkey performed six internally-guided movements with a different sequence (INT 2), and again the cell was inactive. Only when the monkey made the INT 1 sequence for a second time was the cell active. In (b), the premotor neurone was particularly active in the transition phase between visually-guided and internally-guided movements. (From Mushiake, Inase and Tanji, 1991, Figure 9; with permission.)

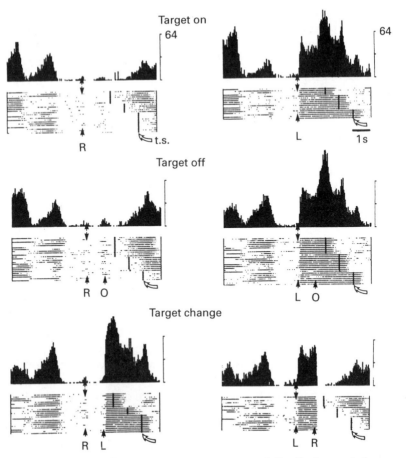

Figure 9.26 Set-related premotor cortex neurone activity. Each panel shows an average histogram of the response of the neurone, with a raster of the response in single trials below. The data is aligned to the onset of the instruction stimulus, which indicated a movement to either the right (R) or left (L). The monkey actually made the movement when the trigger stimulus (t.s.) appeared, up to 3 s later. The top two panels show activity when the instruction stimulus remained on (target on) during the whole of the interval before the trigger stimulus. The cell showed a maintained increase in firing with instructions to the left, and a decrease with instructions to the right. The same occurred if the instruction stimulus was removed before the trigger stimulus was given (middle panels: O indicates the time of removal of the instruction stimulus). The bottom panel shows how the instruction-related activity changed when the direction of the instruction stimulus was changed from right to left or vice versa. (From Wise and Mauritz, 1985, Figure 6; with permission.)

stimulus (cue), and is maintained whether or not the cue remains on or is turned off in the delay period. Many of these cells are directionally specific. For example, a cell may increase its discharge if the instruction stimulus indicates a forthcoming movement to the right, but will decrease its discharge if the movement is to be made to the left. If a rightwards instruction is given, and then replaced 0.5 s later by a leftwards instruction, the cell will initially be excited and then inhibited after the onset of the second instruction. These cells do not fire if no movement is required.

This set-related activity is thought to be involved in the preparation for the upcoming movement. It is not related to the instruction stimulus itself since it is possible to train monkeys to respond in opposite ways to the same signal, or to use different instruction stimuli to signal the same movement. Under these conditions, the set-related activity is dependent only on the movement that is to be made. Precisely what this activity is specifying is unclear at present. However, it seems likely that it is used to preset activity in other systems so that the correct movement is made when the 'go' signal is given. In particular, direction is an important parameter. Set-related activity in premotor cortex does not depend on the amplitude or force of the forthcoming movement. These values must be specified elsewhere (Kurata and Wise, 1988a, 1988b; Riehle and Requin, 1989; Hocherman and Wise, 1991).

Studies of population neuronal discharge in two-dimensional pointing tasks confirm the idea that activity in premotor cortex relates to the direction of a forthcoming movement. First, unlike cells in primary motor cortex, premotor activity is not influenced to any great extent by changing the load against which movements have to be made. Second, in a two-dimensional version of the instructed delay task, the population activity of set-related cells signals the direction of the forthcoming movement even when the spatial location of the cue is separated from that of the movement (Kalaska *et al.*, 1991).

Although set-related activity has been investigated most thoroughly in premotor cortex, it also occurs in the SMA, in a similar proportion of cells. However, there are subtle differences in some tasks which are consistent with the hypothesis that premotor neurones are particularly involved in movement guided by visual stimuli. Monkeys performed two tasks: in one a visual stimulus indicated the direction of the next movement; in the other, the instruction stimulus was irrelevant, and movements always had to be made in the same direction. After a block of 20 trials, the direction was reversed. In both tasks, the movements were made a self-timed interval after presentation of the instruction. In the delay period, set-related activity occurred equally often in premotor as in SMA whether the task was visually instructed or internally generated. However, some 30% of the premotor neurones with set-related activity showed **larger** changes in the visually-instructed than in the internally generated task (Kurata and Wise, 1988a, 1988b).

Bimanual movements

For simple tasks at a single joint, SMA and premotor activity, like that in primary motor cortex is predominantly linked to contralateral movement. However, in more complex tasks, the activity may be bilateral. This is seen, for example, if animals make complex, natural movements of the whole arm, and has been reported for the two-dimensional reaching task used in population vector studies on the premotor area. The vector points in the direction of movement whether the arm is ipsilateral or contralateral to the side of recording. Indeed, when monkeys are trained to perform single digit movements with either left, right or both hands, some cells in the premotor and SMA have activity which is **exclusively** related to only one of the three possible combinations. Thus, the neurones might fire to movement of the left side alone, but might be silent if both hands were used together. This type of activity is never seen in primary motor cortex. If a cell is related to activity of one hand, it will always fire when both hands are used (Tanji *et al.*, 1988).

Lesion studies have also suggested that the SMA may have an important role in bimanual coordination.

After removal of the SMA on one side of the brain, monkeys were left with difficulties in tasks requiring bimanual coordination (Figure 9.27). In retrieving food pellets from a well in a perspex board, monkeys had to push a currant down through the hole with a finger from one hand while they held the other hand cupped underneath to catch it when it

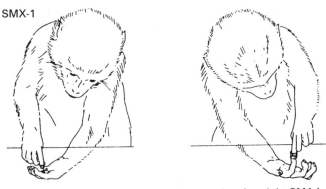

SMX-1

(a) Normal animal   (b) 5 months after right SMA lesion

Figure 9.27 Effect of unilateral SMA lesion on the performance of a complex bimanual task in the monkey. (a) Normal monkey pushing from above to collect a bait lodged in a hole in a Perspex sheet uses the index finger of the preferred hand while the non-preferred hand is cupped underneath, anticipating catch of the falling bait. (b) Reversal of hand position and identical behaviour of both hands five months after ablation of the right SMA. The non-preferred hand is now above the plate, and both hands are used in pushing with the index finger. Even with full visual control, the movements are not corrected. (From Brinkman, 1981, Figure 1; with permission.)

fell. After the lesion, there was a lack of coordination between the hands. Rather than sharing the task between each hand, the monkey might try to use both hands to push the currant out of the hole. It may be that, in the normal animal, the SMA on one side of the brain provides command signals only for the ipsilateral motor cortex. In the lesioned animals, the remaining SMA might influence precentral areas bilaterally and result in mirror movements of the hands (Brinkman, 1984).

Cingulate motor areas

Due to their relatively recent discovery, activity of cells in this area has not been studied extensively. In one experiment, monkeys had to press a keypad either in response to sensory (visual, auditory or tactile) stimuli, or in their own time. Movement-related cells were found in both anterior (CMAr) and posterior (CMAv and CMAd) cingulate regions. There were more neurones selectively active in the self-paced task in the anterior cingulate. In particular, these cells often started to fire a long time (0.5–2 s) before the movement began. Thus, anterior and posterior cingulate may prove to be interesting motor areas, and have different relations even to movements as simple as those used in this study (Shima *et al.*, 1991).

Posterior parietal cortex

Posterior parietal areas of cortex are not traditionally regarded as part of the motor system. However, since lesions in these areas produce characteristic deficits of movement, particularly in control of the hand and arm, they are mentioned briefly here.

The posterior parietal cortex is that part of the parietal lobe which is posterior to the primary sensory area. It is divided into a superior and inferior region by the intraparietal sulcus. In the monkey, the caudal part of the superior parietal lobule is Brodmann's area 5; area 7 occupies the entire inferior lobule. Both areas have been subdivided further into anterior (5a and 7b) and posterior (5b and 7a) regions. The physiological differences between areas 5a and 5b have not been studied in detail, and hence both portions are discussed together. In contrast, areas 7a and 7b are quite distinct. Additionally, those parts of area 7 which lie in the intraparietal sulcus (7ip), and those regions on the medial aspect of the hemisphere (7m) are usually treated separately.

The anterior portions of area 7 (7b) and the majority of area 5 are intimately connected to the sensory and motor systems. They have reciprocal connections between themselves and with premotor and SMA areas of cortex. Area 5 receives strong corticocortical inputs from primary sensory area 2, whereas area 7b receives from the second sensory area (SII). These areas are said to be involved in 'somatomotor' function. The remaining regions of area 7 are much more strongly connected to visually-related areas of the cortex and frontal eye fields. However, there is also a dense projection from area 7 to the pons, and thence to the cerebellum, providing a pathway whereby visual input might influence movement ('visuomotor' function). Stein (1989) has suggested that the combined somatosensory and motor input to area 5 allows this part of the posterior parietal cortex to build up a single picture of the relative position of all parts of the body (and the objects that they contact) with respect to one another. In other words, it describes limb coordinates in egocentric space. Area 7 may perform a similar function for objects in the space which surrounds us.

The separation of posterior parietal cortex into areas involved in somatomotor versus visuomotor function is borne out by single cell recordings. Cells in area 5 respond well to somatosensory input from both superficial and deep structures. They have variable responses with large receptive fields suggesting that the region is involved in 'high level' analysis of sensory input. The discharge of many neurones is enhanced if stimulation is the result of active movement made by the animal. In addition to these somatosensory neurones, Mountcastle and colleagues found a different type of activity in 11% of the cells in the arm region of area 5. This group apparently was not activated by any peripheral sensory inputs, although they discharged in association with active movements made by the animal. These neurones had the remarkable property that their discharge was not tightly linked to a particular movement such as arm extension, but was seen only if given movement occurred in a particular behavioural context. A typical example was the discharge recorded when the animal reached out its arm to obtain food. No discharge was recorded if the same movement was performed spontaneously. A motor role for some of the cells in area 5 is supported by the fact that 14% of cells discharge prior to movement even after deafferentation.

Two hypotheses were put forward to explain the function of these cells. One was that they were 'command' neurones working at a high level in the command process leading to movement. The other was the possibility that the neurones were receiving a corollary discharge from precentral areas. The first possibility might be related to the neglect of contralateral limbs in lesioned animals or man. The second might be related to the phenomenon of astereognosis. The situation is still not clear. Although one third of cells in area 5 discharge before movement, their lead time is still 30 ms or so shorter than the mean for neurones in primary motor cortex, and tends to favour the idea that they receive an efferent copy of the command to move.

Population studies of cell activity during two-dimensional reaching movements have been made in area 5. They show, as in motor cortex, that the population discharge can reflect the direction of movement made by the animal. However, unlike the discharge of motor cortex cells, the discharge of cells in area 5 is not much affected if the manipulandum is loaded in one or other direction (see above). Thus, their activity seems more likely to be coding the abstract direction of movement rather than the forces needed to achieve the movement.

Some cells in area 7b have properties similar to those of area 5. The majority of recordings, though, have been made from other subregions of area 7. In these areas, cells have complex visual and visuomotor activity. Because of their input from both visual and oculomotor centres, some neurones may be able to code the location of visual targets, not in relation to their retinal position, but rather in relation to the position of the head. Such neurones could be a first stage in the representation of visual targets in a body-centred frame of reference. Many area 7 neurones display another property: enhanced activity during directed attention. For example, most area 7 cells have visual receptive fields, but their responses to a stimulus are much enhanced if the stimulus is behaviourally relevant to the animal. Typical experiments might use the stimulus as targets for eye movement, or to signal a forthcoming reward.

In monkeys, combined lesions of both areas 5 and 7 lead to reluctance to use the contralateral arm and neglect of the opposite side of the body. There is misreaching and a failure of the visual and tactile placing reactions. Lesions confined to area 7 produce deficits of visually-guided movements. Reaching for objects in space with the contralateral hand is particularly affected, and there are decreases in the accuracy of eye movements and visual neglect. Lesions of area 5 produce astereognosis (an inability to identify an object, such as a key, through active manipulation with the hand when the eyes are closed), impaired weight discrimination and uncoordinated palpatory movements of the contralateral arm. These deficits seem appropriate for the postulated visuomotor operations performed by area 7, and the somatomotor functions performed by area 5.

In humans, there is some debate over which areas of the cortex are analogous to areas 5 and 7 in the monkey. Somewhat confusingly it appears that human areas 5 and 7 correspond to monkey 5a and 5b (respectively), and that area 39 (the angular gyrus) and area 40 (the supramarginal gyrus) correspond to areas 7a and 7b. Lesions of the posterior parietal cortex in humans produce a fascinating variety of clinical symptoms. As hypothesized by Stein (1989), these can conveniently be divided into those involving disturbance of somatomotor function due to lesions of areas 5 and 7 (equivalent to monkey areas 5a and 5b), such as astereognosis and asomatagnosia (denial or neglect of the contralateral side of the body), and those involving visuomotor functions due to lesions of areas 39 and 40 (equivalent to monkey areas 7a and 7b) such as disorientation and misreaching (optic ataxia). Additionally, in humans there is a clear distinction between symptoms of right and left hemisphere damage.

Right-sided lesions tend to affect the spatial attributes described above, whereas left-sided lesions affect similar abstract temporal, grammatical and logical relations which are described in speech (reviewed by Stein, 1989; Johnson, 1991).

9.7 *Lesions of descending pathways in humans and monkeys*

Capsular lesions

The most common cause of damage to the cortical motor system occurs in capsular hemiplegia. Interruption of cortical efferent fibres in the internal capsule gives rise to a variety of symptoms, the severity of which depends upon the extent and position of damage. For many years, the symptoms were considered to be due to destruction of pyramidal tract fibres, and the symptoms referred to as pyramidal tract symptoms. However, the cortical outputs to red nucleus, pontine nuclei and brainstem tegmental areas are also interrupted and contribute to the symptoms.

Hemiplegia produced by a fairly complete motor stroke in man is characterized by the following sequence of events. Immediately after the stroke there is a period of complete paralysis. The contralateral limbs are flaccid and no reflexes can be obtained. This is the initial or shock stage. The first reflexes reappear after 10–12 hours. The plantar reflex, elicited by firm stroking of the lateral border of the sole of the foot, usually occurs before the others. However, it is now inverted (Babinski's sign), and the big toe dorsiflexes rather than plantarflexes, as is normal (see Chapter 6). A little later, tendon jerks reappear and, over a week or so, gradually become extremely brisk. The patient becomes hyper-reflexive and clonus may appear in some muscle groups.

At the same time, there is a gradual increase in muscle tone which is unevenly distributed. It affects preferentially the extensors of the legs and the flexors of the arms. Such spasticity is accompanied by a clasp-knife reflex (see Chapter 6). Two reflexes do not reappear: the abdominal and cremasteric reflexes. The former consists of contraction of the abdominal muscles during stroking of the abdomen. The latter consists of retraction of the testicle when the inner surface of the ipsilateral thigh is stroked.

In most cases, there is a gradual recovery of voluntary movement. Recovery begins with crude synergic movements of all the muscles in a limb. In the leg these are the extensor and flexor synergies: extension of the knee, for example, is usually accompanied by abduction of the hip and dorsiflexion of the ankle and toes. Flexion of the knee is accompanied by flexion at the hip and plantar flexion of the ankle and toes. In the arm comparable flexor and extensor synergies occur. Independent movements of the digits are the last to appear, if at all.

When maximal strength is tested during a voluntary contraction, muscles contralateral to the lesion are weak, especially distal muscles of the arm and leg. Proximal muscles are less affected, and there may even be a mild decrease in strength ipsilateral to the lesion. This distribution of weakness is consistent with the known large projection from the cortex to distal muscles and the small ipsilateral projection to proximal and axial muscles.

Much of the recovery of function seen after stroke is thought to be due to recovery of transmission in some of the axons which traverse the site of damage. Conduction in these axons may have been blocked by the pressure caused by initial oedema. As the oedema recedes, then the axons recover. However, functional improvement often continues for several months. Other factors must operate over this time scale. One possibility is that some motor functions are taken over by the undamaged ipsilateral hemisphere. Ipsilateral corticospinal pathways account for the relative immunity of face, tongue and jaw movements after stroke, and similar pathways may aid recovery in axial and proximal body muscles. Whether distal movements ever receive a significant contribution from activity in the ipsilateral hemisphere is unknown. One piece of evidence suggests this is possible. Blood flow studies show ipsilateral cortical activity in recovered hemiplegic patients when they move their affected hand (Twitchell, 1951; Colebatch and Gandevia, 1989; Adams, Gandevia and Skuse; Chollet *et al.*, 1991).

The effect of pyramidal tract lesions

Pure lesions of the pyramidal tract in humans are rare since, except in the medullary pyramids, the fibres of the tract are contaminated throughout their length by other systems. Nevertheless, a few cases with lesions in this area have been described. All the patients were described as being weak on the contralateral side, but muscle tone was affected by varying amounts, ranging from flaccid to severe spastic paralysis. Differences between patients were probably caused by varying degrees of involvement of the medial reticular formation and lemniscal systems.

In the late 1950s and 1960s, neurosurgeons experimented with pedunculotomies to relieve abnormal movements, due, for example, to chorea (Chapter 11). They sectioned the middle third of the peduncle, through which it was believed that the majority of pyramidal tract fibres ran. In one patient at autopsy, three and a half years later, it was found that 83% of fibres within the pyramids had degenerated. In contrast to the results above, this patient was said to have retained 'an almost normal volitional control' of the contralateral limbs although the involuntary movements had been abolished. There was no spasticity (Bucy, Keplinger and Sequiera, 1964).

From these reports in man, it would seem as though lesions of the pyramidal tract have very little effect on either movement or tone. More controlled experiments have been conducted on monkeys and chimpanzees, but only recently with trained animals and with quantitative techniques. As in humans, there is surprisingly little gross deficit. Immediately after bilateral pyramidotomy, monkeys can sit, run and climb in their cage with their head erect. However, despite this appearance of normality in such gross movements, careful observation reveals substantial deficits in the control of fine independent finger movement. For example, although the monkeys can use their hands in a power grip to hold onto and swing from the bars of their cages, they cannot pick up food with their hands from the floor. The critical difference in these tasks is that in picking up food, the individual fingers have to move independently, rather than in concert, as in the power grip. Recovery occurs with time, but the animals are left with a deficiency in performing individual movements of the fingers. The best test is to have the animals retrieve food pellets from a shallow depression in a food board. Normally, this is done using the thumb and forefinger together in a 'precision grip', while at the same time, the other fingers are held flexed out of the way. Pyramidotomized monkeys can no longer perform the task in this way. Instead, they use only a 'hooking' movement of all the fingers together, which usually proves inadequate to retrieve the pellet.

Residual muscle weakness is slight and in recent experiments is said to be zero. The only deficit is that the speed of reaction in these monkeys is slower than normal. This is due to a delay and slow rise in the onset of EMG activity in the muscles. The muscle tone of monkeys with sectioned pyramids was never found to be enhanced. In fact, there is a slight hypotonus caused by decreased stretch reflexes. This is probably due to a diminished gamma drive to muscle spindles as a result of interruption of corticofusimotor fibres in the pyramidal tract.

These experiments on animals suggest that few of the symptoms of hemiplegia are caused specifically by damage to corticospinal fibres. The conclusion is that the corticospinal tract superimposes speed and fractionation upon the movements produced by other descending systems. It fits well with the anatomical findings which show an increasing number of corticospinal tract terminals in the spinal motor nuclei, and a gradual increase in importance of the corticomotoneurone connection in monkey, chimpanzee and man. It is speculated that these direct connections from brain to α-motoneurones provide the basis for independent control of finger muscles in higher primates.

A thorny problem has been, and still is, the role of the pyramidal tract in the production of Babinski's sign. Pyramidotomy in chimpanzees produces upgoing big toes after stroking of the plantar surface of the foot, and abolishes the abdominal and cremasteric reflexes. There are no such clear-cut data in man. Reports on the correlation between the presence of pyramidal tract damage, confirmed at autopsy, and the sign of Babinski are contradictory. The most recent suggestion is that the upgoing plantar response probably occurs after damage to the monosynaptic component of the corticospinal projection to the extensor muscles of the toe (Lawrence and Kuypers, 1968; for details about classical studies of pyramidotomy in monkeys).

Effects of lesions of other descending tracts

Lawrence and Kuypers (1968) provided the most complete description of the deficits following transection of descending pathways in the Rhesus monkey. There are no comparable data available in humans. As described in Chapter 7, Lawrence and Kuypers divided the descending pathways into pyramidal tract and brainstem pathways group A and B. Group A comprises the interstitiospinal, tectospinal, vestibulospinal and reticulospinal tracts; group B is mainly the rubrospinal tract. In order to evaluate the contribution of these pathways to movement control, a pyramidotomy was first performed. This removed any pyramidal tract contribution to control of proximal and distal muscles.

Bilateral transection of the brainstem pathways in the upper medullary medial tegmental field (the group A pathways), together with bilateral pyramidotomy produced animals with severe postural deficits. Even after some weeks recovery, animals tended to slump forward when sitting and had great difficulties in avoiding obstacles when walking. There was severe impairment in righting reactions. In contrast to the immobility of axial structures, the hands and distal parts of the limbs were very active. Manipulative ability returned to that seen in pyramidotomized animals; the monkeys could retrieve food from a board with their hands and bring it to their mouth. However, the animals would not reach out to grasp food. They would wait until food was in reach, and bring the food to the mouth mainly by flexion of the elbow.

Animals with pyramidotomy and lesions of the group B descending brainstem pathways looked very different. Their posture appeared normal. They sat up in their cages, walked and climbed the bars. Limbs would be used relatively normally in gross movements of climbing and walking. However, when seated, they held their arms limply from the shoulder. In reaching for food the hand was brought to target by circumduction at the shoulder, with little movement at the elbow. Finger movement was poor, and food usually was obtained by simultaneous flexion of all fingers as part of a total arm movement.

The conclusion from these experiments is that group A pathways are used chiefly in control of gross postural movements. Arm movements only occur as part of a postural synergy. These tracts travel in the ventromedial parts of the spinal cord and terminate on motoneurones of proximal and axial muscles and on long propriospinal neurones which interconnect many segments of cord. The group B pathways terminate on motoneurones innervating distal muscles and on short propriospinal interneurones which interconnect only a small number of spinal segments. These group B pathways appear to give the animals the capability to make arm movements, particularly at the elbow and wrist, which are independent of gross postural synergies. In addition to these methods of control, the corticospinal tract superimposes the ability to fractionate movements even further. This is seen particularly well in finger movements.

These ideas were tested further (without lesioning the pathways themselves) by investigating the control of ipsilateral and contralateral movements of the arm in monkeys with complete commissurotomies. As shown in Figure 9.28, the fibres from one hemisphere project mainly contralaterally to the parts of the spinal grey matter containing motoneurones of the distal extremities. The connections to axial and proximal motoneurones in the ventrolateral grey are mainly bilateral.

Split brain monkeys had one eye covered, and were presented with an object in the nasal hemifield (i.e. the visual input went to the hemisphere ipsilateral to the uncovered eye: the 'seeing' hemisphere). If they

Figure 9.28 (a) The descending connections from the cerebral cortex and the brainstem to the spinal cord in the rhesus monkey. Note that one half of the brain is connected both directly and indirectly to the contralateral intermediate zone and the spinal motoneurones innervating the distal muscles. The connection to the motoneurones in the ventromedial zone which innervate axial and proximal muscles is mainly indirect and bilateral. (b) Drawings from a film showing a split brain monkey retrieving a food morsel from between the examiner's fingers with the hand contralateral (left) and ipsilateral (right) to the 'seeing' hemisphere. The contralateral hand assumes the precision grip posture during the final stages of reaching (top picture), and then seizes the pellet with the index finger and thumb (bottom). The ipsilateral hand does not assume the precision grip at the end of reach. It only assumes this posture after having made contact with the pellet and the examiner's fingers. (From Brinkman and Kuypers, 1973, Figures 1 and 3; with permission.)

used the arm contralateral to the 'seeing' hemisphere, then reaching was accurate, and the fingers and thumb were preformed into a precision grip before the target was reached. If the arm used was ipsilateral to the 'seeing' hemisphere, then although reaching was accurate, the hand was not preformed into the precision grip. Only when the animal touched

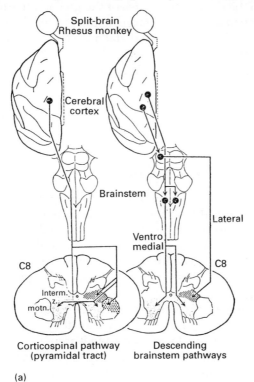

(a)

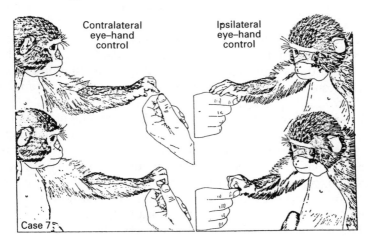

(b)

the object, and obtained somatosensory input, were independent finger movements made (Figure 9.28). The interpretation is that the 'seeing' hemisphere was able to control proximal and axial muscles on both sides via bilateral pathways whereas control of distal muscles was limited to the contralateral arm. Thus the 'seeing' hemisphere had no connection to the distal extremities of the ipsilateral arm and although it could be moved towards the target using proximal and axial muscles, manipulation of the object using distal muscles was only possible when somatosensory input became available and the 'non-seeing' hemisphere could control the hand via crossed pathways.

9.8 *Physiology and pathophysiology of human motor areas*

Stimulation of motor areas during surgery

Electrical stimulation of exposed human cortex is routinely performed during neurosurgical operations which involve removal of diseased areas of cortex. Surface stimulation with 50–60 Hz trains of stimuli is used in order to map out the important areas of motor, sensory, and speech-related cortex, and limit any excision as far as possible to other areas. The classic studies of Penfield and colleagues on the primary motor area have been discussed above. Stimulation in this region produces simple twitches of contralateral body parts and never evokes any complex movements. Patients also report (they are usually conscious, although anaesthetized locally and well-premedicated, during these procedures so that they can identify any sensations evoked) that stimulation prevents them moving voluntarily. The appropriate part of the body may be paralysed while the stimulus is applied, and is no longer available for use by voluntary effort.

Stimulation of the SMA most commonly produces movements of the contralateral arm. In their original studies, Penfield and Welch (1951) concluded that there was no somatotopic representation of the body. More recent work in which stimuli have been applied through arrays of electrodes implanted to monitor EEG activity prior to surgery for epilepsy, have suggested a somatotopic arrangement similar to that reported in the monkey by Woolsey. That is, a rostrocaudal map of face and eyes, then arm and leg. The character of the movement is more complex than is ever seen after stimulation of primary motor cortex. Tonic, sustained postures are characteristic, often involving deviation of the head and eyes to the contralateral side and tonic raising of the contralateral arm. Speech arrest or vocalization may occur, the latter said to be a quite characteristic vowel cry, the 'cri de l'aire motrice supplémentaire'. Epileptic seizures which begin in the SMA involve similar reactions: the arm is elevated and the head and eyes turned in the same direction.

Foerster reported that electrical stimulation of the lateral area 6, in front of primary motor cortex yielded movements very like those seen after SMA stimulation: turning movements of the head, eyes and trunk to the opposite side and abduction and elevation of the arm, elbow flexion, and pronation of the hand. These effects had a much higher threshold than the movements obtained after stimulation of primary motor cortex, and could be seen in patients where area 4 was destroyed on the same side. Again epileptic seizures which originate in this area have a strikingly similar onset.

Transcranial stimulation of the motor cortex

In recent years it has become possible to stimulate the motor cortex of normal, conscious subjects through the intact scalp. The first method to be devised used a single electrical pulse applied through two electrodes fixed to the scalp over the motor strip. Such stimuli produce simple, short-latency, contralateral muscle twitches which are particularly prominent in the muscles of the hand and forearm. They are most likely caused by activiation of the fast-conducting, large-diameter pyramidal tract output from the primary motor cortex. Because the responses occur with remarkably little temporal 'jitter', they are thought to involve the monosynaptic, corticomotoneuronal pathway which is particularly well-developed in man. If stimuli are given during voluntary contraction of a target muscle, the responses are much larger than when subjects are relaxed. This is caused mainly by increased excitability of spinal motoneurones, which are more readily discharged by a given descending volley than in the relaxed state. During voluntary facilitation, short-latency responses can be obtained in virtually every muscle of the body, including the diaphragm, oesophagus, and external anal sphincter. This suggests that the direct corticospinal projection in man is much wider than that in the monkey, where short-latency monosynaptic effects are limited to arm and hand muscles.

Transcranial electrical stimulation employs a short (100 µs or so), but relatively strong (0.5 A) stimulus. The high electrical resistance of the skull and scalp means that only a small fraction of the applied current penetrates into the brain; most travels along the surface where it causes local pain and contraction of scalp muscles. A less uncomfortable method of stimulation uses a magnetic method. A large electrical capacitance is discharged through an insulated coil of wire placed on the scalp. A very large current (several thousand amps) flows transiently (100 µs or so) through the coil and creates a magnetic field perpendicular

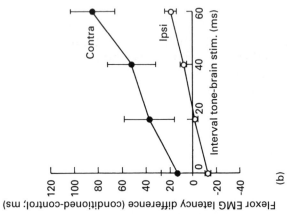

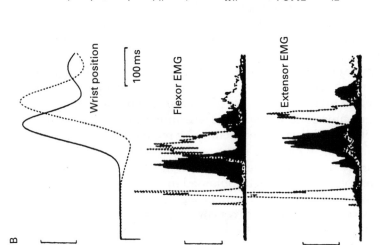

(a)

(b)

to the coil which readily penetrates into the brain. Because the magnetic field changes very rapidly it induces electrical (eddy) currents to flow in the brain. The magnetic field effectively acts as a carrier to make electric current flow in the brain. The method is known as magnetic stimulation, even though it is conventional electric current which provides the stimulus to cortical neurones. Unlike electrical transcranial stimulation, magnetic stimulation is virtually painless, and is now the commonest form of transcranial stimulation.

As reported during neurosurgical procedures, transcranial stimulation not only evokes movement, but can also interfere with voluntary activity. For example, a single stimulus over the motor area, given in the interval between a reaction stimulus and a movement can delay the onset of movement by 100 ms or so (see Figure 9.29). The delay with a single stimulus is not as noticeable as the voluntary paralysis seen with long trains of stimuli to the exposed cortex during surgery. However, it is interesting since with such short delays, the movement which eventually occurs is the same as that seen without stimulation. The implication is that the commands for the movement can be stored during the period of delay so that the movement appears intact. If much smaller stimuli are used, which themselves are subthreshold for producing any movement, then the opposite happens. The reaction time is speeded up. It is as if a small stimulus excites the motor system additively to the voluntary command.

The delaying effect of large stimuli over primary motor cortex seems to apply to all types of movement. In contrast, when large stimuli are applied over SMA, then the disruptive effect is best seen on sequential movements (touching each finger with the thumb in a particular order), consistent with a role of the SMA in complex, internally-guided tasks.

Neither magnetic nor electrical transcranial stimulation can activate

Figure 9.29 Inhibition of cortical function by magnetic transcranial brain stimulation. Subjects were given an auditory reaction tone and required to flex the wrist as rapidly as possible afterwards. (a) Wrist position, and flexor and extensor EMG records show the normal ballistic movement pattern as the filled regions. Superimposed is the response when a magnetic stimulus was given unexpectedly just before the normal time of onset of the movement. In the EMG, the dotted line shows the stimulus artefact, and the direct EMG response to the cortical stimulus. This is followed by a pause, and then the triphasic pattern reappears. All three components of the EMG response (in the agonist, antagonist, and agonist muscles) are delayed by about 50 ms, although their form is unchanged. Graph (b) shows how the delay in onset of movement depends on the timing of the magnetic shock with respect to the auditory reaction tone. This graph illustrates data from an experiment where subjects made a bilateral wrist flexion movement and the stimulus was given only over one side of the head. The major delay was on the side contralateral to the stimulus, although a smaller effect can be observed on the ipsilateral side. (From Day *et al.*, 1989, Figures 1 and 7; with permission.)

motor cortex as focally as an electrode placed on the cortex. Nevertheless, using special designs of stimulator, it is possible to produce maps
of primary motor cortical localization which not only distinguish between the major areas devoted to the leg, trunk, arm and face, but can
also differentiate to a limited extent between the best point to activate,
say, forearm versus hand or upper arm muscles. Changes in the transcranial motor map can be produced by damage to either peripheral or
central structures. Figure 9.30 shows an example of the area on the scalp
from which responses in the left and right deltoid muscles could be
obtained in a patient who seven years earlier had suffered a traumatic
amputation of the right arm below the shoulder. Responses in the right
deltoid after magnetic stimulation of motor cortex were much larger

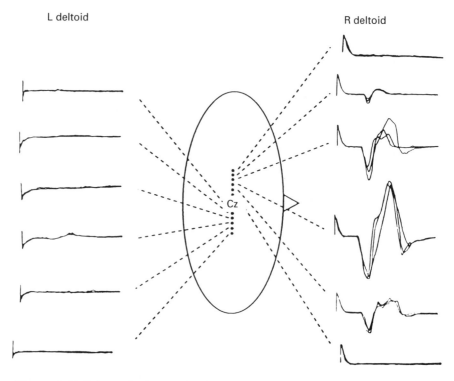

Figure 9.30 Schematic diagram of the top of the head of a patient who had an
amputation of the right arm above the elbow 4 years before. EMG responses were
evoked by stimulation of different scalp positions 1 cm apart along the coronal axis
with equal stimulus intensities. MEPs on the left side of the scalp were recorded
from the right deltoid and those on the right side of the scalp were recorded from
the left deltoid. Note that stimulation of four positions evoked responses from the
deltoid ipsilateral to the stump (right), while stimulation of only one position evoked
responses from the deltoid on the intact side (left). Calibration bar = 5 ms and 100
μV. (From Cohen *et al.*, 1991, Figure 2; with permission.)

and could be obtained from a much wider area than those on the normal side of the body. Similar changes have been observed after traumatic spinal cord injury, and even after transient peripheral anaesthetic block in normal subjects. The results are therefore similar to those reported in rats (see above) after peripheral injury, and are consistent with the idea that the organization of the motor cortex can be changed by alterations in feedback caused either by injury or anaesthesia. However, in humans it is difficult to exclude the possibility that reorganization may have occurred in some subcortical structures such as spinal cord or brainstem. Direct proof that changes in the cortical motor map are due to plasticity at a cortical level is lacking.

In contrast to the plasticity seen after peripheral injury, changes in the cortical motor map after injury to central structures in adults (motor cortex, internal capsule) are very uncommon. Nevertheless, major reorganization can be seen if the damage occurs in infancy or early childhood. Stimulation over the lateral part of the intact hemisphere in patients who underwent hemispherectomy for the relief of intractable epilepsy in childhood produces clear, short-latency movements of both arms. Since bilateral movements are never seen in normal subjects, this indicates that a major reorganization of the corticospinal projection has occurred. Three explanations are possible: (i) the uncrossed portion of the pyramidal tract, which is small and usually distributed to proximal muscles has become much more powerful in these children; (ii) the crossed pyramidal tract projection has sprouted new terminals at a spinal level which have crossed to innervate the side ipsilateral to the intact hemisphere; (iii) in normal development, the crossed pyramidal tract projection may innervate both sides of the spinal cord, but may regress under normal circumstances to supply only the contralateral side. Such **exuberant** projections are common in many parts of the CNS during development. If they were present in the pyramidal tract projection then a failure to regress after hemispherectomy may account for the bilateral corticospinal effects seen in these patients. Bilateral projections have not been reported in most adults after late hemispherectomy or massive unilateral strokes. (Rothwell *et al.*, 1991, reviews details of cortical stimulation in humans.)

Recording cortical activity

EEG activity associated with movement can be recorded readily from conventional electrodes placed on the scalp. However, because movement invariably produces afferent feedback, it is impossible to distinguish between cortical activity related to movement, and cortical activity evoked by sensory input. Because of this, investigations of movement-related activity are confined to the period before the onset

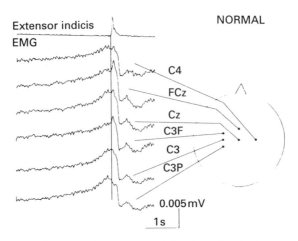

Figure 9.31 Premovement potentials (bereitschaftspotential) recorded simultaneously from several scalp locations during a self-paced voluntary extension movement of the right index finger. The subject was instructed to move in his own time every 5–10 s, and 128 trials were averaged. The EMG responsible for the movement can be seen in the first trace. About 1.5 s before onset of EMG, a slowly rising (negative polarity is upwards) wave can be seen in the EEG records, culminating in a peak just after the EMG has begun. This wave of activity is known as the premovement potential. (Courtesy of Dr J.P.R. Dick.)

of movement. Additionally, in order to avoid any contribution to EEG activity from a sensory triggering signal, the movements are made by subjects in their own time. Under such conditions, a movement-related potential, beginning as much as 1.5 s before the movement can be recorded from wide areas of scalp (Figure 9.31).

This potential is divided into three phases: the initial negative slope (or NS1) from the onset to about 650 ms prior to EMG activity, the second negative slope (NS2), from 650 ms to about 50 ms before EMG onset, and the movement potential (MP), which occurs around the time of the EMG burst. Prior to a unilateral arm movement, the NS1 is symmetrically distributed on both sides of the head, and has a maximum amplitude at the vertex. The NS2 is also bilateral, but is larger on the side contralateral to movement. The MP (which is only a small potential) is strictly contralateral, and localized over the lateral part of the sensorimotor cortex.

It was once believed that, because it was maximum at the vertex, the NS1 reflected preparatory activity in SMA prior to movement, whilst the NS2 and MP indicated progressively greater activation of the primary motor cortex. Recently it has become common to implant arrays of recording electrodes on the surface of the brain in patients with epilepsy prior to surgery. This allows for a much more detailed localization of the focus of their seizures than can be obtained from scalp recordings.

In some of these patients, arrays have been implanted over lateral or medial aspects of the brain and recordings have been made directly from the surface of primary motor cortex or SMA during voluntary movement. They indicate that although the SMA is one source of the NS1, bilateral activity in the primary motor area also occurs at the same time (see also blood flow studies below). This activity summates at the vertex, and hence the NS1 appears larger there than on either hemisphere alone. During the NS2 and MP, primary motor cortex activity becomes lateralized; activity in SMA also continues.

Movement-related potentials are larger prior to complex than simple movements (e.g. sequential wrist and elbow movement versus either movement alone), presumably reflecting increased preparatory activity before more difficult tasks. Movement-related potentials have also been investigated in disease. The NS1 component is decreased in Parkinson's disease, perhaps as a result of a decreased contribution from SMA activity in this condition (see Chapter 11). The NS2 component is absent in certain types of cerebellar disease and has led to the idea that cerebellar activation of motor cortex may be an important contributor to this part of the movement-related potential (Ikeda and Shibasaki, 1992).

Perhaps the most interesting fact about the premovement potentials is that they appear to start before one is aware of taking the decision to move. In a fascinating experiment, Libet *et al.* (1983) showed that if subjects estimated the time at which they made the decision to move, this was always some 0.5 s later than the onset of the premovement potential. Estimation of the time of other types of past event such as sensory stimulation was, in contrast, quite accurate. The reason for the interest in this fact is purely philosophical. It means that consciousness must be secondary to activity of cerebral neurones. The idea that humans decide to move and then somehow set in train the physiological processes to accomplish the task is quite wrong. The decision to move is made first: only later do we become consciously aware of it (Libet *et al.*, 1983).

Epilepsy and myoclonus

Jacksonian seizures

The movements produced by abnormal cell activity in supplementary and premotor areas during the onset of epileptic seizures are described above. In contrast with such complex synergies involving the activation of many muscles in different parts of the body, seizures with onset in primary motor cortex have much simpler characteristics. They were first described by Hughlings Jackson several years before the discovery of the electrical excitability of the cerebral cortex. Such fits, now known as Jacksonian seizures, begin with focal twitching of a small group of muscles in one part of the body. Usually, this is in the hands or the feet or the corner of the mouth. From this point, the twitching spreads proximally to involve more and more muscle groups, over the space of several seconds. For example, it may progress from one hand to the forearm, elbow, shoulder on the same side; finally a full tonic–clonic convulsion of the whole body may ensue.

At cortical level, the seizure begins as focal activity in one part of the motor cortex, and because of their relatively large representation this frequently involves the hand, foot or mouth. Activity spreads to neighbouring areas of cortex, to recruit neurones projecting to neighbouring muscle groups, according to the somatotopic cortical representation. A full grand mal convulsion follows when (and if) the seizure spreads beyond the motor strip to the rest of the cerebral cortex.

Cortical myoclonus

A less devastating form of increased cortical excitability can be seen in patients with cortical myoclonus. Myoclonus refers to muscular twitching in any part of the body and may involve one or many muscles of the body. Hypnic jerks that are experienced by many people on falling off to sleep are a form of physiological myoclonus.

In contrast to these whole body jerks, cortical myoclonus usually involves muscles in only one part of the body. The jerks are brief, and look very like the muscle twitch produced by a single electrical stimulus to a peripheral nerve. The jerks may occur at rest, during voluntary activation of the affected part, or may be provoked by sensory stimuli, particularly from the affected body part. Jerks which occur at rest and are repetitive, regular and persist continuously for many days, disappearing only during sleep are known as **epilepsia partialis continua**. In some patients, epilepsia partialis continua is associated with Jacksonian seizures if the activity in the cortical focus spreads to other areas.

The reflex variety of cortical myoclonus has been studied in some detail. It is the result of hyperactivity in the cortical reflex loops from peripheral skin or muscle receptors to motor cortex and back down to muscle (Figure 9.32). This can be demonstrated in many patients quite easily by recording the muscle and brain activity which is produced by the somatosensory stimuli that evoke myoclonic jerks. Brain activity in humans cannot be recorded via intracortical microelectrodes, as in a monkey. Instead, distant electrodes have to be attached to the surface of the scalp over the area of brain of interest. Electrodes placed over the sensorimotor areas normally record small waves of activity with onset

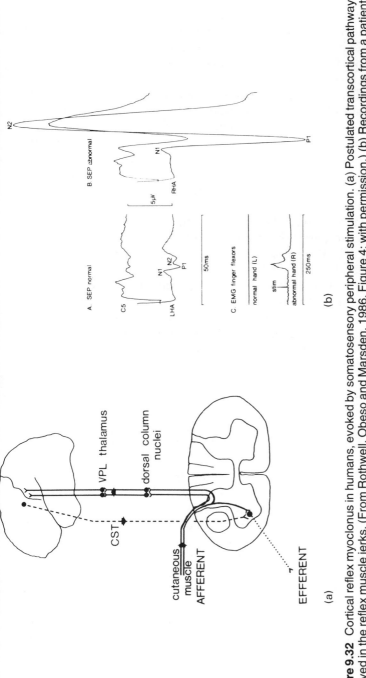

Figure 9.32 Cortical reflex myoclonus in humans, evoked by somatosensory peripheral stimulation. (a) Postulated transcortical pathway involved in the reflex muscle jerks. (From Rothwell, Obeso and Marsden, 1986, Figure 4; with permission.) (b) Recordings from a patient with reflex myoclonus of the right hand. A and B: evoked potentials to electrical stimulation of the median nerve at the wrist recorded from electrodes over the cervical spinal cord (C5) and the somatosensory hand area of cerebral cortex (LHA, RHA). Stimulation of the normal (right) hand gives the usual complex of evoked responses at the spinal cord and scalp (N1, P1, N2; negative deflection upwards). Stimulation of the abnormal (left) hand gives normal early responses over the cervical cord and a normal N1 response at the scalp. However, the later P1, N2 responses are greatly enlarged. (The enlarged later waves of the C5 lead are due to pick-up of remote activity from the scalp, and not to activity in the spinal cord.) C: EMG recordings from the first dorsal interosseous muscle on a longer time sweep. Stimulation of the normal hand produces no EMG response on that side. Stimulation of the abnormal hand generates a myoclonic jerk, seen in the EMG as increased activity with a latency of about 50 ms after stimulation. (From Rothwell, Obeso and Marsden, 1984, Figure 1; with permission.)

18–20 ms after electrical stimulation of the median nerve at the wrist. These potentials are known as somatosensory evoked potentials (SEPs).

Figure 9.32 shows results from a patient who had reflex myoclonus of the left hand. Electrical stimulation of the median nerve at the wrist on the right (normal) side evoked normal scalp potentials and no myoclonus. Electrical stimulation of the left side produced myoclonic jerks which had a latency of about 50 ms in the first dorsal interosseous muscle. In addition, it produced giant evoked potentials over the contralateral central region of scalp. The initial part (N1) of this potential was normal, but the later components (from P1 onwards, at 25 ms) were greatly enlarged. The hypothesis is that in this patient, a normal afferent volley reached somatosensory cortex, but that later stages in processing were abnormal. Increased excitability produced an abnormal SEP which may have driven cells in motor cortex to discharge and produce a visible myoclonic jerk. The latency between onset of the enlarged SEP and the muscle EMG response is the same as the corticospinal conduction time to hand muscles in humans.

Patients with spontaneous jerks can be studied in the same way. In this case, spike activity over contralateral pre/postcentral areas can be recorded some 20 ms before spontaneous muscle jerks in the arm. In such cases, it may be that an excitatory focus is discharging spontaneously in the cortex. (Obeso, Rothwell and Marsden, 1985, give further details of cortical myoclonus.)

Blood flow

When an area of brain is active, the flow of blood increases locally to supply the extra energy that is needed. The major energy-requiring process is synaptic transmission, not the firing of action potentials, so that changes in flow actually occur in the grey, not the white matter. It should always be remembered that flow will increase whether the synaptic activity is excitatory or inhibitory. Three techniques are available to monitor blood flow in the brain: positron emission tomography (PET), single photon emission computed tomography (SPECT), and magnetic resonance imaging (MRI). In the first two, radioactive tracers are injected into the blood or, if they are gases, they can be inhaled. Tomographic scanning machines then monitor the amount of radioactivity in three dimensions within the brain, and from this, an estimate of blood flow can be calculated. Estimation of blood flow using MRI is in the early stages, but has the advantage that no radiation need be given to subjects, so that the procedure is considerably safer, and can be repeated as often as necessary.

The time resolution of any technique which relies on blood flow changes is necessarily limited by the time taken for blood vessels to react to changes in demand. This is of the order of 0.5 s or so. In fact, both PET and SPECT have a much coarser time resolution. In PET, scans take of the order of 2 minutes to complete. The precise time resolution of SPECT scans is difficult to estimate. With SPECT, a single bolus of tracer is injected. This is taken up by the endothelial lining of brain capillaries by an amount proportional to the flow in the vessel. The tracer has a long half-life and remains in the cells for some time. When scans are taken, they effectively provide a snapshot of blood flow in the brain at about the time the bolus was injected.

Because movement generates afferent input, blood flow changes occur in both pre- and postcentral areas of cortex. With present scanners, it is not possible to distinguish reliably between these two regions, and therefore most studies simply refer to changes in sensorimotor activity. Some authors try to overcome this problem by performing two scans: one during active movement, and one during passive movement by the experimenter of the same part of the body when the subject is relaxed. Subtracting the flow in the passive from the active condition produces a signal more nearly related to motor activity, although this can never be completely accurate since the sensory input produced during active movement is never the same as that in the passive condition (e.g. spindle and tendon organ input will be completely different in the two states).

In the sensorimotor cortex, changes in blood flow during movement of different parts of the body reflects the topographic sensorimotor map; leg movements produce the greatest flow changes in the medial part of the sensorimotor strip; arm movements are more lateral, and face movements more lateral still. Within the arm area, shoulder movements produce flow that is significantly more medial than finger movements. Elevation of the arm at the shoulder is accompanied by increased blood flow on the ipsilateral as well as contralateral motor (and premotor) cortex. This may reflect involvement of ipsilateral projections to proximal and axial muscles involved in stabilizing the trunk during movement. It may also be related to the ipsilateral activity recorded in BP studies noted above (see illustration on front cover).

In the first studies of cerebral blood flow during different types of movement, it was reported that activity in primary motor cortex increased during all types of movement, whilst that in SMA and premotor cortex increased only when the subject performed complex sequences of finger and thumb movements. This was consistent with the notion that premotor and SMA were 'higher' in the notional hierarchy of movement control than the primary motor area. However, recent work suggests that this may be an oversimplification. Blood flow increases in both premotor and supplementary motor area even during simple finger or arm movements, and is not confined to complex tasks. The result is compatible with more recent ideas that there is no strict segregation of activities between motor areas, but that many functions are carried out in parallel in all regions.

Despite the occurrence of parallel processing in cortical motor areas, differences between activities of SMA and primary motor cortex are evident in some tasks. The clearest difference was reported by Roland and colleagues (1980). They asked subjects to imagine performing a complex sequence of finger movements without any actual movement occurring. In such conditions, there was no change in blood flow in primary motor cortex, but there was a significant increase in the SMA, consistent with a role in planning internally-guided movements. A similar increase in SMA activity (and in many other areas of the frontal cortex) is seen in tasks where subjects themselves have to choose randomly to make one of four different movements rather than being instructed which one to perform on the basis of an auditory cue.

Blood flow studies have also been performed on visually-guided movements. When subjects track a target on a screen by moving their hand, activity increases in several areas of the brain, over and above that seen when subjects simply follow the target by moving their eyes. Some of this activity reflects the instructions for movement, and some the afferent feedback produced by movement. However, if subjects are given a visual cue as to which direction, and how far, the target will move as it jumps across the screen, then there is additional flow bilaterally in the dorsal parietal cortex. Since the movement of the arm (and hence the somatosensory feedback) is the same in each case, the extra activity in dorsal parietal cortex must indicate that this area is important in interpreting visual cues for movement. (Articles by Colebatch *et al.*, 1991; Dieber *et al.*, 1991; Grafton *et al.*, 1992; give further references to PET studies.)

Lesions of motor areas of cortex

Lesions of primary motor cortex produce effects similar to those seen after damage to the descending pathways in the internal capsule. Clinically the two are distinguished by the extent and the severity of the symptoms. Capsular lesions, because of the bunching of fibres as they pass through the capsule, generally affect a larger part of the body and are more severe than cortical strokes. A very large cortical lesion is needed to produce combined motor and sensory symptoms, whereas this is common after capsular lesions. Similarly, a cortical stroke may often produce weakness of only a limited part of the body, whilst a capsular stroke usually affects the whole of the contralateral half of the body. Finally, cortical, but not capsular, lesions, may be accompanied by 'high level' deficits such as dysphasia or apraxia, due to damage of neighbouring areas of cortex, or interruption of corticocortical connections.

For many years, it was not clear whether lesions of premotor cortex could be distinguished from the effects of damage to the primary motor cortex. Several authors emphasized that premotor lesions produced a degree of 'clumsiness' or 'awkwardness' of movement that was greater than expected from the degree of muscle weakness which also occurred. Unfortunately, slow, clumsy movements are also seen in many other conditions, such as Parkinson's disease, cerebellar disease and recovered hemiparesis. More precise descriptions of premotor abnormalities have been given by Freund and colleagues (1985). They emphasize two forms of deficit. One group of patients had a peculiar distribution of muscle weakness, involving all the hip muscles, and the muscles at the shoulder involved in shoulder elevation and abduction. Such a distribution would be very uncommon in lesions of the primary motor cortex. Since the trunk representation lies between the shoulder and hip muscles, a primary motor cortex lesion would be expected to affect *all* shoulder, hip and axial muscles. In addition to the distribution of weakness, the patients had particular difficulty in making movements involving both arms, or in certain tasks involving the coordination of the whole arm and hand. Thus, they could not perform 'windmilling' movements of the arms together (rotating the arms about the shoulder), although each arm separately could make the movement. Similarly, although subjects could move their arms under visual guidance, they were grossly impaired in making rapid movements, such as catching a ball. The problem was not simply due to weakness at the shoulder, but to a disorder in the timing of muscle activity, which led to incoordination and clumsiness. The manual movements of these patients, made with supported arms were completely normal. It was proposed that the problems of these patients arose because of interruption of premotor connections to reticular centres controlling girdle muscles. The particular deficits in shoulder elevation and abduction were seen as indicating a role for this area in pointing movements of the whole arm.

More complex tasks, yielded further deficits after premotor lesions. Halsband and Freund (1990) devised a task similar to that used in monkeys to investigate the role of premotor cortex in the selection of movements of the basis of external cues. These studies showed that patients had no difficulty in learning a non-motor task in which an arbitrary visual, tactile or auditory cue had to be associated with different spatial locations on a display. The same patients, in contrast, could not learn a motor task in which the **same** stimuli had to be associated with different hand movements.

Lesions of the supplementary motor area are rare. The commonest sympton is difficulty in initiating or maintaining speech. Although patients may be able to repeat phrases and provide short answers to questions, they have great difficulty in spontaneous conversation. Same patients are said to exhibit the 'alien hand' sign, a condition in which the hand contralateral to the lesion behaves independently of the subject's will, and reaches out to grasp objects in the environment. The latter is interpreted as a possible uncoupling of the externally guided premotor system from the internally guided SMA system.

Of particular interest are the descriptions of patients in whom the SMA has been removed because of intractable epilepsy. Laplane and colleagues described three patients with unilateral lesions. All three patients had an initial three-week period of contralateral akinesia and paucity of verbal expression, although there was no weakness of the muscles. The face was motionless and there was difficulty in lateral deviation of gaze. Such akinesia is remarkably similar to that seen in parkinsonism, and it is possible that it is a consequence of lesioning one of the main targets of basal ganglia outflow via VL_m and VL_o thalamus to SMA. Recovery was almost complete, except that the patients were left with a permanent reduction in the speed at which they could perform rapid alternating movements of the hands. As with the blood flow monitoring this sequence of events suggests a role in 'higher level' control of movement (Laplane *et al.*, 1977).

Grasp reflexes and spasticity

The grasp reflex and spasticity are two common features of lesions affecting motor areas of cortex. However, these symptoms are not specific to lesions of any one particular area of cortex: they arise after damage to many different areas. A grasp reflex is seen in normal human infants, but disappears within a few months of birth. It may reappear after lesions of area 6 in posthemiplegic patients. There are three different types of grasp reflex and, in any one patient, one or all three of them may be present.

The classical grasp reflex of neurology consists of involuntary prehension elicited by a tactile stimulus to the palm. It is usually accompanied by a traction response, which is a heightened stretch reflex in the finger flexor muscles, such that attempts to remove the object from the hand by extending the flexor tendons results in an increased force of grip. This traction response is known sometimes as forced grasping, to indicate the difficulty which the patient or animal has in releasing his grip on an object under these conditions. The traction response can be influenced by the posture of the patient. When lying on one side, the reflex is heightened in the upper limbs, and decreased on the opposite side.

The final version of the grasp reflex is rather more complex, and known as the instinctive grasp reflex. It is an involuntary closure of the hand elicited by tactile stimulation of any part of the hand or fingers.

The hand is first oriented towards the object so as to bring it into the palm. This stimulus then secondarily elicits a tactile grasp reflex to close the fingers on the object. The initial exploratory orientation of the hand in the instinctive grasp reflex was considered by Denny-Brown (1966) to be the opposite of the tactile avoiding reaction, in which the hand is withdrawn from a stimulus. Lesions involving the premotor cortex produce the instinctive grasp, whereas lesions of the parietal cortex produce the avoiding reaction.

Unfortunately, despite the wealth of clinical description of the grasp reflex, it is not known which is the precise region of cortex which must be lost in order for the reflex to appear. A recent study concluded that grasping was most often seen after lesions affecting the medial areas of the frontal cortex including the cingulate gyrus and SMA, although in some patients, grasping was seen after lateral lesions of the premotor area (De Renzi and Barbieri, 1992).

There is also disagreement about the role of premotor cortex lesions in production of spasticity. The old idea was that there was a 'suppressor' strip just anterior to area 4 ('strip area 4 s') which had to be lesioned for spasticity to occur. This idea is now discredited. It is probably safest to conclude that if a lesion is made in area 6 (and this includes both premotor and supplementary motor areas), then the larger it is, and especially if accompanied by lesions of area 4, the more likely spasticity is to ensue (Wiesendanger, 1981).

Apraxia

Lesions of the cerebral cortex do not only produce the primary disorders of movement described above, that is, a disorder of the effector regions of motor cortex, but also can produce secondary disorders of movement in which there is no weakness or damage to the primary motor areas. Such secondary disorders of movement provide some insight into the mechanisms used to control the final output of motor cortex. Diseases of the cerebellum or basal ganglia provide good examples of secondary disorders of movement, and are described in the following chapters. At a cortical level, an example of this type of disturbance is **apraxia**.

Apraxia (Geschwind and Damasio, 1985) may be defined as an impairment in producing an appropriate movement in response to visual or auditory instruction. For true apraxia to be considered, the incorrect movement cannot be due to weakness (that is, damage to effector structures), or to akinesia or bradykinesia (as in basal ganglia damage), ataxia (as in cerebellar damage), or to impairment in comprehension or sensation. Furthermore, the patient must be attentive and cooperative. The reason for this fairly strict definition is to define a syndrome in which there is a disorder of movement control in the absence of deficits in either effector or perceptual mechanisms.

The most common form of apraxia is **ideomotor** apraxia. If a patient is asked to hold out his tongue, he may be unable to do so, or will make an inappropriate movement such as chomping the teeth. However, although unable to use the tongue on command, the same patient may be able to move it quite normally in automatic movements of licking his lips or speaking. While ideomotor apraxia refers to the inability to perform single movements, **ideational** apraxia (less common) refers to an inability to perform correctly a sequence of movements, such as taking a cigarette, lighting it and using an ashtray. Each movement may be performed well, but, for example, in the wrong order.

Apraxia usually affects movements of the face or limbs. Axial apraxia is rare. An example of apraxia of facial movements is an inability to blow out a match. Patients may fail to round their lips to perform the task, and may instead shake the match in their hands to put it out. This type of error clearly shows that the patient knows what to do; he is simply unable to transform the idea of the movement into the appropriate action.

The question is, where in the brain does this transformation occur? The answer is not yet clear. Apraxia is a symptom of lesions in several parts of the brain. It is particularly common in patients with left hemisphere lesions, who often have aphasia as well, and in patients with lesions of the anterior part of the corpus callosum. Both types of lesions have been interpreted as producing 'disconnection syndromes', disconnecting areas of the brain which receive the instruction to move or formulate the idea of the movement, from the final effector areas of the motor system.

For example, one particular patient with a lesion of the anterior part of the corpus callosum could use both legs normally in walking, turning and so on. Yet when asked to demonstrate how he would kick a ball, he would only use his right leg; he was unable to demonstrate any movement with his left leg. If the patient was asked to imitate a visual instruction, performance was normal with both legs. The interpretation is as follows. The left side of the brain was able to interpret the auditory command and to feed this forward to premotor centres on the left so that movements of the right side were normal. However, the lesion of the corpus callosum disconnected the frontal areas on the right side of the brain, making them unable to receive information about the verbal command. Hence, movements on the left were apraxic. In contrast, visual information to the right side of the brain was sufficient to produce appropriate movements on the left.

With left hemisphere strokes, there may be paralysis of the right limbs and apraxia of the left, often associated with aphasia. The explanation is the same as above. There is damage to the areas of the left hemisphere which send commands to the right side of the brain. This also explains why axial apraxia is rare. Cortical control of axial muscles is mainly via brainstem reticular pathways, which are bilaterally organized.

Is there, then, a 'centre' in the left hemisphere responsible for transforming commands into appropriate plans for movement? One clue to the answer comes from the observation that lesions of the lower parts of posterior parietal areas 5 and 7 in the left hemisphere can lead to pronounced bilateral apraxia of the limbs (but not face). Geschwind and Damasio (1985) have suggested that this may be a 'master control system' receiving inputs from all sensory modalities and projecting to frontal motor areas on both sides of the brain.

Bibliography

Review articles and books

Brooks, V. B. (ed) (1981) *Handbook of Physiology*, sect. 1, vol.2, part 2, Williams and Wilkins, Baltimore.

Caminiti, R., Johnson, P. B. and Burnod, Y. (eds) *Control of Arm Movement in Space*, Springer-Verlag, Berlin.

Georgopoulos, A. P., Kalaska, J. F., Crutcher, M. D. *et al.* (1984) The representation of movement direction in the motor cortex: single cell and population studies, in G. M. Edelman, W. E. Gall and W. M. Cowan (eds), *Dynamic Aspects of Neocortical Function*, John Wiley and Sons, New York, pp. 501–524.

Humphrey D. R., and Freund, H. J. (1991) *Motor control: concepts and issues*, John Wiley, Chichester.

Ito, M. (ed.) (1989) *Neural Programming*, Karger, Basel.

Lemon, R. N. (1990) Mapping the output functions of the motor cortex, in G. M. Edelman, W. E. Gall, W. N. Cowan (eds.), *Signal and Sense*, Wiley-Liss, New York, pp. 315–355.

Passingham, R. E. (1987) Two cortical systems for directing movement, in *Motor Areas of the Cerebral Cortex*, Ciba Foundation Symposium 132, John Wiley, Chichester, pp. 151–164.

Phillips, C. G. and Porter, R. (1977) *Corticospinal neurones. Their role in movement*, Academic Press, London.

Rothwell, J. C., Thompson, P. D., Day B. L. *et al.* (1991) Stimulation of the human motor cortex through the scalp, *Exp. Physiol.*, **76**, 159–200.

Stein, J. F. (1989) Representation of egocentric space in the posterior parietal cortex, *Q. J. Exp. Physiol.*, **74**, 583–606.

Original papers

Adams, R. W., Gandevia, S. C. and Skuse, N. F. (1990) The distribution of muscle weakness in upper motoneurone lesions affecting the lower limbs, *Brain*, **113**, 1459–1476.

Andersen, P., Hagan, P. J., Phillips, C. G. and Powell, T. T. S. (1975) Mapping by microstimulation of overlapping projections from area 4 to motor units of the baboon's hand, *Proc. Roy. Soc. Lond. (Biol)*, **188**, 31–60.

Asanuma, H. (1981) The pyramidal tract, in V. B. Brooks (eds), *Handbook of Physiology*, sect. 1, vol. 2, part 2, Williams and Wilkins, Baltimore, pp. 703–733.

Brinkman, C. (1981) Lesions in supplementary motor area interfere with a monkey's performance of a bimanual coordination task, *Neurosci. Lett.*, **27**, 267–270.

Brinkman, C. (1984) Supplementary motor area of the monkey's cerebral cortex: short-and long-term deficits after unilateral ablation and the effects of subsequent collosal section, *J. Neurosci.*, **4**, 918–929.

Brinkman, J. and Kuypers, H. G. J. M. (1973) Cerebral control of contralateral and ipsilateral arm, hand and finger movements in the split-brain rhesus monkey, *Brain*, **96**, 653–674.

Brodal, A. (1981) *Neurological Anatomy in Relation to Clinical Medicine*, Oxford University Press, Oxford

Bucy, P. C., Keplinger, J. E. and Sequiera E. B. (1964) Destruction of the pyramidal tract in man, *J. Neurosurg.*, **21**, 385–398.

Butler, E. G., Horne, M. K. and Rawson, J. A. (1992) Sensory characteristics of monkey thalamic and motor cortex neurones, *J. Physiol.* **445**, 1–24.

Buys, E. J., Lemon, R. N., Mantel, G. W. H. and Muir, R. B. (1986) Selective facilitation of different hand muscles by single corticospinal neurones in the conscious monkey, *J. Physiol.*, **381**, 529–549.

Cheney, P. D. and Fetz, E. E. (1980) Functional classes of primate corticomotoneuronal cells and their relation to active force, *J. Neurophysiol.*, **44**, 773–791.

Cheney, P. D. and Fetz, E. E. (1984) Corticomotoneuronal cells contribute to long-latency stretch reflexes in the rhesus monkey, *J. Physiol.*, **349**, 249–272.

Cheney, P. D. and Fetz, E. E. (1985) Comparable patterns of muscle facilitation evoked by individual corticomotoneuronal (CM) cells and by single intracortical microstimuli in primates: evidence for functional groups of CM cells, *J. Neurophysiol.*, **53**, 786–804; 805–820.

Chollet, F., DiPierot V., Wise, R. J. S. *et al.* (1991) The functional anatomy of motor recovery after stroke in humans: a study with positron emission tomography, *Ann. Neurol.*, **29**, 63–71.

Cohen. L. G., Bandinelli, S., Findley, T. W. and Mattett, M. (1991), Motor reorganisation after upper limb amputation in man, *Brain* **114**, 615–627.

Colebatch, J. G. and Gandevia, S. C. (1989) The distribution of muscular weakness in upper motoneurone lesions affecting the arms, *Brain*, **112**, 749–763.

Colebatch, J. G., Deiber M.-P., Passingham R. E. *et al.* (1991) Regional cerebral blood flow during voluntary arm and hand movements in human subjects, *J. Neurophysiol.*, **65**, 1392–1401.

Day, B. L., Rothwell, J. C., Thompson, P. D. *et al.* (1989) Delay in the execution of voluntary movement by electrical or magnetic brain stimulation in intact man: evidence for storage and motor programmes in the brain, *Brain,* **112,** 649–663.

Deiber, M. P., Passingham, R. E., Colebatch, J. G. *et al.* (1991) Cortical areas and the selection of movement: a study with positron emission tomography, *Exp. Brain Res.,* **84,** 393–402.

Denny-Brown, D. (1966) *The cerebral control of movement,* University Press, Liverpool.

DeRenzi, E. and Barbieri, C. (1992) The incidence of the grasp reflex following hemispheric lesion and its relation to frontal damage, *Brain* **115,** 293–313.

Dum, R. P. and Strick, P. L. (1991) The origin of corticospinal projections from the premotor areas in the frontal lobes, *J. Neurosci.,* **11,** 667–689.

Evarts, E. V. (1966) Pyramidal tract activity associated with a conditioned hand movement in the monkey, *J. Neurophysiol.,* **29,** 1011–1027.

Evarts, E. V. (1968) Relation of pyramidal tract activity to force exerted during voluntary movement, *J. Neurophysiol.,* **31,** 14–27.

Evarts, E. V. (1972) Pre-and post-central neuronal discharge in relation to learned movement, in T. Frigyesi, E. Rinvik and M. D. Yahr (eds), *Cortico-thalamic Projections and Sensorimotor Activities,* Raven Press, New York, pp. 449–458.

Evarts, E. V. (1973) Motor cortex reflexes associated with learned movement, *Science,* **179,** 501–503.

Evarts, E. V. (1981) Role of motor cortex in voluntary movements in primates, in V. B. Brook (ed.), *Handbook of Physiology,* sect. 1, vol. 2, part 2, Williams and Wilkins, Baltimore, pp. 1083–1120.

Evarts, E. V., Fromm, C., Kroller J. and Jennings, V. A. (1983) Motor cortex control of finely graded forces, *J. Neurophysiol.,* **49,** 1199–1215.

Fetz, E. E., Cheney, P. D., Mewes, K. and Palmer, S. (1989) Control of forelimb activity by populations of corticomotoneuronal and rubromotoneuronal cells, *Prog. Brain Res.,* **80,** 437–449.

Friedman, D. P. and Jones, E. G. (1981) Thalamic input to areas 3a and 2 in monkeys, *J. Neurophysiol.* **45,** 59–85.

Gentilucci, M., Forgassi, L., Luppino, G. *et al.* (1989) Somatotopic representation in inferior area 6 of the macaque monkey, *Brain Behav. Evol.,* **33,** 118–121.

Georgopoulos, A. P. (1989) Motor cortex and reaching, in M. Ito (ed.) *Neural programming,* Karger, Basel, pp. 3–12.

Georgopoulos, A. P., Kalaska, J. F., Caminiti, R. and Massey, J. T. (1982) On the relations between the direction of two dimensional arm movements and cell discharge in primate motor cortex, *J. Neurosci.,* **2,** 1527–1537.

Georgopoulos, A. P., Crutcher, M. D. and Schwartz A. B (1989a) Cognitive spatial-motor processes. 3. Motor cortical prediction of movement direction during an instructed delay period, *Exp. Brain Res.,* **75,** 183–194.

Georgopoulos, A. P., Lurito, Petrides, S. M. *et al.* (1989b) Mental rotation of the neuronal population vector, *Science,* **243,** 234–236.

Georgopoulos, A. P., Ashe J., Smyrnis N. and Taira, M. (1992) The motor cortex and the coding of force, *Science* **256**, 1692–1695.

Geschwind, N. and Damasio, A. R. (1985) Apraxia, in J. A. M Fredericks (ed.) *Handbook of Clinical Neurology*, vol. 1 (45), Elsevier, Amsterdam, pp. 423–432.

Grafton, S. T., Mazziotta, J. C., Woods, R. P. and Phelps, M. E. (1992) Human functional anatomy of visually-guided finger movements, *Brain*, **115**, 565–587.

Halsband, U. and Freund, H. -J. (1990) Premotor cortex and conditional motor learning in man, *Brain*, **103**, 207–222.

Hocherman, S. and Wise, S. P. (1991) Effects of hand movement path on motor cortical activity in awake, behaving rhesus monkeys, *Exp. Brain Res.* **83**, 285–302.

Humphrey, D. R. (1986) Representation of movements and muscles within the primate precentral motor cortex: historical and current perspectives, *Fed. Proc.* **45**, 2687–2699.

Humphrey, D. R. and Tanji, J. (1991) What features of voluntary motor control are incoded in the neuronal discharge of different cortical motor areas? in D. R. Humphrey and H. J. Freund (eds.) *Motor Control: Concepts and Issues*, John Wiley, Chichester, pp. 413–444.

Ikeda, A. and Shibasaki, H. (1992) Invasive recording of movement-related cortical potentials in humans, *J. Clin. Neurophysiol.*, **9**, 509–520.

Johnson, P. B. (1991) Toward an understanding of the cerebral cortex and reaching movements: a review of recent approaches, in R. Caminiti, P. B. Johnson and Y. Burnod (eds.), *Control of Arm Movement in Space*, Springer Verlag, Berlin, pp. 199–261.

Kalaska, J. F., Crammond D. J., Cowan D. A. D. *et al.* (1991) Comparison of cell discharge in motor, pre-motor and parietal cortex during reaching, in R. Caminiti, P. B. Johnson and Y Burnod (eds) *Control of Arm Movement in Space*, Springer Verlag, Berlin, pp. 129–146.

Kasser, R. J., and Cheney, P. D. (1985) Characteristics of corticomotoneuronal post-spike facilitation and reciprocal suppression of EMG activity in the monkey, *J. Neurophysiol.*, **53**, 959–978.

Kurata, K. and Wise, S. P. (1988a) Premotor cortex of rhesus monkeys: set-related activity during two conditional motor tasks, *Exp. Brain. Res.*, **69**, 327–343.

Kurata, K. and Wise, S. P. (1988b) Premotor and supplementary motor cortex in rhesus monkeys: neuronal activity during externally- and internally-instructed motor tasks, *Exp. Brain Res.*, **72**, 237–248.

Kwan, H. C., MacKay, W. A., Murphy, J. T. and Wong, Y. C. (1978) Spatial organisation of precentral cortex in awake primates. II. Motor outputs, *J. Neurophysiol.*, **41**,, 1120–1131.

Laplane, D., Talairach, J., Meininger, V. *et al.* (1977) Clinical consequences of corticectomies involving the supplementary motor area in man, *J. Neurol. Sci.* **34**, 301–314.

Lawrence, D. G. and Kuypers, H. G. J. M. (1968) The functional organisation of the motor system in the monkey, Parts 1 and 2, *Brain*, **91**, 1–14; 15–36.

Lemon, R. N. (1988) The output map of the primate motor cortex. *Trends Neurosci.* **11**, 501–506.

Libet, B., Gleason, C. A., Wright, E. W. and Pearl, D. K. (1983) Time of conscious-intention to act in relation to onset of cerebral activity (readiness potential), *Brain*, **106**, 623–642.

Matelli, M. and Luppino, G. (1991) Anatomo-functional parcelation of the agranular frontal cortex, in R. Caminiti, P. D. Johnson and Y. Burnod (eds) *Control of Arm Movement in Space*, Springer Verlag, Berlin, pp. 85–101.

Mitz, A. R. and Wise, S. P. (1987) The somatotopic organisation of the supplementary motor area: intracortical microstimulation mapping, *J. Neurosci.* **7**, 1010–1021.

Muir R. B. and Lemon, R. N. (1983) Corticospinal neurones with a special role in precision grip, *Brain Res.*, **261**, 312–316.

Mushiake, H., Inase, M. and Tanji, J. (1991) Neuronal activity in the primate premotor, supplementary, and precentral motor cortex during visually-guided and internally-determined sequential movement, *J. Neurophysiol.*, **66**, 705–718.

Obeso, J. A., Rothwell, J. C. and Marsden, C. D. (1985) The spectrum of cortical myoclonus, *Brain*, **108**, 193–224.

Penfield, W. and Jasper, H. H. (1954) *Epilepsy and the functional anatomy of the human brain*, Little Brown, Boston.

Penfield, W. and Welch, K. (1951) Supplementary motor area of the cerebral cortex, *Arch. Neurol. Psychiatr.*, **66**, 289–317.

Riehle, A. and Requin, J. (1989) Monkey primary motor and premotor cortex: single-cell activity related to prior information about direction and extent of an intended movement, *J. Neurophysiol.*, **61**, 534–549.

Rosen, I. and Asanuma, H. (1972) Peripheral afferents to the forelimb area of the monkey motor cortex: input-output relation, *Exp. Brain Res.*, **14**, 257–273; 243–256.

Rothwell, J. C., Obeso, J. A. and Marsden, C. D. (1984) On the significance of giant somatosensory evoked potentials in cortical myoclonus, *J. Neurol. Neurosurg. Psychiatr.*, **47**, 33–42.

Rothwell, J. C., Obeso, J. A. and Marsden, C. D. (1986) Electrophysiology of somatosensory reflex myoclonus, *Adv. Neurol.*, **43**, 353–366.

Ruch, T. C. and Fetz, E. E. (1979) The cerebral cortex: its structure and motor functions, in T. Ruch and H. D. Patton (eds.) *Physiology and Biophysics*, vol. II, W. B. Saunders, Philadelphia, pp. 53–122.

Sanes, J. N. and Donoghue, J. P. (1991) Organisation and adaptability of muscle representations in primary motor cortex, in R. Caminiti, P. B. Johnson and Y. Burnod (eds), *Control of Arm Movement in Space*, Springer Verlag, Berlin, pp. 103–127.

Schell, G. R. and Strick, P. L. (1984) The origin of thalamic inputs to the arcuate premotor and supplementary motor area, *J. Neurosci.*, **4**, 539–560.

Schiebel, M. H., Davies, T. L., Lindsay, R. D. *et al.* (1974) Basilar dendrite bundles of giant pyramidal cells, *Exp. Neurol.* **42**, 307–319.

Shima, K., Iya, K., Mushiake, H. *et al.* (1991) Two movement-related foci in the primate cingulate cortex observed in signal-triggered and self-paced following movement, *J. Neurophysiol.*, **65**, 188–202.

Shinoda, Y., Yokota, J. I. and Futani, T. (1981) Diverse projection of individual corticospinal axons to motoneurones of multiple muscles in the monkeys, *Neurosci. Lett.*, **23**, 7–12.

Strick, P. L. and Preston, J. B. (1982) Two representations of the hand in area 4 of a primate. I. Motor output organisation, *J. Neurophysiol.*, **48**, 139–149.

Tanji, J. and Evarts, E. V. (1976) Anticipatory activity of motor cortex neurones in relation to direction of an intended movement, *J. Neurophysiol.*, **39**, 1063–1068.

Tanji, J., Okano, K. and Sato, K. C. (1988) Neuronal activity in cortical motor areas related to ipsilateral, contralateral, bilateral digit movements of the monkey, *J. Neurophysiol.*, **60**, 325–343.

Twitchell, T. E. (1951) The restoration of motor function following hemiplegia in man, *Brain.*, **74**, 443–480.

Wiesendanger, M. (1981) Organisation of secondary motor areas of cerebral cortex, in V. B. Brooks (ed.) *Handbook of Physiology*, sect. 1, vol. 2 part 2, Williams and Wilkins, Baltimore, pp. 1121–1147.

Wiesendanger, M. and Miles, T. S. (1982) Ascending pathway of low threshold muscle afferents to the cerebral cortex and its possible role in motor control, *Physiol. Rev.*, **62**, 1234–1270.

Wise, S. P. and Mauritz, K.-H. (1985) Set-related neuronal activity in the premotor cortex of rhesus monkeys: effects of changes in motor set, *Proc. Roy. Soc. B.*, **223**, 331–354.

Wise, S. P., Alexander, G. E., Altman *et al.* (1991) What are the specific functions of the different motor areas? D. R. Humphrey and H. J. Freund (eds) *Motor Control: Concepts and Issues*, John Wiley, Chichester, pp. 463–485.

Woolsey, C. N., Settlage, P. H., Meyer, D. R. *et al.* (1952) Patterns of localisation in precentral and supplementary motor areas and their relation to the concept of the premotor area, *Res. Pub. Ass. Nerv. Ment. Dis.*, **30**, 238–264.

The cerebellum 10

There are more neurones in the cerebellum than there are in the whole of the rest of the brain, yet surprisingly little is known about the function of this remarkable organ. Lesions of the cerebellum do not produce muscle weakness nor disorders of perception. However, they do produce disturbances of coordination in limb and eye movements, as well as disorders of posture and muscle tone. Despite the lack of any direct efferent connections between cerebellum and spinal or brainstem motor nuclei, it is a structure which is undoubtedly concerned with the control of movement.

10.1 Cerebellar anatomy

The cerebellum consists of a convoluted outer mantle of grey matter covering an inner core of white matter. The structure is folded over upon itself so that the anterior and posterior ends meet beneath, above the surface of the fourth ventricle. The conspicuous features of the human cerebellum are two lateral hemispheres which lie either side of a narrow midline ridge known as the vermis (Figure 10.1). Deep within this folded mass are the three pairs of nuclei, arranged either side of the midline. In animals they are known as the medial, interposed (subdivided into an anterior and posterior part), and lateral nuclei. In man, they have different names: fastigial, globose and emboliform, and dentate. The globose and emboliform are sometimes spoken of together as the interposed nucleus, as in animals.

The input and output of the cerebellum travels via three pairs of large fibre tracts which connect it to the brainstem: the superior, middle and inferior cerebellar peduncles, which are known also as the brachium conjunctivum, the brachium pontis and restiform body, respectively. There are far more fibres entering the cerebellum than there are leaving it. The ratio of afferents to efferents has been put as high as 40:1. Most of the efferent fibres leave the cerebellum through the superior peduncle

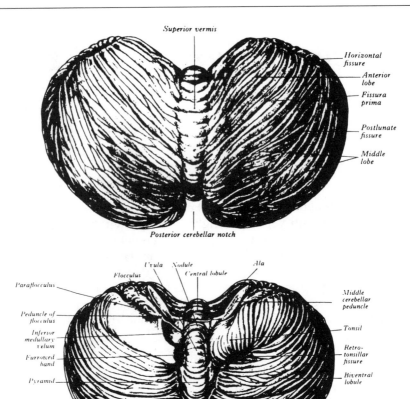

Figure 10.1 Surface views of the superior (upper) and inferior (lower) aspects of the human cerebellum. In the lower diagram, the tonsil and the adjoining part of the biventral lobule on the right side have been removed. (From Williams and Warwick, 1975, Figures 7.72 and 7.75; with permission.)

(0.8 million fibres) and pass to the red nucleus, thalamus and brainstem.

Subdivisions of the cerebellum

The cerebellar cortex has basically the same neuronal organization throughout (see below), so that cytoarchitectonics, which were so useful in describing the anatomy and functional subdivisions of the cerebral cortex, cannot be applied here.

The oldest scheme for describing the structure is based simply upon the appearance of the cerebellum when dissected. A series of ridges or folia, separated by deep fissures run transversely across the cerebellar surface. Two of the fissures are deeper than the rest and divide the cerebellum into three **lobes**. The **anterior** and **posterior** lobes are separated by the **primary fissure**. Together, these lobes form the mass of the cerebellum and are therefore sometimes known as the **corpus cerebelli**. The **postero-lateral fissure** separates the **flocculonodular lobe** (consisting of the flocculus and nodule) from the posterior lobe. The anterior and posterior lobes are further subdivided by smaller fissures into a series of **lobules**. In animals, these lobules are known either by a series of names, or by a set of ten roman numerals shown in Figure 10.2. Yet another, different, set of names is sometimes used to describe these lobules in man.

Comparative anatomical studies have also divided the cerebellum into three parts. The archicerebellum, paleocerebellum and neocerebellum. The phylogenetically oldest region (the archicerebellum) is the flocculonodular lobe. The paleocerebellum comprises the vermis of the anterior lobe plus the pyramis, paraflocculus and uvula. The neocerebellum, which is by far the largest subdivision in man consists of the remaining portions of the hemispheres and the middle portion (lobules VI and VII) of the vermis.

Roughly speaking, these phylogenetic subdivisions reflect the pattern of afferent terminations in the cerebellum. Vestibular afferents terminate in the archicerebellum; direct projections from spinal cord terminate in the paleocerebellum; and pontine inputs terminate in the

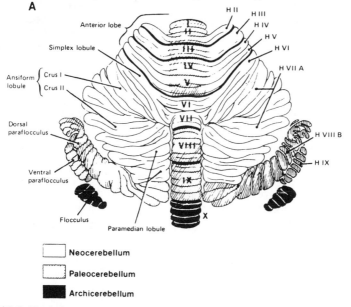

Figure 10.2 Evolutionary subdivisions of the monkey cerebellum. (From Gilman, Bloedel and Lechtenburg, 1981, Figure 2A, with permission.)

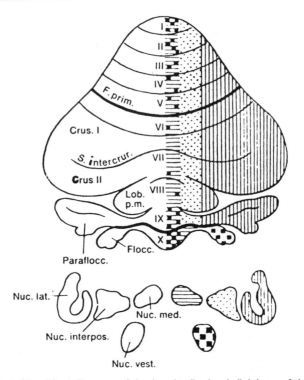

Figure 10.3 Simplified diagram of the longitudinal subdivisions of the cerebellum based on the pattern of efferent projections to the cerebellar and vestibular nuclei. Similar shading of cerebellar cortex and nuclei indicates an efferent connection between the two. F. prim, primary fissure; Flocc., flocculus; Lob. p.m., paramedian lobule; Nuc. lat., lateral nucleus; Nuc. interpos., interpositus nucleus; Nuc. med., medial nucleus; Nuc. vest., vestibular nucleus. (From Jansen and Brodal, 1958, Figure 183; with permission.)

neocerebellum. Because of this, these subdivisions are referred to as vestibulocerebellum, spinocerebellum and pontocerebellum respectively. However, although useful, this subdivision is a little misleading. The sites of termination in practice overlap quite considerably.

Perhaps the most useful concept of cerebellar organization depends upon the pattern of efferent projections from the cerebellar cortex. Three main longitudinal divisions can be recognized: a medial zone (vermis) which sends fibres to the fastigial nucleus, an intermediate zone which sends fibres to the interposed nuclei, and a lateral zone which sends fibres to the dentate nucleus (Figure 10.3). The vestibular nuclei also receive cerebellar output. This comes from the flocculonodular lobe and from regions in the vermis and anterior and posterior lobes.

These three main longitudinal zones have been further subdivided by both anatomical and physiological techniques into a series of parallel

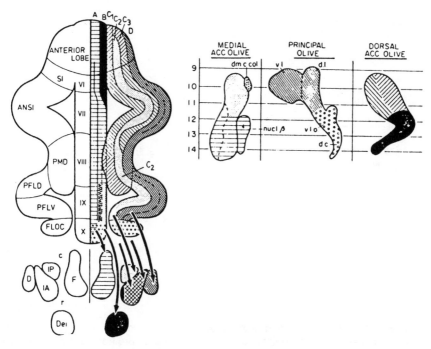

Figure 10.4 Fine subdivision of longitudinal and zones of cat cerebellum according to the pattern of olivocerebellar projections and the projections from cerebellar cortex onto the cerebellar nuclei. As indicated by the different shading, different regions of the inferior olivary complex (top right) project to longitudinal strips of cerebellar cortex. These strips in turn project in an orderly fashion onto the cerebellar nuclei. F, fastigial nucleus; IA, IP, anterior and posterior interposed nuclei; D, dentate nucleus; Dei, lateral vestibular nucleus of Deiters. (From Groenewegen, Voogd and Freedman, 1979, plate 16; with permission.)

strips which run longitudinally for almost the whole length of the cerebellar cortex (Figure 10.4). These strips have a strikingly precise pattern of input and output connections related to the inferior olive. Olivary cells which project to the cerebellum are arranged into small groups. Cells within each group all send axons to innervate the same longitudinal zone of the cerebellum. In addition, they send axonal branches to clusters of cells within the cerebellar nuclei. The efferent fibres from the longitudinal zones then complete the circuit by projecting to those nuclear clusters which receive input from the same olivary group of cells.

Afferent connections of the cerebellum

Most of the afferents to the cerebellum travel in the middle cerebellar peduncle (20 million fibres), and come from the pontine nuclei. The

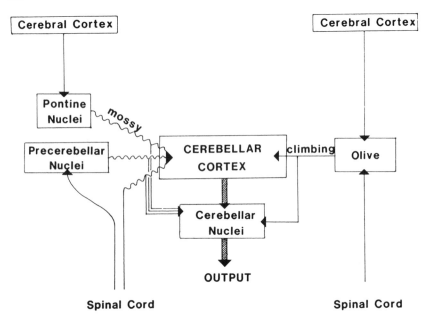

Figure 10.5 Highly simplified diagram showing main sources of afferent input to the cerebellar cortex and nuclei. Note that both mossy fibre (left) and climbing fibre (right) inputs project to both cortex and nuclei.

other afferents which come from the olive and via direct spinocerebellar pathways are found mainly in the inferior peduncle (0.5 million fibres), with a small contribution in the superior peduncle. There are also inputs from visual and vestibular systems.

Figure 10.5 gives a very simplified diagram of the main sources of cerebellar afferents. It is thought that all afferents to the cerebellum terminate both in the cortex and the cerebellar nuclei. This arrangement has given rise to the idea that the cerebellar cortex functions as a modulatory 'side-loop' in a direct pathway from afferent input to nuclear output. However, the relative strength and importance of the inputs to cortex and nuclei, are in most cases, not known. (Bloedel and Courville, 1981, is a major review of afferent cerebellar connections, Gilman, 1992, is a condensed version.)

Direct input
Direct input to the cerebellum comes from two sources: spinal cord and vestibular system. That from the spinal cord travels in the spinocerebellar tracts. The dorsal spinocerebellar tract and its forelimb analogue, the cuneocerebellar tract, both transmit input from large diameter sensory fibres with restricted peripheral receptive fields. In both cases, although

input from cutaneous and muscle receptors remains separate within the tracts, their terminations appear to overlap within the cerebellar cortex. These occupy areas which correspond approximately with the hindlimb and forelimb representations of the two somatotopic body maps (see below) in the anterior and posterior lobes. The ventral and rostral spinocerebellar tracts contain much smaller diameter fibres, and determinate more diffusely and bilaterally. Direct spinal inputs are also found in the cerebellar nuclei.

Primary vestibular afferents terminate in the nodulus, flocculus, and the ventral part of the uvula (the vestibulocerebellum). A smaller number of fibres supply the whole of the vermis. The fibres also terminate in the fastigial nucleus.

Inputs from precerebellar reticular nuclei
There are three precerebellar reticular nuclei which are usually considered separately from the pontine nuclei referred to below. They are the lateral reticular nucleus, which lies just lateral to the inferior olive, the nucleus reticularis tegmenti pontis, which lies dorsal to the pontine nuclei proper, and the paramedian reticular nucleus in the medulla.

Inputs from the pontine nuclei
The pontine nuclei are the main relay station for cortical inputs to the cerebellum. The bulk of the fibres arise in the ipsilateral frontal and parietal lobe. Most of them are not collaterals of pyramidal tract fibres but constitute a specific corticopontine projection. Other inputs are from the cerebellar nuclei and the colliculi. The pontocerebellar projection is massive, and forms the bulk of the middle cerebellar peduncle. Terminals are found in almost all parts of the contralateral cerebellum (except in the nodulus and ventral uvula), as well as in the interposed and dentate nuclei.

Inputs from the inferior olivary nuclei
The inferior olive is a relay station for a large variety of different convergent inputs. These arise in the spinal cord (travelling via the spino-olivary tract), motor cortex, superior colliculus, vestibular nuclei, trigeminal nuclei and pretectum.

Olivary afferents travel in the inferior cerebellar peduncle to end in all regions of the contralateral cerebellar cortex and nuclei. Those from the principal olive terminate mainly in the lateral cerebellum and those from the dorsal accessory olive in the medial and intermediate zones. In the cortex, they terminate as climbing fibres. Each Purkinje cell receives one climbing fibre input (see below) from a single olivary neurone. Since there are more Purkinje cells than olivary axon, this means that each axon contacts 7–15 Purkinje cells.

Aminergic afferent projections
There is a diffuse aminergic input to all regions of the cerebellar cortex
and nuclei which is quite separate from the systems discussed above.
The locus coeruleus supplies noradrenergic fibres; the raphe nuclei
supplies serotonergic fibres and the ventral mesencephalic tegmental
area supplies dopaminergic fibres.

Somatotopy (Figure 10.6)

Anaesthetized
Afferent input from all parts of the body reaches the cerebellar cortex.
In anaesthetized preparations, two complete somatotopic representa-
tions of the body can be found, one in the anterior lobe and one in the

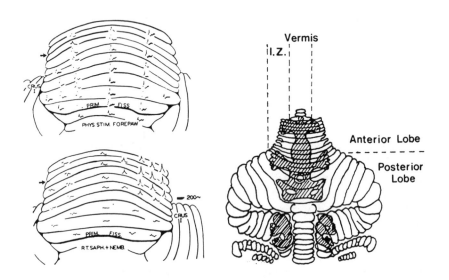

Figure 10.6 Differences in cerebellar somatotopy in awake and anaesthetized cats.
On the right are the three classical somatotopic representations produced by
electrical stimulation of various parts of the body in anaesthetized animals. The
study on the bottom left appears to confirm this. The diagram shows the size and
distribution of responses in the anterior lobe of an anaesthetized animal after
electrical stimulation of the saphenous nerve in the right leg. The potentials are
localized in the ipsilateral 'leg' area of the anterior lobe. In contrast, the upper
diagram shows the distribution and size of evoked responses after natural
stimulation of the forepaw in an unanaesthetized decerebrate cat. Similar findings
of widespread, bilateral activity also have been seen in unanaesthetized intact
animals. (From Combs, 1954, Figures 8 and 9; with permission.)

paramedian lobule and pyramis of the posterior lobe. The representations are inverted with respect to one another, and have axial structures represented in the midline. In the middle part of the posterior lobe vermis are found visual and auditory inputs. Overlying the peripheral somatotopic maps is another input from cerebral cortex. Stimulation of somatosensory cortex gives a topographic map on the surface of the cerebellar cortex in which stimulation of, for example, leg areas of the cerebral cortex evokes activity in the same cerebellar regions as receive direct spinal input from the leg.

Awake
Somatotopy disappears in awake animals. Stimulation of one arm can give responses over a wide area of cerebellar cortex. Such divergence can be seen at the level of single cells. For example, peripheral stimuli may arrive at a Purkinje cell via mossy fibres from one part of the body and via climbing fibres from another. The same can be seen when cerebral cortex is stimulated.

The reason for the differences is that in anaesthetized animals, only the most direct pathways remain active. The more indirect routes, containing more synapses, are more likely to be affected by anaesthesia. Thus input from the leg may arrive directly in the dorsal spinocerebellar tract. However, information from the same part of the body may also reach cerebellar cortex in ventral spinocerebellar tract and via relays in the olive and precerebellar reticular nuclei. These latter pathways, which are active in the awake animal probably are responsible for obscuring the somatotopy seen in anaesthetized preparations (Bloedel and Courville, 1981, Gilman, Bloedel and Lechtenberg, 1981).

10.2 *Circuitry of the cerebellar cortex (Figures 10.7 and 10.8)*

Anatomy

Because of its many convolutions, the cerebellar cortex of man covers a large surface area: it has a total anteroposterior length of over one metre. Yet throughout this area it maintains the same basic neuronal structure. Viewed in sections cut vertical to the surface, three neuronal layers can be distinguished. From the surface down these are the molecular layer, the Purkinje layer and the granule layer. The only output cell of the cortex is the Purkinje cell; the majority of cortical inputs terminate either as mossy or climbing fibres. There are a small number of aminergic terminals.

Purkinje cells are the main neuronal elements of the middle layer of

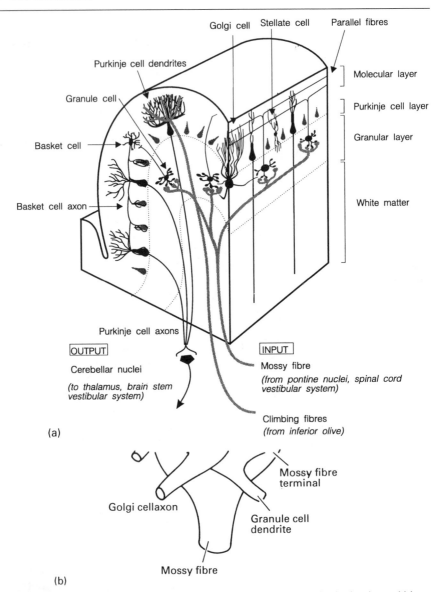

Golgi cell Stellate cell Parallel fibres

Purkinje cell dendrites

Granule cell

Basket cell

Basket cell axon

Molecular layer

Purkinje cell layer

Granular layer

White matter

Purkinje cell axons

OUTPUT

Cerebellar nuclei

(to thalamus, brain stem vestibular system)

INPUT

Mossy fibre

(from pontine nuclei, spinal cord vestibular system)

Climbing fibres

(from inferior olive)

(a)

Mossy fibre terminal

Golgi cellaxon

Granule cell dendrite

Mossy fibre

(b)

Figure 10.7 (a) Three-dimensional diagram of the principal circuitry within a cerebellar folium. Input enters as mossy and climbing fibres: the only output is via Purkinje cell axons. (Courtesy of Dr P.Thompson.) (b) Structure of the glomerular junction between mossy fibre axons, granule cell dendrites and Golgi cell axons. (From Ghez and Fahn, 1985, Figure 39.3; with permission.)

the cerebellar cortex. These flask-shaped cells are the largest neurones in the brain and form a nodal point in cerebellar circuitry. They are GABAergic cells and their axons leave the cortex to form inhibitory

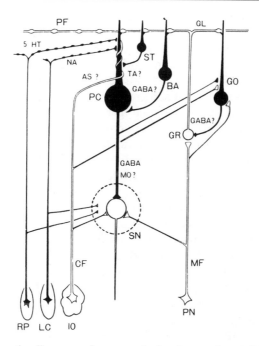

Figure 10.8 Schematic diagram of neuronal circuitry and possible synaptic transmitters in the cerebellar cortex. Inhibitory cells and synapses are black; open cells and synapses are excitatory. The major efferents are excitatory: the climbing fibres (CF) from the inferior olive (IO) and the mossy fibres (MF) from the pontine nuclei (PN) and elsewhere. Inhibitory inputs are very much smaller in number and arise in the locus coeruleus (LC) and raphe nucleus (RP). The dotted circle represents the cerebellar nuclei, which receive inputs from all classes of afferent. PF, parallel fibres; GO, Golgi cell; GR, granule cell; BA, basket cell; PC, Purkinje cell; ST, stellate cell. NA, noradrenaline; AS, aspartate; TA, tachykinin. (From Ito, 1984, Figure 3; with permission.)

synapses in the cerebellar or vestibular nuclei. Their dendrites stretch upward throughout the depth of the molecular layer (400 μm) forming a dendritic tree with a total length of some 4 mm per cell. The arrangement of the dendrites is extremely precise. They are oriented in a single narrow plane perpendicular to the long axis of the folia. Rows of Purkinje cells appear to have their dendritic trees stacked on top of each other like plates.

The two types of input to the cortex show extremes of convergence and divergence in their contacts with the Purkinje cells. Climbing fibres, which originate in the inferior olive, synapse directly with Purkinje cells. Each Purkinje cell is contacted only by one climbing fibre, which forms 150–200 synaptic contacts on the proximal dendrites. This synaptic arrangement is very powerful: a single impulse in the preterminal axon always produces firing of the Purkinje cell (see below). Each climbing fibre may, however, contact 7–15 Purkinje cells.

Mossy fibres, which do not contact the Purkinje cells directly, come from many sources and have axons of larger diameter than those of the climbing fibres. Their axons end in the granule layer. This is made up of an enormous number of small densely packed cells (10^{10} to 10^{11}) with very scanty cytoplasm. Each granule cell has four or five small dendrites which end in clawlike expansions known as rosettes. Rosettes from up to 28 different granule cells are clustered together to form a structure known as a glomerulus. It is at these sites that synapses are formed with the incoming mossy fibres and also with axons of Golgi cells (see below). A single mossy fibre may branch to innervate many glomeruli in more than one cerebellar lobule. Since the mossy fibres contact many granule cells in one glomerulus, it has been calculated that each mossy fibre may contact on average over 400 separate granule cells.

The axons of the granule cells ascend vertically through the Purkinje layer to the molecular layer, where they bifurcate in a 'T' junction. The branch on each of the arms of the 'T' is known as a parallel fibre, and runs along the axis of the folia so as to contact the maximum number of Purkinje cells, whose planar dendritic trees are stacked perpendicular to their course. The parallel fibres are longer than was once thought. In the monkey, they average about 6 mm, which is roughly one third of the width of the macaque's cerebellar hemisphere. Along their length, therefore, each parallel fibre will pass through the dendritic trees of several hundred thousand Purkinje cells (Mugnanini, 1983). One, or at most two, synapses are formed with any Purkinje cell, and frequently none at all. A Purkinje cell has up to 80 000 parallel fibre synapses on its distal dendritic tree. It is the end point of a remarkable degree of divergence from a single mossy fibre input.

Both parallel and climbing fibre inputs are excitatory to Purkinje cells, and synapse at special sites known as dendritic spines. The parallel fibre synapse is probably glutamatergic, whereas the climbing fibre transmitter is thought to be aspartate. The other inputs are from cortical interneurones or aminergic fibres. They are inhibitory and contact the smooth surface of the cell soma or dendrites.

There are three types of cortical interneurone. Two of them, the stellate and basket cells have cell bodies in the molecular layer. These cells receive input from parallel fibres and direct their inhibitory output to the Purkinje cells. Stellate cells synapse with the middle and distal parts of the dendritic tree, while the basket cells have axon terminals which interweave, basket-like, around the soma and proximal dendrites of the Purkinje cell. Each Purkinje cell may receive input from 20–30 basket cells. Basket cells inhibit Purkinje cells which lie 'off the beam' of the parallel fibres which excite them.

The final interneurone of the cerebellar cortex is the Golgi cell, which has its cell body in the granule layer. Its dendrites ramify up into the molecular layer, where they receive input from parallel fibres. In addition, they receive input from mossy fibres at the glomeruli in the granule layer. Unlike the Purkinje, stellate and basket cells, their dendrites are distributed to form a cylinder rather than a plane. They inhibit granule cells, and may perhaps be involved in a feedback inhibition of the mossy fibre input.

Electrophysiology

Because of its size, the Purkinje cell has been intensively investigated with electrophysiological techniques. It contains the usual voltage sensitive Na^+ and K^+ conductances in the soma and axon membranes which are responsible for the production of rapid action potentials. In addition, there are other voltage sensitive channels which give the cell some unusual firing characteristics.

If the soma is depolarized by passing a just threshold steady current through an intracellular microelectrode, then the cell will fire a repetitive series of impulses. Increasing the intensity of the depolarizing current produces a different behaviour. The onset of repetitive firing occurs earlier and, rather than giving a continuous train of impulses, rhythmic bursts of impulses are seen. The end of each burst is signalled by a reduction in the amplitude of the individual spikes of the burst.

Two main conductances underly this behaviour. A non-inactivating Na^+ conductance in the soma membrane is responsible for the repetitive firing of the cell during prolonged depolarization, A voltage-sensitive Ca^{2+} conductance in the dendritic membrane is responsible for the bursting behaviour. This latter can be demonstrated by blocking the Na^+ channels with tetrodotoxin (TTX), which abolishes the fast spikes, leaving a late, slow-rising burst potential intact. This potential is blocked by removing Ca^{2+} ions and replacing them with cobalt or manganese ions.

Under normal circumstances, the Purkinje cell is activated either by parallel or climbing fibre input. Activation of the climbing fibre input generates a very large (25 mV) EPSP in the Purkinje cell. It is the most powerful synapse in the CNS, and is a consequence of the large number of synaptic contacts made by each climbing fibre. It causes a burst of spikes followed by a long repolarization phase, which may have additional spikes superimposed upon it. This type of activity is produced by activation of both the dendritic Ca^{2+} conductance and the Na^+ and K^+ conductances of the soma and axon. The spikes are known as **complex** spikes. A single action potential in a climbing fibre always produces a complex spike in a Purkinje cell. It is an obligatory synapse with all-or-none properties.

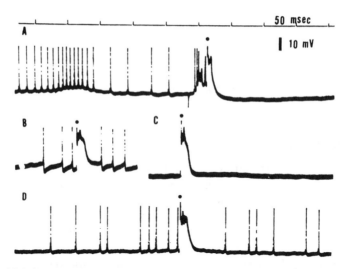

Figure 10.9 Intracellular potentials recorded from cerebellar Purkinje cells. Dots mark responses evoked by climbing fibre connections: complex spikes. The shorter action potentials are simple spikes caused by mossy fibre input. (From Martinez, Crill and Kennedy, 1971, Figure 1A–D; with permission.)

Parallel fibre input is much weaker. It generates a graded EPSP in the Purkinje cell which may or may not generate a single rapid action potential. These are known as **simple** spikes and probably do not involve activation of the dendritic Ca^{2+} conductance. Due to its different types of spiking, the Purkinje cell is one of the few places in the CNS where the morphology of the response indicates the mode of afferent excitation (Figure 10.9). Only if exceptionally synchronous activation

Figure 10.10 Changes in the response of a Purkinje cell to mossy fibre input after a period of conjunctive mossy and climbing fibre stimulation. (a) Experimental arrangement. A Purkinje cell was recorded in the flocculus and three of its inputs were stimulated (S_1, S_2, S_3) in the ipsilateral and contralateral vestibular nuclei (iVN and cVN) and in the dorsal cap of the inferior olive (DC). Mossy fibres (MF), climbing fibres (CF) and cerebellar cortical interneurones are shown. (b) Records of simple spike Purkinje cell discharge. *A*, Part of a continuous recording period. Over the period marked by dashed lines (*b* and *c*) stimuli were given at 2 Hz to the ipsilateral vestibular nerve. Each stimulus produced, in *b*, a short duration increase in spike discharge which is shown more clearly in *B*. *B*, A time histogram of the number of simple spikes seen in the Purkinje cell following a single stimulus to the vestibular nerve (at arrow). On this short time scale, an early increase in excitability of the Purkinje cell is quite obvious. During the period marked by the solid horizontal line in *A*, there was conjunctive stimulation of the ipsilateral vestibular nerve at 20 Hz and the olive at 4 Hz. This gave a greatly increased rate of simple spike firing. After this period, the spike rate decreased again but became much more variable. The response to the ipsilateral vestibular nerve stimulation at 2 Hz over time period *c* was now considerably diminished (see histogram *C*). (From Ito, Sakurai and Tongroach, 1982, Figures 1 and 5; with permission.)

occurs via the parallel fibre system can complex spikes be produced. (Ito, 1984, gives a detailed survey of cerebellar circuitry.)

Synaptic 'learning'

It has often been speculated that the cerebellum participates in one way

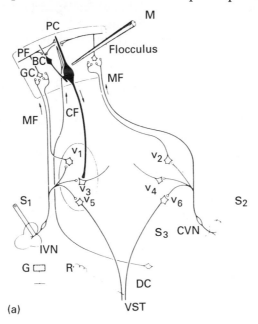

(a)

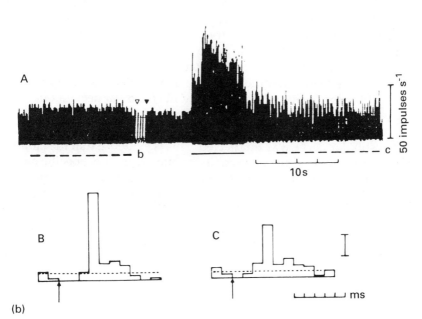

(b)

or another in motor learning. As described below, the cerebellum might adjust the gain of certain reflexes, time the order of contraction of muscles in a complex task, or coordinate the action of many different muscles in a multi-joint task. The basic postulate of many of these theories is that the activity in climbing fibres can modulate the effectiveness of the parallel fibre–Purkinje cell synapse. In effect, the climbing fibre input is thought to 'teach' the cerebellar cortex how to respond to particular patterns of parallel fibre inputs.

At a cellular level, the essential component is a protein kinase molecule in Purkinje cells which, when activated, can desensitize the glutamate receptors of the parallel fibre synapse. The kinase is activated by (i) a cofactor (diacylglycerol), which is formed whenever a synapse is activated and (ii) a high concentration of calcium ions. Climbing fibre input opens a voltage-dependent calcium channels in the dendrites and therefore is a powerful method of providing elevated calcium levels. In addition, concurrent activation of parallel fibre synapses may even facilitate this calcium influx.

Thus, if a Purkinje cell is activated at approximately the same time by climbing fibre and parallel fibre inputs, the cofactor will be formed and a high level of calcium will occur. These two phenomena will induce protein kinase, and the efficiency of the glutamatergic synapse between parallel fibre and Purkinje cell would be decreased. This effect may form the basis of cerebellar learning.

Some evidence that this change in synaptic function actually occurs comes from experiments performed on the vestibular input to cat cerebellum (Figure 10.10). A Purkinje cell was recorded while its input from three separate sources was stimulated: (i) mossy fibre input from the left vestibular nerve; (ii) mossy fibre input from the right vestibular nerve; (iii) climbing fibre input from the olive. If the climbing fibre input was stimulated at 4 Hz at the same time as stimuli were given to the mossy fibre input from the right (but not the left) vestibular nerve, then later tests showed a specific influence on the effectiveness of the right mossy fibre input. Stimulation of the right vestibular nerve alone (without any concomitant climbing fibre input) evoked a **reduced** simple spike discharge in the Purkinje cell. The response to stimulation of the left vestibular nerve was unaffected. This is evidence that the climbing fibre inputs can produce changes in the effectiveness of specific parallel fibre inputs to Purkinje cells. Whether such changes occur under more physiological conditions, in which climbing fibre discharges are much less regular and frequent than that used in these experiments is still under investigation (see behavioural studies detailed below) (Ito, Sakurai and Tongroach, 1982; Ito, 1984).

10.3 *Efferent pathways of the cerebellum*

The Purkinje cells are the only output cells of the cerebellar cortex. They project to the cerebellar nuclei and the vestibular nuclei and in all cases their action is inhibitory. The efferent fibres from these nuclei are the only route whereby the cerebellum can influence other parts of the brain and spinal cord (Asanuma, Thach and Jones 1983; Rispal-Padel, Harnors and Troiani, 1987; Gilman, 1992).

As already discussed above, the three main longitudinal zones of the cerebellar cortex project to the three ipsilateral cerebellar nuclei. In addition, the lateral vestibular (Deiter's) nucleus receives a direct projection from the lateral portion of the anterior lobe. Other regions of the vestibular nuclei which are involved mainly in the control of eye movements receive afferents indirectly by the flocculonodular node and the uvula. Although the projection from the cerebellar cortex to the nuclei is topographically quite specific, there is still a certain amount of divergence in this pathway. Each nuclear cell, for example, can be contacted by up to 800 different Purkinje cells.

Efferent pathways of the cerebellar nuclei

All three nuclei send projections via the thalamus to cerebral cortex in addition to their terminations in various midbrain and brainstem nuclei, and even spinal cord (from the interpositus and fastigial nuclei). The fastigial nucleus provides a minority of cerebellar output. Only a small proportion ends in the thalamus; the majority travels in the superior and inferior cerebellar peduncles to terminate in the vestibular nuclei, various regions of the reticular formation and the spinal cord.

The bulk of the cerebellar output is provided by the interpositus and dentate nuclei. Efferents from these nuclei leave the cerebellum in the ipsilateral superior peduncle, and cross to the contralateral midbrain tegmentum. At this point the tract divides into an ascending and descending component, with many of the fibres branching to provide axons in both components (Figure 10.11). The descending axons innervate nuclei in the pontine and medullary reticular formation. The ascending axons innervate the contralateral red nucleus, and what has been termed the 'cerebellar' thalamus. This area of the ventrolateral thalamus is virtually devoid of terminals from either spinal or basal ganglia afferents and consists of the following anatomically-defined nuclei: ventroposterior lateral pars oralis (VPLo), ventrolateral pars postrema (VLps), ventrolateral pars caudalis (VLc) and nucleus X.

The two portions of the red nucleus receive fibres from different parts of the cerebellar nuclei. The anterior interposed nucleus sends fibres

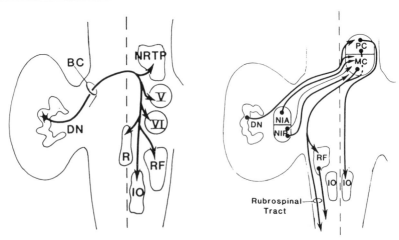

Figure 10.11 Diagram of the non-thalamic outputs of the dentate and interposed nuclei. On the left are the descending projections (only the dentate is shown here, for clarity: the interpositus connections are the same) to the nucleus reticularis tegmenti pontis (NRTP), the V and VIth cranial nerve nuclei, the raphe nuclei (R), reticular formation (RF) and inferior olive (IO). On the right are the ascending projections to the red nucleus. The dentate innervates the rostral and anterior caudal parts of the nucleus: the interposed (mainly the anterior region) projects to the caudal part. The rostral (or parvocellular (PC)) red nucleus projects to the ipsilateral inferior olive. In the monkey, the caudal (or magnocellular (MC)) region is the origin of the rubrospinal tract, whereas in the cat, some contribution is also seen from the rostral part (dotted line). (From Gilman, Bloedel and Lechtenberg, 1981, Figures 3 and 4, with permission.)

somatotopically to the caudal, or magnocellular, portion of the red nucleus. Thus this part of the cerebellar nuclei has direct access to the rubrospinal tract which has its origin in the large cells of the red nucleus (in the monkey). There is also a small projection from posterior interposed to the medial portion of the red nucleus. Fibres from the dentate enter the rostral, or parvocellular part of the red nucleus. In man this region is much larger than the magnocellular part. From here fibres descend in the ventral tegmental tract to innervate the ipsilateral inferior olive. The olivocerebellar projection completes a cerebello-rubro-olivo-cerebellar loop.

The cerebellar projection to the thalamus travels through the red nucleus and hence is difficult to study in isolation. All three nuclei contribute fibres, and their terminals form interdigitating patches or rods within at least three of the 'cerebellar' thalamic nuclei (VPLo, VLc, VLps). From here, fibres project to the primary motor cortex (Brodmann's area 4). Nucleus X may be slightly different. Its main input comes from the dentate, possibly from a region caudal and ventral to the projection to the rest of the thalamus. In addition, there is a contribution from posterior interpositus. Nucleus X projects to the arcuate premotor area (lateral Brodmann's area 6). Although cortical areas 4 and 6 are the main sites of termination of cerebellar efferents, the most recent evidence suggests that cerebellar output may influence wider areas of cortex including prefrontal and parietal regions. The projections from the cerebellar nuclei to cerebral cortex are somatotopically arranged so that, for example, the arm area of the motor cortex receives input separately from all three nuclei. Thach and colleagues (1992) have suggested that this allows the three subdivisions of the cerebellum to contribute simultaneously to different aspects of movement control (Schell and Strick, 1984; Orioli and Strick, 1989).

In addition to these connections with other areas of the brain, there are also projections from all three cerebellar nuclei back on to the cerebellar cortex. In general, these connections reciprocate the corticonuclear projections and end as mossy fibres. An exception to this rule in primates is that from the dentate, which sends fibres to the vermis as well as to the lateral areas of the cerebellar cortex.

10.4 *Electrophysiological studies of the cerebellum*

Background levels of discharge

At rest in either decerebrate, lightly anaesthetized or awake animals, Purkinje and nuclear cells both discharge at high rates of between 40 and 80 Hz. This is far higher, for example than the maximum 20 Hz seen in pyramidal tract cells of relaxed animals and suggests that the cerebellum, even at rest, may be providing some tonic input to other structures. Indeed, removal of the cerebellum can produce a loss of muscle tone (see below) which would be consistent with this idea. The Purkinje cells and probably the nuclear cells as well are driven to fire at such high frequencies by sustained mossy fibre input. Recordings of complex spike activity show that climbing fibre input is infrequent, having a discharge rate of only about 1 Hz. Even during active movement, when Purkinje rates may exceed 400 Hz, climbing fibre input remains at approximately the same rate.

Despite the low level of activity in the climbing fibre system, the climbing fibres are essential for proper operation of the cerebellum. Removal of the inferior olive, with subsequent degeneration of the climbing fibres produces symptoms in experimental animals that are indistinguishable from removal of the cerebellum itself. Climbing fibre input, although infrequent, and with little immediate effect on Purkinje cell discharge rates, must have some vital long-term influence on cerebellar function.

In general, nuclear and Purkinje cells change firing frequency in phase with each other. For example, increases in Purkinje cell discharge usually are accompanied by increases in nuclear cell firing. This is consistent with the anatomical findings that input to cerebellum projects both to the intracerebellar nuclei and to the cortex. Since the output of cerebellar cortex is then mainly to the nuclei, it may be that the cortical discharge **modifies** the nuclear firing. One idea is that the cortical output may **sculpt** the firing pattern of the nuclei. This might happen either in terms of space or time. Thus the inhibition from the cerebellar cortex may allow only certain groups of nuclear cells to respond to mossy fibre input. Alternatively, the inhibitory cortical output might cut short the nuclear discharge and sharpen the temporal firing pattern of the cells.

Evidence for this comes from studies in which afferent input has been used to modify cerebellar firing rates. For example, many inputs to the cerebellum produce an initial excitation followed by a longer period of inhibition in the neurones of the cerebellar nuclei. The inhibition is abolished by cooling the cerebellar cortex, suggesting that it is produced by activity in Purkinje cells excited by the same afferent systems as project to the nuclei. These ideas on the reciprocal relationship between cortical and nuclear firing have mostly come from studies comparing population discharge in the two areas. At the level of individual corticonuclear cell pairings there is no correlation between discharge rate in a Purkinje cell and its target cells in the cerebellar nuclei. This implies that the activity of a single nuclear neurone is not influenced noticeably by the firing of one Purkinje cell. In fact, since up to 800 Purkinje cells converge onto a nuclear cell, this emphasizes the importance of population discharge in determining the final cerebellar output (McDevitt, Ebner and Bloedel, 1987). Most studies on movement-related firing of cerebellar cells have, however, concentrated on nuclear, rather than Purkinje cells.

Activity of the cerebellar neurones during active movement

Many of the neurones of the cerebellar cortex and deep nuclei change their discharge rate in association with movements of the body. Unlike the cells of the motor cortex, cerebellar neurones have no direct connections with either spinal or bulbar motoneurones, so that their discharge is less likely to be related to specific patterns of muscle activity. Instead, their discharge may provide insight into how a different level of the CNS is involved in the control of movement (Brooks and Thach, 1981; Gilman, 1992; Thach, Goodkin and Keating, 1992).

Timing of discharge
The timing of changes in discharge in the cerebellar nuclei has been

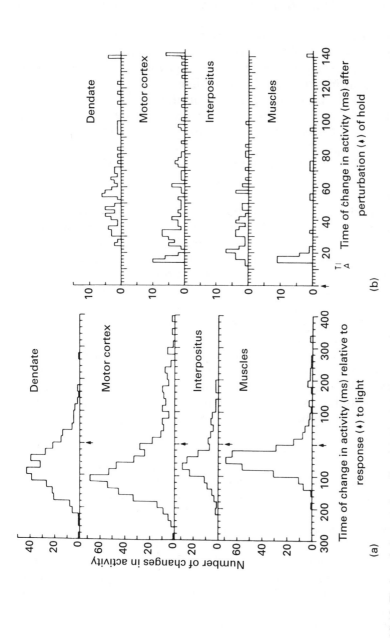

Figure 10.12 Timing of changes in cell activity in dentate and interpositus nuclei, motor cortex and muscle. (a) Responses prior to and during a visual reaction task. The histograms show the number of cells responding at different times relative to arm movement (arrow at $t = 0$ ms). (b) The monkey was instructed to maintain a constant position of his wrist in a manipulandum while receiving randomly timed perturbations from a torque motor (arrow at $t = 0$ ms). Prior to voluntary (reaction time) movement, cells start to fire first in dentate and motor cortex. Activity occurs slightly later in interpositus and muscle. Following a wrist perturbation, muscles respond first (tendon jerk component of a stretch reflex), followed by interpositus, motor cortex and then dentate. (From Thach, 1978, Figure 16; with permission.)

examined in two situations: in relation to the onset of voluntary movement, and in relation to perturbations given during the course of a movement.

In monkeys flexing and extending the wrist in response to a visual signal, there is considerable overlap in the onset of changes in discharge rate in the cerebellar nuclei. However, on average, dentate neurones change their firing at about the same time as motor cortex cells in the same task. Both occur before activity in the interpositus (see Figure 10.12). Fastigial neurones usually discharge after interpositus in single joint movements. In contrast, when monkeys perform whole arm reaching movements, fastigial activity occurs first, even before that in the dentate. In those studies in which they were recorded, Purkinje cells often changed firing at the same time as the nuclear cells to which they projected.

The hypothesis from such experiments is that firing of dentate neurones is related to the initiation of a voluntary response, whilst interpositus neurones fire mostly in relationship to the actual movement being executed. The variation in timing of fastigial activity may be related to whether postural control is important in the task.

Experiments in which perturbations have been given confirm this general impression of the timing of nuclear activity. Rather than have a monkey flex its elbow in response to a target light, monkeys were trained to hold their wrist in a constant position (Figure 10.12b; Figure 10.13). At random intervals, the elbow was extended or flexed by application of an external force. The monkeys could be instructed in advance on how to react to the disturbance. On some occasions the animal was told to assist the disturbance by moving the elbow rapidly in the direction of the force. In others, the monkey had to assist the perturbation.

The EMG response consisted of any early reflex component followed by a later voluntary response which was dependent upon the prior instruction. In these circumstances, interpositus neurones changed their firing pattern before dentate neurones. The effect on the discharge was related to the direction of the perturbation. The discharge of dentate cells was more closely related to the direction of the voluntary response and could occur even when there was no preceding elbow perturbation. Again, dentate neurones appear to play a role in the initiation of the voluntary component of the movement, whereas interpositus neurones may have been more involved in the automatic, reflex, control of such movements.

Another piece of evidence favouring these arguments is that during locomotion on a treadmill in cats, the discharge of interpositus neurones is highly correlated with EMG changes occurring during the step cycle. In contrast, dentate nuclear discharge is not closely coupled with

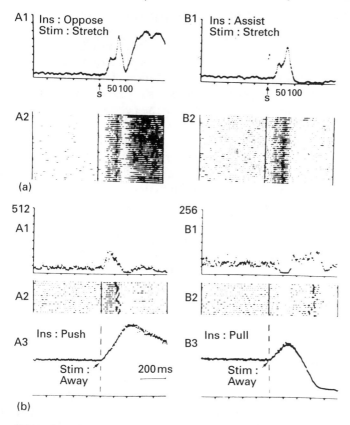

Figure 10.13 Responses of human muscle and neurones in the dentate nucleus of a monkey to perturbations of the arm and their dependence on prior instruction. (a) Muscle responses recorded from a **human** subject. The task was to maintain a constant wrist position against a torque motor which could apply random supination/pronation perturbations. Imposed pronation stretched the biceps. In some trials the subject had to oppose the displacement and in others he had to assist it. The responses from biceps are shown in two ways: in the upper traces is an average rectified EMG, in the lower traces are raster displays of the responses recorded on each single trial. On the left, the biceps was stretched and the instruction was to oppose the stretch. In the EMG there is a stretch reflex starting about 20 ms after stretch (short-latency response) with a later component just visible (long-latency response) starting at about 50 ms. Following these reflex responses is the voluntary muscle activity at 100 ms. On the right, the same stretch was given but the instruction was to assist the disturbance. In this case, the stretch reflex response remains, but the voluntary activity disappears. (b) Responses of a dentate neurone recorded in an awake **monkey** are shown in the same way. In this instance, the monkey held a push/pull handle within a target zone. The handle could be moved by a motor so as to extend triceps (Stim: Away) by moving away from the monkey. When the instruction calls for the monkey to assist the perturbation by pushing (left), there is short-latency increase in dentate activity at about 10 ms. When instructed to resist the disturbance (right, Pull), this activity is not present and there is a reduction in dentate firing. Corresponding position records from the handle are shown below. (From Strick, 1978, Figure 1; with permission.)

stepping at all, but is very responsive to perturbations (e.g. slowing the treadmill) involving arrest and resumption of locomotion. It is as if the dentate activity was related to restarting the locomotor pattern.

Finally, there is one piece of evidence that dentate activity may sometimes contribute directly to the initiation of movement. Cooling of the dentate nucleus in a monkey increases the reaction time in visually-triggered wrist movements and delays the occurence of motor cortical discharge. The motor cortical delay was not due to removal of background facilitation provided by the dentate, since cooling did not appreciably change the resting discharge in motor cortex cells.

Other experiments suggest that the situation is more complicated. For example, cooling the ventrolateral thalamus, which is thought to be the synaptic relay from dentate to motor cortex, does not produce any delay in reaction time. Similarly, cooling of the motor cortex itself has little effect on reaction time in the monkey in comparison with dentate cooling. The movements are much smaller but they appear to begin at the same short latency. A possible explanation of this is that output from dentate can in certain circumstances produce movement via subcortical pathways (for example, its projection to the reticular formation), although this may not be the normal mode of activity.

The relative contributions of dentate and interpositus to different phases of the movement is consistent with the anatomical relationships of the two nuclei. The dentate receives its major input from two sources, the overlying lateral cerebellar cortex and from direct pontine afferents. Both regions are supplied primarily by cerebral cortical fibres, perhaps providing input concerned with the command to move. The interpositus nucleus receives input from intermediate cerebellar cortex and from spinocerebellar and pontine afferents. It is thus receiving both peripheral and central input, and it has been suggested that such a combination would allow comparison of the central command to move with the present limb position and provide a possible basis on which to make corrective adjustments to the movement in progress.

Relation of discharge to movement parameters

The studies above show that the discharge of cerebellar nucleus neurones changes at specific times during specific types of movement. To what aspect of the movement is this discharge related? There have been many experiments in which possible relationships between force, velocity, amplitude and direction of the movement have been examined. The best summary of the data is that if the relationships exist, then they are not very clear (Mink and Thach, 1991; Thach, Goodkin and Keating, 1992).

The majority of experiments have been conducted on relatively fast wrist or elbow movements, with the monkey moving in either flexion or extension at different speeds to different targets against different loads in order to separate out possible correlations with the parameters of movement. In several studies, interpositus neurones discharged in relation to muscle force or EMG with weaker correlations with direction or velocity of movement. The discharge of dentate neurones was not clearly correlated with any particular aspect of the task, although several groups reported that the direction of movement was an important factor. A tentative conclusion from this might be that, given the relative timing of the discharges, dentate activity may be signalling that a movement is to occur, without specifying its precise characteristics (except perhaps direction). Interpositus activity would be involved in control of the muscle activity needed for the movement underway (e.g. Thach, 1978; MacKay, 1988).

Unfortunately, there are three sets of results which complicate the picture.

(a) *Population coding.* The techniques of Georgopoulos and colleagues (see Chapter 9) on population coding have been applied to the firing of cerebellar Purkinje and nuclear cells. Monkeys were trained to point from a central target to one of eight different targets located on a surrounding concentric ring. The discharge pattern for many neurones was then recorded for each of the movements. When averaged over all eight movements, a large majority of the neurones (84%) had a preferred direction of movement for which their discharge was highest. In the model of Georgopoulos, the neurones are said to be coding the direction of their preferred movement. If a cell discharges little during any movement, then this indicates that the movement made was in some direction distant from the neurone's preferred direction. If the discharge is high, then the actual direction of movement is presumed to be similar to the preferred direction of that cell.

For any one cell in the cerebellum, the relation between direction moved and discharge rate is usually very noisy. Because of this it is not possible to predict a direction of movement from the firing pattern of any one cell. However, by studying a large population of cerebellar neurones it is possible to show that the average movement predicted by all cells is a very good indicator of the direction actually moved. This averaging process is called vector summation (Figure 10.14).

The net result of these experiments indicates that the population discharge of Purkinje cells, interpositus and dentate cells is highly correlated with the direction of movement. Thus, despite the fact that single cell discharge was found in several studies to be independent of movement direction, this is not so when the population is examined. The implication is that the relationship of cerebellar discharge to parameters of movement may have been inadequately studied in the past.

(b) *Relationship of cerebellar firing to activity in antagonist muscles.* So far,

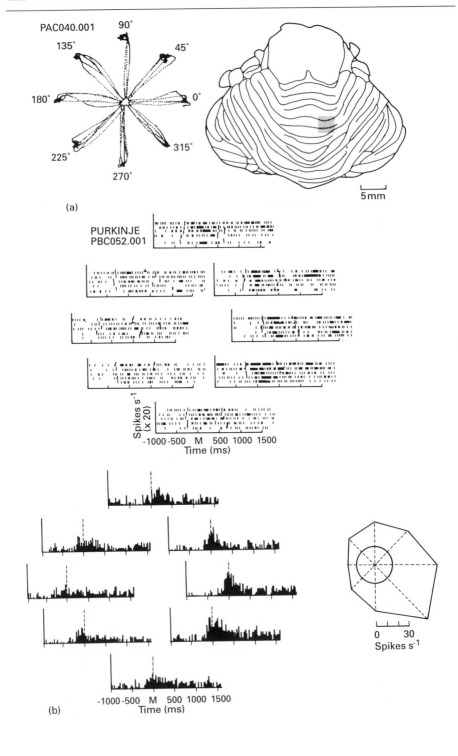

(a)

(b)

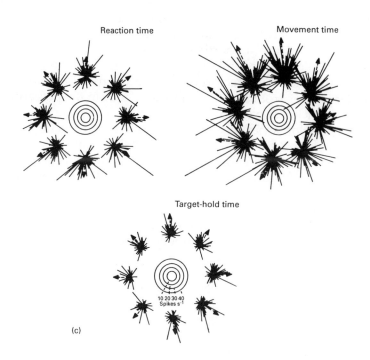

Figure 10.14 Population coding of neural discharge in cerebellar reaching movements. Monkeys were trained to make arm reaching movements in one of eight different directions from a central point (a; trajectories), whilst neurones were recorded in the shoulder representation area of the anterior lobe of the cerebellum (a, right). (b) The typical pattern of cerebellar neurone activity during performance of each of eight different directions of movement. The upper panel shows the discharge as a raster-plot aligned to the onset of movement (M), whilst the lower panel shows the same data in histogram form. The panel on the right is a polar-plot of the data. The circle in the middle of the polar-plot represents mean activity of the cell during the hold time. The length of each axis represents the mean neuronal activity at the onset of movement. This cell responded best during movement towards the lower right quadrant. (c) A vector representation of cerebellar population discharge during three portions of the movement: the reaction time (before motion of the arm had begun), the movement time (during movement towards the arm from the centre to the target position), and the target hold-time (the hold period at the end of the movement). The eight groups of vectors represent the activities during the eight directions of reaching. The position of a group of vectors relative to the centre of the array represents the direction of movement, and each vector within a group represents mean activity and preferred direction of a single neurone during this movement. The heavy arrows represent the vector sum for each group of vectors. Note that the population vectors point in the approximate direction of movement in the reaction time and target hold-time. The vectors are not quite so clearly related to the direction of movement during the movement time. (From Fortier, Kalaska and Smith, 1989, Figures 2, 3, 6 and 11; with permission.)

it has been considered that cerebellar activity might be related to the control of the prime mover, or agonist, muscle. However, there is no reason why the cerebellum might not be involved in control of antagonist muscles. Evidence supporting this possibility comes from studies in which monkeys were trained to co-contract antagonist muscles, rather than to use them reciprocally as in the usual type of experiment. Recordings from identified Purkinje cells in the posterior part of the anterior lobe revealed that about 70% of cells decreased their firing rate during an isometric grip of thumb and forefinger which involved co-contraction of antagonist wrist flexor and extensor muscles in the forearm. In contrast, almost all of these cells showed a reciprocal activation to isometric flexion or extension of the wrist which did not involve co-contraction of antagonists. When recordings were made from dentate or interpositus nuclei, many cells were found to behave in a similar fashion except that these neurones increased their firing during co-contraction, rather than decreasing it as did the Purkinje cells.

The hypothesis is that one role of Purkinje cell discharge is to control activity of antagonist muscles. In situations requiring reciprocal activation of antagonist muscles, increased Purkinje cell discharge might cause inhibition of antagonists. Conversely, in situations in which co-contraction is required, a decrease in Purkinje cell firing might cause an increase in antagonist activity.

The interest in these studies is the finding that there is a reciprocal relationship between the discharge of Purkinje and nuclear cells. Although it is not proven, it may be that the decrease in Purkinje output during co-contraction reduces inhibition of nuclear cells and causes an increase in nuclear firing. Additionally, it has also been found that other inhibitory interneurones of the cerebellar cortex actually increase their discharge during co-contraction. Such discharge may then be directly responsible for inhibition of Purkinje cells (Brooks and Thach, 1981; Wetts, Kalaska and Smith, 1985).

(c) *Activity of cerebellar neurones during slow movements.* Slow movements do not have a precisely defined time of onset, so that it is not possible to investigate time relations between cerebellar activity and movement. However, it is possible to investigate the activity of cells during the movement. Unlike the reciprocal activity of many cells which is seen during rapid flexion and extension tasks, almost all task-related cells in dentate and interpositus discharge equally well in slow movements in either direction. They are said to be **bidirectional** neurones. Many of the cells are the same as those which have unidirectional discharge during rapid movements, or in response to torque perturbations. In monkeys that have been examined in this way, there was no antagonist co-contraction or excessive postural muscle activity which might explain such relatively non-specific responses, so that other explanations have been sought.

One hypothesis which has been put forward is that the activity of these cerebellar neurones is more related to the activity of γ-motoneurones than α-motoneurones. It is further suggested that in slow tracking tasks, gamma discharge to both agonist and antagonist muscles is increased, making the spindles more sensitive in both muscles. In accordance with this idea, spindle recordings in the monkeys did show an increase in spindle activity from both agonist and antagonist muscles at the onset of movement. The increased antagonist activity persisted even during slight unloading of the muscle. At least in this instance, gamma discharge to the antagonist increased even though there was no sign of activity in the extrafusal muscle.

This is an example of α–γ dissociation which was discussed in more detail in Chapter 6. However, it should be remembered that α–γ dissociation has not yet been documented in humans, particularly during similar slow movements of the wrist. The idea that the cerebellum may be involved in control of γ-motoneurones is supported by evidence from lesion studies (see below) (Brooks and Thach, 1981).

10.5 *Effects of cerebellar lesions*

Recordings from single neurones do not give a very clear idea of which particular aspects of movements are coded by cerebellar neurones. Studies of the effects of cerebellar ablation give more information of its possible role.

Muscle tone

The cerebellum can influence muscle tone by changing the balance of α and γ activity to muscles and muscle spindles. Thus, the background level of excitability in spinal motoneurones can be increased, so that they discharge more readily even at rest, or the level of fusimotor drive can be changed so that spindle sensitivity is altered.

Cerebellar lesions in cats and dogs, unlike those in primates (see below) produce **increased** muscle tone. This is due primarily to increased excitability of α-motoneurones. The earliest studies were made on decerebrate animals. In these preparations, removal of the cerebellum increases extensor tone still further in all four limbs and the neck (opisthotonus). The effect can be produced by ablation of the anterior lobe of the cerebellum alone, indicating that this is the principal structure responsible for increasing the hypertonus (see Figure 10.15, in

Figure 10.15 Extreme extensor rigidity in all four limbs and neck of a decerebrate cat in which the anterior lobe of the cerebellum also had been inactivated by ligation of the internal carotoid arteries. (From Pollock and Davis, 1927, Figure 3; with permission.)

which ligation of the internal carotid arteries has produced both decerebration and inactivation of the anterior lobe of the cerebellum). The rigidity does not depend upon stretch reflex activity since it is not abolished (as is decerebrate rigidity) by sectioning the dorsal roots of the spinal cord. It is said to be an **alpha** rigidity, produced by direct activation of α-motoneurones in the spinal cord from the vestibulospinal tract.

The mechanisms responsible for this effect are as follows. The anterior lobe of the cerebellum can influence activity in the vestibulospinal tract, which projects mainly to the extensor muscles, in two ways: a direct inhibitory projection to the lateral vestibular (Deiter's) nucleus by Purkinje cells in the anterior lobe, and an indirect net excitatory input via the fastigial nucleus. Removal of the anterior lobe abolishes the inhibitory input and vestibulospinal activity increases. This can be reduced either by ablation of the fastigial nucleus, or by section of the vestibular nerve.

Although α effects predominate, cerebellectomy in cats and dogs also produces effects on fusimotor discharge. Recordings from γ-motoneurones show that in contrast to the excitation of α-motoneurones there is a decrease in fusimotor activity and a decrease in sensitivity of the stretch reflex similar to that seen in primates (see below). This decrease in fusimotor activity is due to the removal of tonic input from all three cerebellar nuclei. Some days after cerebellar ablation (the time depends on the species of animal involved) there is a gradual decrease in extensor tonus, accompanied by a gradual increase in spindle sensitivity to normal values.

In primates, the situation is quite different. The main cerebellar influence on muscle tone is thought to be primarily via an action on the fusimotor system. The lack of direct, α effects is probably due to the smaller cerebellar output to Deiter's nucleus in primates and the increasing dominance of the cerebellar hemispheres. Gilman's (1969) classic studies in the monkey showed that cerebellectomy produces a *hypotonia* of all muscles which recovers only very slowly, if at all. More pronounced deficits are in the rate and force of voluntary muscle contractions together with tremor (especially if the cerebellar nuclei are damaged) (see sections below on pathophysiology). Limbs tend to be held in flexion and there is nystagmus. All deficits are seen especially well if movements are made which require postural support.

The decreased tone is accompanied by a decrease in sensitivity of primary spindle afferents to both phasic and sustained stretch. The sensitivity of secondary spindle endings is unchanged. It is thought that these effects are caused by underactivity of the fusimotor system.

The idea that one function of the cerebellum is to regulate the balance of activity in α and γ motoneurones is consistent with the observations of Thach, Goodin and Keating (1992) discussed above who concluded that under certain circumstances dentate and interpositus discharge was related to fusimotor activity in wrist flexion/extension tasks. However, a more recent experiment has cast doubt on the importance of any changes in fusimotor function caused by cerebellar dysfunction. Unlike the experiments discussed so far, which were carried out on anaesthetized preparations, the new experiments were performed on conscious, freely moving animals so that the discharge of muscle spindles could be examined under more natural conditions. Neurones of the dentate or interpositus nucleus of cats were inactivated temporarily by small injections of the local anaesthetic lidocaine. Recordings of muscle spindle activity were then made in the dorsal roots of the spinal cord whilst the cat walked around the laboratory (Gorassini, Prochazka and Taylor, 1992).

Inactivation of dentate failed to produce any observable effect in either the cat's behaviour, or the discharge of its spindles. In contrast, inactivation of the interpositus produced marked ataxia of gait, with prominent hyperflexion of the hindlimb during the swing phase. Despite this change in behaviour there was no change in the discharge pattern of muscle spindles. Thus, cerebellar symptoms occurred in the absence of changes in receptor sensitivity.

Why, then, are these results different to those described previously? One important difference is that during normal walking, fusimotor output changes a great deal during different phases of the step cycle. In the conscious cat, changes of two to five times the resting discharge rate are common. This is much larger than the reductions in fusimotor discharge reported for anaesthetized decerebellate animals, and hence it may be that these small effects were effectively concealed by the large fluctuations which accompany natural movement. In other words, the cerebellar influence of fusimotor discharge in these conditions is much smaller than the contribution to spindle sensitivity provided by other structures.

Effect of cerebellar lesions on movement control

Single joint movements

Most of these experiments have used the technique of reversible inactivation of the cerebellar nuclei with local cooling probes. (Brooks and Thach, 1981). Combined cooling of dentate and interpositus of monkeys produces deficits of movement very similar to those seen in human patients with cerebellar disease (see illustrations in following sections). In the movements made about a single joint such as the elbow or wrist:

(1)   Reaction times are increased.
(2)   If the movement is rapid and self-terminated (rather than being made up to a fixed end-stop), then the duration of the initial burst of agonist EMG activity is longer but smaller in average amplitude than normal, and the onset of antagonist activity (with respect to the peak velocity of movement) is delayed. This leads to overshoot of the target.
(3)   In rapid, repetitive movements between two end stops, the amount of time spent at each turn-around point is increased and the rhythm of the movement is interrupted. The critical deficit is delay in the onset of antagonist activity needed to provide rapid reversal of movement.
(4)   Smooth, slow movements break up into jerky intermittent segments which occur at about 3–5 Hz. Records from motor cortical neurones also reveal a predominant oscillation at about 3 Hz.

A further deficit in monkeys has been examined in more detail than that in humans. This involves 'set-dependent' bursts of EMG activity which occur when movements are made under predictable conditions.

In normal, well practised monkeys, short-lasting perturbations evoke a stretch reflex in the agonist muscle (which helps to restore the wrist towards its start position), followed by a burst of EMG activity in the antagonist muscle. This latter helps to damp out the mechanical oscillations of the wrist which follow a pulsatile torque disturbance. Such antagonist activity is not a reflex response to muscle stretch since it occurs before the wrist begins to return towards its original start position (when stretch of the antagonist muscle will occur) (see Figure 10.16). It is thought to be a 'set-dependent' response triggered by the

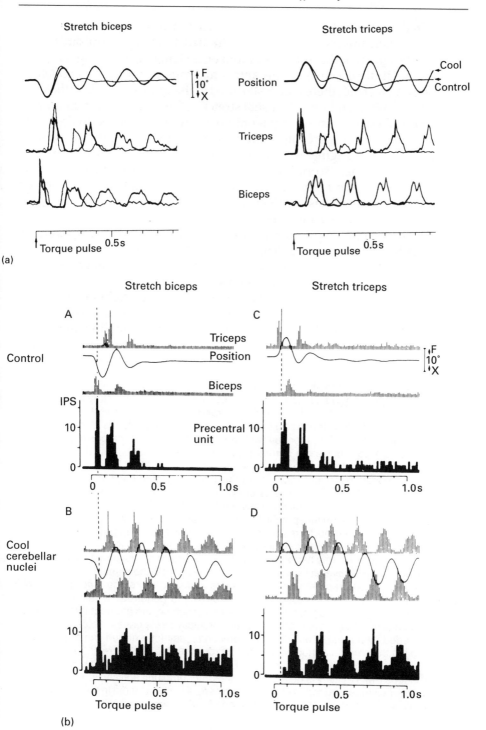

initial stretch of the agonist muscle. That is, a response which occurs only when the monkey has been trained to expect a pulsatile wrist displacement. If the 'set' of the monkey is changed, by giving long-lasting wrist displacements, this antagonist muscle response is no longer present. Instead, there is an EMG burst in the agonist muscle at the same latency, which follows the usual stretch reflex response in that muscle. Under these conditions, antagonist activity is not needed to damp out oscillations. It is more important that extra agonist activity be called up to restore the original wrist position against the maintained load.

The responses to pulsatile disturbances have been investigated in some detail (see Figure 10.16). Cooling of the cerebellar dentate and interposed nuclei has no effect on normal stretch reflex responses of the agonist muscle. It is the 'set-dependent' EMG activity of the antagonist which disappears, to be replaced by later activity driven purely by stretch reflex pathways. This stretch reflex antagonist activity now occurs later than usual, and no longer helps to dampen out oscillations in limb movement. It may even make them worse.

In the normal animal, 'set-dependent' activity in the antagonist muscle is preceded by changes in motor cortical activity in neurones linked to the contraction of that muscle. During cerebellar cooling, this linked precentral activity also becomes delayed and no longer occurs before

Figure 10.16 Cerebellar control of 'set-dependent' EMG activity in antagonist muscles, possibly via a precentral motor cortical pathway. A monkey was trained to hold its elbow in a fixed position and at random times it received brief, pulsatile disturbances which stretched either biceps or triceps and took the elbow away from the target position. The monkey was well trained to restore the position as rapidly as possible. (a) shows the normal EMG responses in triceps (middle traces) and biceps (bottom traces) to this type of disturbance (thin lines), together with superimposed responses recorded during cooling of interposed and dentate nuclei (thick lines). When the biceps is stretched (left panel), there is a stretch reflex response in biceps. This helps to restore the original elbow position. It is followed by a burst of activity in triceps which in the normal condition (thin lines) begins before the elbow has started to return to the control position. This predictive activity in triceps helps damp out the oscillation in the elbow position produced by the disturbance. The same happens when triceps is stretched, but now it is biceps which shows the 'predictive' activity. (b) Another experiment in the same monkey, in which a biceps-linked precentral cell also was recorded at the same time as the EMG activity. In the control state (A), stretch of biceps generates a reflex EMG plus short-latency excitation of the precentral unit which occurs at about the same time. Stretch of triceps (C) generates the usual 'predictive' responses in biceps which are preceded by a burst of activity in the precentral cell (see dotted line at $t = 50$ ms). Cooling the cerebellar nuclei does not affect either the initial stretch reflex EMG or precentral response to biceps stretch (B). However, when triceps is stretched, onset of biceps EMG activity is delayed as is the precentral unit activity D. The precentral activity no longer precedes this burst of biceps activity, as it did in the intact animal. (From Vilis and Hore, 1980, Figures 4 and 5; with permission.)

the EMG onset. It appears that the cerebellum may drive these 'set-dependent' EMG responses via its connections with precentral motor cortex. Such observations after cooling suggest that the tremor seen after cerebellar lesions may be due to the operation of stretch reflex mechanisms which are no longer supplemented by the predictive 'set-dependent' activity of the cerebellum.

Tracking moving visual targets with the hand
Several studies have shown that slow visual tracking movements tend to break-up during cooling of the cerebellar nuclei. If normal monkeys track a moving visual target by moving a joystick with their hand, then they usually match the peak velocity of their movements to the peak velocity of the target. In addition, the animals make catch-up movements when they begin to lag behind the target. The amplitude of these corrections is again best related to the velocity of target movement. This use of velocity information allows the animals to make more accurate tracking movements since it allows them to predict the future position of the target more effectively.

During cooling of the interpositus nucleus, the behaviour of the ipsilateral tracking arm becomes jerky, and the velocity of the corrective movements becomes better related to the position than the velocity of the target. Thus, cerebellar inactivation seems to have impaired the ability of the animal to use velocity information to increase the accuracy of its corrective movements (Miall, Weir and Stein, 1987).

Single joint or coordination of many joints?
The majority of studies have described deficits in simple, single joint movements which were performed by monkeys restrained in a primate chair. However, much more dramatic changes in performance are seen when animals perform unrestrained, multi-joint movements in a natural setting.

Only one experiment (Thach, Goodin and Keating, 1992) has been specifically designed to compare the performance of a simple, re-strained movement with that of normal, multi-joint movement in a group of animals in which the fastigial, dentate or interpositus nuclei were inactivated by local injection of either muscimol (temporary inactivation) or kainate (permanent inactivation). The effects on simple movement were very similar to those described above. That is, small, rather subtle changes in performance which in no case ever resulted in the animal being unable to produce the required movement. In contrast, in complex tasks, clear abnormalities were seen after inactivation, with different nuclei being associated with different deficits. Fastigial inactivation prevented sitting, standing or walking with frequent falls to the side of the lesion. Interposed inactivation produced a severe 3–5 Hz

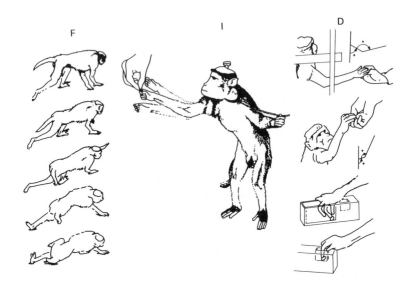

Figure 10.17 Major deficits in monkey movements produced by microinjection of muscimol into the individual cerebellar nuclei. Column F represents problems with stance after fastigial injections of muscimol. The figures represent sequential video frames of a monkey falling towards the side of the lesion. Column I shows prominent arm tremor during reaching after injection of muscimol into the interpositus nucleus. Column D shows deficits in reaching and pinching after dentate injection of muscimol. (From Thach, Goodin and Keating, 1992, Figure 2; with permission.)

action tremor which affected reaching movements of the arm, but not standing, sitting or walking. Dentate inactivation resulted in hypometria (overshooting) on reaching to a target and a tendency to use only one finger to retrieve food from a well in a board in front of the monkey instead of a precision grip (Figure 10.17).

The important conclusion from this study is that the cerebellum may be much more concerned with the coordination of activity at many joints rather than with the performance of movement about one joint. If this is true, then it may explain why previous studies of single joint tasks failed to show a clear relationship between nucleus and discharge and any parameters of movement.

10.6 *Adaptation and learning*

When we imagine what is meant by motor learning, examples such as riding a bike or playing the piano spring to mind. Unfortunately, these are extraordinarily complex tasks and as such are very difficult to analyse in detail. However, it is possible to make one point about motor learning with these examples before going on to consider simpler problems. That is, the memory that we have of motor tasks is something quite separate from the memories that we have of other events in life. One way of demonstrating this is to ask an expert typist to draw a diagram of a typewriter keyboard, indicating the correct position of all the letters. When my secretary did this, there was a lot of phantom practising with her fingers on the top of her desk, and it took several minutes to produce the final picture. In fact there were still some mistakes on it, all from a person who could type 'the quick brown fox jumped over the lazy dog' faultlessly in about 20 seconds!

When physiologists examine motor learning, they are usually tackling something a little more prosaic. Very often, this involves adaptation of motor commands to small changes in the environment. We make such adjustments to our movements hundreds of times of day in normal life. An example would be the adaptation we make in our hand movements to control a cursor on a computer screen using a mouse. The amount we move the mouse is considerably less than the amount of movement of the cursor, but we adapt almost immediately to the change in gain. The examples that follow in animals implicate a role of the cerebellum in three different forms of learning and adaptation: (i) changing the gain of reflex responses; (ii) classical learning in a Pavlovian conditioning task; (iii) 'recalibration' of voluntary arm movement to changing task conditions. The role of the cerebellum is inferred from the effects of lesions on the learned or adapted response (Ito, 1989).

(1) The vestibular–ocular reflex in the external ocular muscles maintains the axis of the eyeball in a constant position with respect to the visual scene during movements of the head. An afferent signal from the vestibular apparatus is analysed to provide information about head position velocity. This is then used to stabilize the eyes on an object in space by producing movements of the eyeball which compensate for the change in head position. The reflex is extremely important for stabilized vision and can be modulated in various ways. For example, if reversing prisms are worn as spectacles so that the directions of left and right are reversed in the horizontal plane, the reflex gradually decreases in effectiveness over 4–5 days until it almost disappears. Further training may even result in reversal of the response in order to compensate for the spectacles. In animal experiments, this compensation is abolished by ablation of the flocculonodular lobe of the cerebellum.

(2) A puff of air on the cornea of a rabbit causes the nictitating membrane or third eyelid to sweep across the eyeball. At the same

time there is usually closure of the external eyelids. In a classical conditioning experiment, the air puff (unconditioned) stimulus is given in association with an acoustic (conditioning) stimulus, and after a number of pairings, the noise alone will elicit a blink in the absence of the air puff. Rabbits or cats can learn and retain the unconditioned response normally after removal of the hippocampus, neocortex or all tissue above the level of the thalamus. However, this is prevented by lesions of the lateral cerebellum, dentate or interpositus.

(3) A final example involves calibration of eye/wrist movement in a visual tracking task. Monkeys are trained to flex or extend their wrist to move a tracking spot on a screen before them in order to catch a target which moves randomly left or right. Adaptation has to occur when the gain of the hand coupling to movement of the cursor is unexpectedly changed. For example, after the change the monkey may need to move the wrist much further to produce the same displacement of a tracking spot on a screen. Effectively, the animal has to 'rescale' his wrist movement to the new gain. This rescaling takes place in a roughly exponential fashion over ten to twelve trials, and is abolished by injections of small quantities of muscimol into a localized region of the lateral zone of the cerebellar hemisphere. The injections only affect rescaling, but have no noticeable effect on the performance of the movement itself.

Cell Activity During Learning

In all studies of motor learning, there has been debate over whether the cerebellum is the primary site of learning or whether the cerebellum is necessary for learning to occur whilst the 'memory' is stored elsewhere. In electrophysiological terms, the argument boils down to whether or not the synaptic changes which occur in learning are in the cerebellum.

For those authors who champion a cerebellar site, one model of cerebellar learning involves climbing fibre activity modifying the effectiveness of the parallel fibre–Purkinje cell synapses. Thus, during learning, we should expect to see a transient increase in climbing fibre activity during the learning period, and this should decline after learning is complete. Over the same period, the number of simple spikes (from parallel fibre activity) in the Purkinje cells should change (it should decline if climbing fibres **reduce** synaptic efficiency), and, most importantly would remain at a new value even after climbing fibre activity had returned to normal. A long-term change in the synaptic effectiveness of parallel fibres would have been produced by a short-term change in climbing fibre discharge.

Such a pattern has been described during wrist movements in the

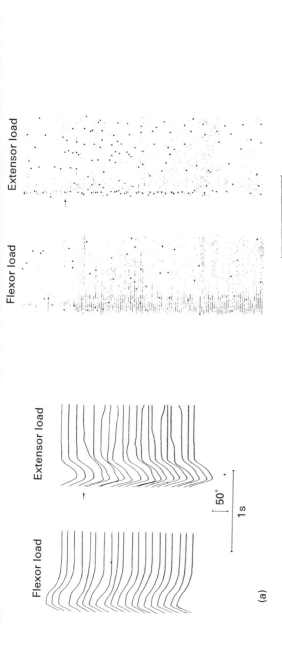

Figure 10.18 Possible role of climbing fibre discharges in learning new limb movements. A monkey was trained to hold his wrist in a constant stationary position in the face of alternating flexor and extensor loads (lasting 300 ms) which moved the handle away from target. With practice, when the same flexor and extensor loads were given each time, the monkey was able to restore the target position rapidly and repeatedly. (a) The performance of the monkey in terms of records of wrist position. The traces should be read from left to right and top to bottom. The first trace shows the response to the known flexor load. There is a pause and then a known extensor load is given, and so on: both sets of responses are very repeatable. At the arrow, the extensor load was changed from 300 **g** to 450 **g**. The monkey performed badly at first in response to this novel load, although his responses to the flexor load (which had not been changed) were relatively unaffected. Over several trials, performance and repeatability increased with the new extensor load. (b) Recordings from a single Purkinje cell in the cerebellar cortex for a similar series of trials. The discharge is shown as raster displays of simple spike (small dots) and complex spike (large dots) activity. This Purkinje cell was load-related, with a higher simple spike frequency during the flexor response. At the arrow, the known extensor load was unexpectedly changed from 300 to 450 **g**, while the flexor load remained constant. After the load change, there was a greatly increased frequency of complex spike activity in the extensor task, although the corresponding activity in the flexor task persisted for about 70 trials and was unchanged. This increased activity in the flexor task persisted for about 70 trials and was accompanied by changes in simple spike activity. (From Gilbert and Thach, 1977, Figures 1 and 3; with permission.)

monkey. Recordings were made from Purkinje cells, so that the frequency of both complex (climbing fibre) and simple (parallel fibre) spikes could be monitored while a monkey maintained a constant wrist position. At unpredictable intervals, a known flexor or extensor load was applied alternately and the monkey had to restore the original wrist position as rapidly as possible (Figure 10.18).

When the task had been learned, the appearance of complex spikes was irregular and infrequent as usual. At this stage, the extensor load was suddenly changed to a new value while the flexor load remained the same. The wrist movements after the introduction of the new extensor load were initially inaccurate, but improved over the succeeding 40 to 60 trials as the monkey learned the new task. After the appearance of the new load, the number of complex spikes following the extensor, but not the flexor, load increased, and the number of simple spikes decreased. As performance improved, the number of complex spikes gradually reverted back to its original value, but the simple spikes remained at the new lower frequency. The suggestion is that first, the cerebellum played a role in adapting the movement to the novel task. Second, that the climbing fibre activity accompanying the novel stimulus decreased the response of the Purkinje cells to mossy fibre input (Gilbert and Thach, 1977).

A similar result also has been described during adaptation to changes in the gain of visual feedback in task (3) above. These experiments have recently been repeated by Ojakangas and Ebner (1992). They recorded complex and simple spike discharges from Purkinje cells during adaptation to changes in gain of the visual feedback in task (3) above. As in the previous experiments, there was a transient increase in complex spike discharge during the learning period which then returned to baseline. Simple spike discharge also changed, but (i) some Purkinje cells showed increased firing whilst others showed decreased firing, and (ii) changes in simple spike discharge occurred in some Purkinje cells in which there had been no change in complex spike activity. The authors suggested that although complex spike firing may be involved in motor learning, it is not necessary to produce sustained changes in simple spike firing.

One question which arises in such experiments is which factor is responsible for producing the transient change in climbing fibre discharge in learning which is not present at other times during the task? That is, what mechanism allows a climbing fibre activity to serve as an error detector? One hypothesis is that error detection occurs in the olive. The olive receives descending projections from cortex which would inform it of motor commands from higher centres. In addition there is input from the periphery via the spino-olivary tract. By somehow comparing these inputs, the olive might generate a signal related to the mismatch between the expected result of a motor command and the movement actually produced.

Finally, it should be pointed out that there are examples of behaviour in which climbing fibre activity does **not** produce the changes of simple spike behaviour seen above. Thus, over short time periods, climbing fibre discharge can **increase** the responsiveness of Purkinje cells to other inputs. Natural stimulation of cat forelimb normally evokes a simple spike discharge in some Purkinje cells. If the stimulus is given 20–70 ms after a complex spike in the cell, the simple spike response is augmented. Clearly, the mechanism of cerebellar involvement in learning is not yet completely understood.

Visually triggered movements

The role of the cerebellum can also be seen during visually triggered movements. It takes many days before animals first learn to associate the movement of the arm with the onset of a visual signal, and many more days before they learn to react as rapidly as possible to the appearance of the signal. Recordings in motor cortex show that the gradual decrease of reaction time which is seen in such experiments is associated with the appearance of surface positive – depth negative field potentials. These potentials represent EPSP activity in superficial dendrites of pyramidal cells, generated by thalamocortical input to these layers, and they increase in size as reaction time becomes shorter. This input is produced by activity in cerebellar efferent fibres, and is abolished by removal of the cerebellum. This procedure also increases the latency of reaction time. It is thought that this is another reflection of the role of the cerebellum in the initiation of movements via motor cortex (Sasaki, Genba and Mizuno, 1982).

10.7 *Theories of cerebellar function*

Despite the large mass of data on the anatomy and physiology of the cerebellum, there is little agreement on the precise function of the cerebellum in control of movement. Three main functions have been attributed to the cerebellum in different theories: (i) the cerebellum has been supposed to work as a timing device; (ii) to be involved in learning new movements; and (iii) to coordinate movements by transforming the idea of where to move into the appropriate changes in joint angle necessary to make the movement. It is not known whether any of these theories is correct. Perhaps all of them are.

Cerebellum as a timing device

This was one of the earliest theories of cerebellar function. It proposes that the cerebellum is used to time the duration of agonist muscle activity and the latency of antagonist activity, so that any movement is halted at the correct point. Hypermetria on this theory is simply an inability to stop a movement at the correct point in space due to a defect in timing of agonist and antagonist muscle activity. As already seen, there is some evidence from the study of rapid limb movements that suggests that defects in timing of EMG activity do exist in animals with cerebellar deficits.

In detail, the theory suggests the following mechanism. The cerebral cortex would initiate activity in the agonist muscle to start the limb moving, and it would be the function of the cerebellum to stop the movement at the correct point (by agonist inhibition and antagonist excitation). Mossy fibre input would be activated by a cortical control signal and produce a pattern of activity on the surface of the cerebellum corresponding to the desired end position of the limb in space. In contrast, the pattern of climbing fibre input would reflect its actual position. Thus at the beginning of a movement, the two patterns would represent the start and stop positions of the limb.

It was suggested that the time taken for impulses set up by mossy fibre activity (that is, in the 'stop' position) to travel down the parallel fibre system to the area of climbing fibre input would be related to the timing of EMG activity needed in the movement. The larger the distance between the mossy and climbing fibre patterns of activity, the longer the EMG burst would last. Conjunction of parallel and climbing fibre activity in a Purkinje cell would signal the command to halt.

Cerebellum as a learning device

The principal features of the postulated learning mechanism are described in detail above. All that remains to be mentioned here is that although there is good evidence that the cerebellum is involved in some aspects of learning, and evidence that the parallel fibre – Purkinje cell synapses are, under some conditions, modifiable, there is some debate of how much of a learned 'programme' actually resides in the cerebellum and how much is distributed to other brainstem and spinal centres. This is a question which may take some time to resolve.

Cerebellum as a coordinator

This is the most recent theory, and incorporates, to some extent, both of the previous hypotheses. The basic idea is that the cerebellum is involved in 'open-loop' movements; that is, movements in which muscle commands are calculated centrally in advance, rather than being driven through reflex arcs from sense organs. In such movements, the cerebellum might be involved in coordinating muscle activity at several joints and in adapting that activity to cope with different external conditions. For example, in a pointing task involving the shoulder, elbow and wrist, there will be instructions to move the shoulder, extend the elbow and rotate the wrist. The cerebellum may scale the amplitude of these components so that the movement is accomplished as desired by the subject. If external conditions are changed, for example, if we use a pencil to point at the same object, then the cerebellum may be involved in adapting the muscle commands to what is effectively an instantaneous change in the length of our arm.

The theme of coordination can also be expanded to control of reflex pathways. As described in Chapter 8, reflexes in leg muscles during standing are not the same if disturbances are produced by moving the support surface backwards or forwards (translation) compared with rotation of the surface about the ankle joint. The cerebellum, with its wide range of inputs, may be in a good position to compare all these inputs during disturbance of support and adapt reflexes over the course of several trials to produce an appropriately coordinated response which maintains balance.

In this type of scheme, coordination between different joints might be achieved by the long parallel fibres traversing the body maps in the cerebellar cortex, linking together areas controlling separate joints. The strength of the parallel fibre–Purkinje cell synapses over different portions of this length might be varied and thereby change the effectiveness of coupling between joints.

10.8 *Studies of cerebellar dysfunction in humans*

Clinical symptoms

The principal symptom of cerebellar disease in humans is **ataxia** of voluntary movement. Ataxia is a general term which is used to describe decomposition of movement. A typical example of an ataxic arm movement is shown in Figure 10.19. A patient with injury to the left cerebellum was asked to pull on two springs of equal length by extending his arms, and then to maintain a final end position against the force that they supplied. The normal (right) arm starts promptly and moves quickly and smoothly to the end point where it is maintained with a few minor corrections. The ataxic (left) arm starts late and moves slowly and jerkily towards the target. Toward the midpoint of the movement a

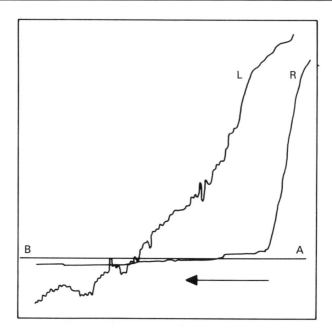

Figure 10.19 Smoked drum recordings of arm movements in a patient with a left cerebellar injury. The recordings are to be read from the right to the left side. The patient was asked to extend springs of equal strength with his two arms simultaneously after a 'go' signal, and was required to maintain the stretch at the level indicated by the line A–B. Note the slowness of the left arm in starting, the irregular tremulous character of the movement and the failure to maintain the final position. (From Holmes, 1939, Figure 9; with permission.)

tremor appears, the target is overshot, and further corrections of limb position occur.

There are specific terms to describe some of these abnormalities. **Kinetic tremor** refers to the oscillation seen during the course of movement. **Intention tremor** refers to the (usually larger) oscillations which occur around the target at the end of movement. There may also be a **postural tremor** which appears when the patient holds his arms out in one position. **Dysmetria** describes the inaccuracy in achieving a final end position (hypermetria equals overshoot; hypometria equals undershoot). **Dysdiadochokinesia** refers to the inability to perform movements of constant force and rhythm. This is appreciated by asking patients to tap their thigh alternately with the dorsum and palm of the hand. The rhythm is poor and the force of the tap varies throughout the sequence (Figure 10.20). Ataxia of gait is a common feature of cerebellar disease. Patients are unsteady and stabilize themselves by walking with a broad-based gait. At first sight they often appear as if they were drunk.

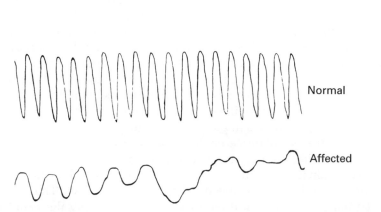

Figure 10.20 Tracings of rapid pronation–supination in a patient with unilateral cerebellar symptoms. The movements of the affected arm (below) were for a time regular though slower and of smaller amplitude than normal, but grew irregular and the arm became more or less fixed in supination. Muscular contractions appeared to spread aimlessly over the whole of the affected limb including unwanted shoulder movements and finger movements. One patient commented to Holmes, 'The movements of my left arm are done subconsciously, but I have to think out each movement of the right (affected) arm. I come to a dead stop on turning and I have to think before I start again.' (From Holmes, 1939, Figure 16; with permission.)

Walking heel to toe is very difficult, if not impossible. Running the heel of one leg down the shin of the other whilst seated or lying (heel–shin test) also is very poorly performed. Cerebellar deficits can be seen in bulbar muscles. This leads to slurring of speech and to a coarse nystagmus of the eyes, especially prominent at the extremes of gaze.

There are two other manifestations of cerebellar disease which are a little more difficult to define. These are hypotonia and asynergia. As pointed out in Chapter 6, two mechanisms contribute to the resistance felt during passive manipulation of the limbs: the inherent viscoelastic properties of the muscle itself and the tension set up in the muscle by reflex contraction caused by muscle stretch. In normal subjects at complete rest, the stretch reflex component of tone is not present. This is because of the very low level or absent background fusimotor discharge of relaxed muscle and hence the afferent response to relatively slow movements of muscle is small. From this, it might be assumed that for hypotonus to occur, there must be some fundamental change in the intrinsic mechanical properties of muscle. However, this is probably a relatively rare occurrence. Under most conditions, it is extremely difficult to ensure that muscles are completely relaxed during manipulation, and a small degree of contraction probably results in some stretch reflex contribution to muscle tone. This is the normal state of patients during clinical examination, especially if they are at all anxious. Patients with detectable hypotonus are either perfectly relaxed (and hence normal) or, more commonly, they have a diminished stretch reflex contribution to their tone. Three other phenomena are associated with hypotonia (see description of physiology below): (i) pendular tendon reflexes; (ii) reduction in check of a limb after sudden release of an isometric force; (iii) reduced tonic vibration reflexes. As in the monkey model, all forms of hypotonia are more pronounced after acute cerebellar injury such as those described by Holmes in patients with gunshot wounds. Chronic degenerative diseases of the cerebellum rarely produce changes in tone.

Strictly speaking, asynergia describes the inability of cerebellar patients to coordinate movement at more than one joint. The difficulty, however, is whether such a specific deficit actually exists. That is, are the irregularities seen in multi-joint movements simply the result of combining the errors in each constituent single joint movement, or is the incoordination greater than expected from the analysis of the movement at each joint alone? If the latter, then this extra problem should be termed asynergia. No one seems to have measured this quantitatively in humans, so the problem is unresolved. Nevertheless, the experiments using reversible inactivation of the cerebellar nuclei of monkeys suggest that true asynergia may be an important fundamental deficit.

Not all patients with cerebellar lesions have all these symptoms, although most have some form of gait ataxia. From the clinical viewpoint, the most useful way of interpreting cerebellar symptoms is on the sagittal subdivision of the structure into three zones projecting to the three cerebellar nuclei. Damage to the medial zone (anterior and posterior vermis and fastigial nucleus) produces disorders of stance and gait. The arms are relatively spared. In man there are no descriptions of patients with lesions limited to the intermediate zone (paravermal cortex and interposed nuclei), so that the symptoms are unknown. However, abnormalities of the lateral zone (majority of the cerebellar hemispheres and dentate nucleus) are common and lead chiefly to ataxia of the arms, with some difficulties of gait.

A cerebellar 'picture' can be seen in other conditions which affect the afferent or efferent connections of the cerebellum. The spinocerebellar ataxias are a group of diseases with a number of common clinical features. The most common, although still exceedingly rare, is Friedreich's ataxia. The disease produces degeneration of the spinocerebellar tracts and dorsal columns, with some involvement of the pyramidal tract and dorsal roots. Multiple sclerosis also may produce a cerebellar picture by interfering with cerebellar connections.

Physiological investigations of cerebellar patients (e.g. reviews by Diener and Dichgans, 1992; Gilman, 1992; Thompson and Day, 1993)

Stretch reflexes

Detection of hypotonus in cerebellar disease is difficult because passive manipulation of the joints in totally relaxed normal individuals meets with relatively little resistance to movement. Indeed, such normal subjects would be described as hypotonic on clinical criteria (see Chapter 6). Another sign of cerebellar hypotonia is the pendular knee jerk (Chapter 6), which is seen following the initial extension of the leg after tapping the patellar tendon. Undamped pendular decay of the movement is due to lack of resistance by the thigh muscles to extension at the knee joint.

As with manual assessment of muscle tone, pendular knee jerks are in practice rather difficult to spot. Other objective methods for detecting hypotonia are said to be more accurate. The tonic vibration reflex (TVR, Chapter 6) is decreased in cerebellar disease as can be seen clearly in patients with unilateral symptoms. The reflex builds up steadily on the non-affected side but is much smaller or absent on the other. Lack of a TVR cannot be due to the lack of gamma drive to the muscle spindles which has been demonstrated in cerebellectomized animals. This is because vibration is such a powerful stimulus that it can 'drive' the spindle output in the absence of any tonic fusimotor input. During vibration spindle discharge is thus believed to be maximal. Cerebellar hypotonia detected in this way must therefore be due to central mechanisms. One possibility is that cerebellar disease produces increased presynaptic inhibition of spindle Ia afferent fibres, although there are as yet no comparable animal data that indicate that this would be possible.

A further manifestation of hypotonia is thought to be the reduced check of suddenly released limb contraction. The usual demonstration of this involves the patient flexing his elbow against resistance. On releasing the forearm abruptly, the patient may strike his chest with his wrist because of failure to halt the movement in sufficient time. Again, this may be due to lack of reflex resistance from the stretched triceps muscle.

Long latency stretch reflexes have not been examined extensively in cerebellar disease, but in the cases so far examined the reflex is of normal latency but decreased in size and duration, consistent with the clinical observations of hypotonia.

Ballistic limb movements

These are voluntary movements of a limb made as rapidly as possible from one target position to another. The movement is made quite freely without end stops in the apparatus, so that the subject has to reach and

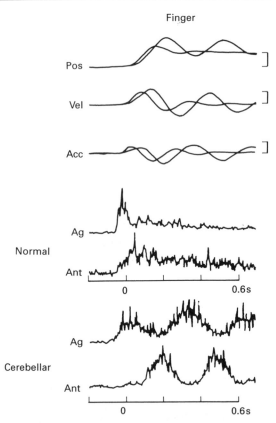

Figure 10.21 Typical ballistic finger flexion movements made on the normal and the affected side of a patient tested three years after infarction of the left cerebellar hemisphere. The records are averages of 15–20 movements in each case and show the position (Pos), velocity (Vel) and acceleration (Acc) of the finger movement together with the agonist (Ag) and antagonist (Ant) EMG activity. Note the prolonged first agonist burst on the cerebellar side and the late activity in the antagonist followed by repeating cycles of tremor. This is reflected in the position records which show overshoot followed by terminal oscillation. In addition, the velocity profile is asymmetric: the time taken to reach peak velocity on the cerebellar side is longer than that to return to zero velocity. The initial acceleration is also smaller on the affected side. (From Hore, Wild and Diener, 1991, Figure 8; with permission.)

maintain the final end position himself. The movements are completed in 200 ms or less, which is so short that the subject is unable to make any voluntary corrections to the ongoing movement. Corrections can only be made on completion of the initial movement. The movements are made with a relatively stereotyped pattern of muscle activity. This begins with a burst of activity in the agonist muscle which lasts for 100 ms or so. It is followed by a burst of activity in the antagonist and then by a small second burst in the agonist. The agonist activity provides the initial impulsive force for the movement, while the antagonist activity brakes the motion. The function of the second agonist burst of activity is unknown.

This pattern of EMG activity has been studied extensively since it represents a package of nervous commands which can be 'preprogrammed' in advance by the central nervous system. In other words, the size and timing of the agonist and antagonist bursts of activity are decided upon prior to the initiation of movement and then released upon receiving a 'go' signal. That such a pattern of activity can be seen in patients with no afferent input from their limbs (due to peripheral neuropathies of various causes), emphasizes their central origin. Despite the fact that the ballistic movement pattern can be generated centrally, this does not mean that commands cannot be **modulated** under normal conditions by peripheral feedback mechanisms. There is, in fact, good evidence that perturbations to a ballistic movement can result in reflex modulation of the size and timing of the antagonist and second agonist burst at very short latencies.

There are a number of abnormalities in the ballistic movements of patients with cerebellar disease, all of which involve errors in the timing of muscle contraction (Figure 10.21). These are seen for movements of both proximal and distal joints.

1. If the movements are made in response to a visual reaction stimulus, the reaction time is some 70 ms or so longer than normal. This is the same result as seen in monkeys with cooling probes implanted in their cerebellar nuclei, and illustrates the importance of the cerebellum in initiating movements.
2. The movements tend to overshoot their target, as expected from the clinical observations of hypermetria.
3. Consistent with the overshoot in the movement, the duration of the first burst of agonist EMG activity is longer in patients with cerebellar disease than normal. This presumably results in increased muscle force and contributes to hypermetria. Interestingly, the patients seem unable to compensate for this disturbance by aiming, for example, at an imaginary target slightly nearer than that indicated.
4. Cerebellar patients show a disorder of reciprocal inhibition of antagonist muscles. In normal subjects, if there is any prior activity in the antagonist muscle before a ballistic movement, then this is silenced before activation of the agonist muscle. This prevents the antagonist from opposing the action of the agonist during initiation of the movement. In cerebellar disease, antagonist activity sometimes persists until after EMG onset in the agonist, a phenomenon never seen in normal subjects.

5.   There is an asymmetry in the acceleration:deceleration profile of the movement. The movements of cerebellar patients have lower peak accelerations and higher peak decelerations than normal. This gives rise to an asymmetrical velocity profile in which the time taken to reach peak velocity is longer than that to return to zero velocity. The longer period of acceleration is probably caused by the longer first agonist EMG burst. The larger peak deceleration could be due either to a larger antagonist burst, or a late second agonist burst, which may under normal circumstances limit the deceleratory action of the antagonist. (Note that this velocity profile asymmetry is seen only in moderately affected patients. A reverse asymmetry has been observed in patients with very mild symptoms.)

6.   The latency to onset of the antagonist EMG burst is longer than normal. This is not just a consequence of the longer first agonist action: it is delayed even when measured relative to the time of peak movement velocity. This delay presumably contributes to hypermetria.

7.   In normal subjects, the joint quickly stabilizes at the end position after one or two small oscillations. In cerebellar patients the oscillations are much larger and continue for some time, causing an obvious tremor.

To summarize, the main errors in performing ballistic movements at one joint can be related to problems in timing: the EMG bursts are long and delayed. These timing errors either reflect changes in central timing mechanisms involved in 'preprogramming' movement, or they represent inappropriately timed use of peripheral feedback which modulates the central programme. At present, it is not possible to say which of these possibilities is correct, although relatively minor changes in simple reflexes which have been described in cerebellar disease lead most workers to favour a disorder of central timing. Changes in burst timing lead to incoordination of agonist and antagonist muscles, resulting in overshoot of the target. Continued oscillation at the end of the movement may, as suggested in animal experiments, be due to failure to time corrective bursts of activity which normally stabilize the limbs (Brown *et al.*, 1991; Hallett *et al.*, 1991; Hore, Wild and Diener; 1991).

Timing studies
The question of whether the cerebellum is involved in central timing mechanisms has been investigated in several other ways. One set of experiments investigated the preparatory process involved in timing of sequence of finger movements. Subjects performed three sets of reaction time trials. In the first, a single key-press had to be made with the index

finger. In the second, the index movement was followed by a key-press made with the ring finger. In the third, a sequence of key-presses was made in the order index, ring then middle. In normal subjects, two effects are related to sequence length. First, reaction time increases as the length of the sequence increases. This is known as the **sequence effect** and is thought to occur because more time is needed to set-up or retreive a complicated set of motor instructions than it is to organize a simple movement. The second effect is that the interval between successive key-presses in a sequence is less than the initial reaction time. This is called the **position effect** and may be a beneficial result of setting up the sequence in advance before movement begins. Cerebellar patients, particularly those with moderate disability, have a much smaller sequence effect than normals: their initial reaction time to a single movement is long, but does not increase greatly as the length of the sequence increases. Perhaps as a result of this, these patients fail to show any position effects. That is, the inter-key press time is the same as the initial reaction time, rather than being shorter. One hypothesis is that such patients do not set up the instructions to time the whole sequence in advance. Each element of the sequence is therefore performed as if it were a single separate movement.

A second set of experiments investigated timing problems during the execution of a movement. In this task, subjects were asked to tap in rhythm to an auditory signal. After a few beats, the tone was stopped and subjects continued to tap on their own at the same frequency. Wing and Kristofferson (1973) proposed that after removal of the tone, variability in the timing between successive taps was due to two processes: (i) variability in a central timing system which estimates a required inter-beat interval and (ii) variability in the effector mechanism which produces the tapping movement. In the latter case, one can imagine that a central command to tap occurs at the correct time, but there is variability in the amount of time taken to transform that command into movement. Analysis of the variation of tap-to-tap timing can give an estimate of the relative role of these two phenomena. Patients with diseases primarily affecting lateral cerebellum have a disorder of the **central** rhythm generator, whereas patients with midline cerebellar deficits (gait and posture abnormalities) had a primary disorder of the **effector** system (Ivry, Keele and Diener, 1988; Inhoff *et al.*, 1989).

Slow movements and visuomotor tracking
A slow movement can be defined as a movement made at a speed which allows the subject time to make feedback corrections to the trajectory using visual or somaesthetic information. The movement is said to be under feedback control. Such slow movements are more variable than rapid ones and have for that reason been investigated in less detail. One

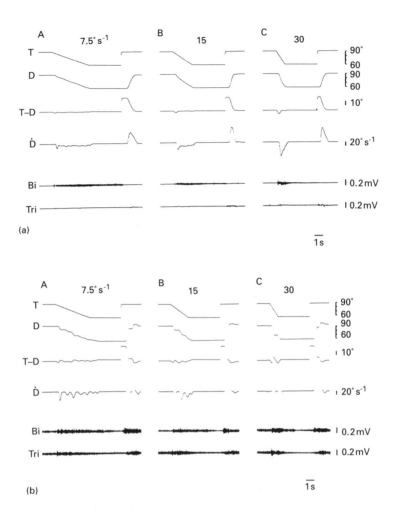

Figure 10.22 Slow tracking movements in cerebellar ataxia. Subjects had to track a slowly moving target by flexing their elbow at the appropriate rate. Both target and elbow position were displayed to the subjects on an oscilloscope screen. Targets were moved at 7.5, 15 or $30°s^{-1}$, and began to move at random times. The performance of a normal subject is shown in (a). Shortly after the target (T) moves, the subject makes a rapid 'catch-up' movement (trace D) and then continues to track the target perfectly well until the end of the movement. The error between target and actual movement (T–D) and the velocity of elbow movement (D) are shown, together with EMG records from biceps (Bi) and triceps (Tri). Performance of a cerebellar patient is shown in (b). The initial 'catch-up' movement is delayed and is bigger than necessary (overshoot). The smooth phase of tracking is broken up into a series of steps. (From Beppu, Suda and Tanaka, 1984, Figures 2 and 3; with permission.)

obvious clinical feature is that unlike the hypermetria seen in rapid movements, slow tracking movements are often hypometric.

The usual experiment is to ask subjects to track a visual target on a screen before them by moving the elbow or wrist (Figure 10.22). In the simplest task, subjects start at rest and track a target which begins to move at a steady speed, but at an unpredictable time, across the display screen. In normal individuals, there is an initial pause before the subject moves due to the finite reaction time. The movement begins with an initial 'catch-up' phase which corrects for the lag introduced by the pause. After this, it continues steadily at a speed equal to that of the target.

In cerebellar disease, there is, as usual, an increase in the reaction time to the onset of movement. The 'catch-up' phase which follows is not scaled appropriately to the lag which has evolved. Thus, patients over-shoot or undershoot the target during this phase. Finally, and most dramatically, patients cannot then switch into the smooth tracking mode seen in normal subjects. The rest of the tracking movement is made up in a series of 'saccadic' steps: rapid movements interrupted by pauses, as if the patient was making a series of corrections throughout the movement, rather than tracking the target smoothly. There are delays also in the termination of movement, as expected from problems in controlling antagonist muscle activation.

The discontinuities in smooth movement are not due to a superim-posed tremor. The movement is quite clearly 'stepwise' (that is, including definite pauses) and not oscillating, as expected from tremor. Neither was the breakdown of attempted smooth movements into 'saccadic' tracking due to a fundamental inability of the movement control system in cerebellar patients to produce such movements. Pa-tients could perform smooth, continuous movements if they moved their arms as they wanted in their own time, rather than following a moving visual target. The defect seems to be a problem in moving from the 'catch-up' (or truly saccadic) phase to the smooth pursuit phase of movement. Such discontinuous movements are similar to those seen in monkeys with cooling probes implanted into the interposed and den-tate nuclei. Both have a similar frequency of 1–3 Hz (Beppu, Sudu and Tanaka, 1984; Beppu, Nagoka and Tanaka, 1987).

Abnormal use of visual feedback is suggested by other experiments. Studies of free-arm ball throwing experiments using modern 3D anal-ysis techniques show that throwing by cerebellar patients is remarkably normal. The coordination between elbow and wrist is preserved, as is the timing of ball release relative to the maximum velocity of movement. The major deficit is that throws are much more variable than normal because the patients cannot reproduce the same hand direction from trial to trial. The suggestion is that this variability occurs because patients cannot transform their perception of visual target position into the appropriate motor commands (Becker *et al.*, 1990).

Posture
Unsteadiness of balance is one of the most common features of cerebellar disease. Accordingly, there have been a number of investigations into the possible abnormalities which might be responsible for this. Unfortunately, there are little data from animal experiments (there are no other **bipedal** mammals) to compare with the description of human cerebellar patients. Because of this, the mechanisms of the reflexes described below and the precise role of the cerebellum in each of them is unknown.

Postural reflexes in normals are summarized in Chapter 8. Cerebellar patients with postural instability show deficiencies in three types of reflex testing.

1) When normal subjects stand on a movable platform, the responses to horizontal translation consist of a stereotyped sequence of muscle contractions. For example, if the platform is moved backwards, the body sways forwards and EMG responses are evoked first in the triceps surae muscles (after about 100 ms), and these are followed by activity in hamstrings and paraspinal muscles.

   In cerebellar patients, these responses are disrupted. The order of muscle activation varies considerably from the stereotyped pattern seen in normals. In addition, the relative amount of activity in the calf and thigh muscles is also variable and different from normals. Thus, these reflex responses are affected in the same way as simple voluntary movements: there are errors in timing and force of muscular contraction.

2) Besides moving the platform horizontally, rotation about the axis of the ankle also has been used to investigate postural reflex function. Rotation of the supporting surface poses an interesting problem for the body. Consider dorsiflexion of the foot in the standing subject. This will induce backwards sway of the trunk. However, the motion at the ankle will initially produce stretch of the triceps surae muscles. If this stretch evokes a stretch reflex, contraction of the triceps surae will tend to pull the subject backwards and hence further off balance, rather than correcting for the disturbance. In normal subjects, a strong triceps surae stretch reflex can be recorded on the first occasion that platform dorsiflexion is given. It is a destabilizing reflex, which gradually adapts if the same stimulus is given again and again. However, in cerebellar patients, the response never adapts. It seems that there is a failure to learn the new situation.

Platform rotation at the ankle also produces very strong responses in muscles shortened by the disturbance. Following platform dorsiflexion, the pretibial muscles show a very strong reflex contraction with a latency of about 130 ms. This is the response which actually **compensates** for the body sway. Patients with atrophy of the cerebellar anterior lobe have enlarged responses in the tibialis anterior muscles following dorsiflexion of the ankles. These overcorrect for the displacement by bringing the body too far forward of the centre of foot pressure, and set up stretch reflexes in the triceps surae. Repetitive cycles of anterior tibial and calf muscle activity produce a postural tremor at about 3 Hz. Other categories of cerebellar patient showed little abnormality in these tests, except for a group with Friedreich's ataxia. In this group, the responses were delayed. This suggests that the responses may have been mediated by the spinocerebellar or pyramidal tracts which are involved in this disease.

3) A different method of producing postural disturbances involves applying forces directly to the trunk or arms, and then measuring the reflex responses in the muscles of the legs. In normal subjects, a small pull forwards to the arm at the wrist can evoke EMG responses in the triceps surae muscles at a latency of 80–100 ms. These responses are in such a direction to compensate for any induced sway of the body. However, the responses are actually initiated **before** the body sway reaches the leg. They appear to be leg muscle responses which are 'driven' by input from the moving arm. These reflexes are absent or decreased in cerebellar patients with postural instability.

Thus, unsteadiness of cerebellar patients is reflected in a number of abnormal postural reflexes. There may be a lack of organization or timing of the muscle responses, lack of adaptation of responses to new situations or even absence of responses altogether. The mechanism of the responses produced by the different types of disturbance is unknown. It is likely that they involve some supraspinal mechanisms. If so, the cerebellar influence may act directly on the reflex pathway or indirectly via its influence on other brain structures.

In addition to reflex deficits in posture, cerebellar patients also have a disturbance of **preparatory** postural activity which precedes voluntary movement. For example, if standing subjects are instructed to rise onto tip-toe, the main contraction of the calf muscles is preceded by activity in tibialis anterior. This pulls the body forward slightly to reduce the backward shift of the centre of gravity on contraction of the calf. In cerebellar patients, the basic preparatory pattern is present, but the timing is delayed. In addition, tibialis contraction often fails to turn

(a)

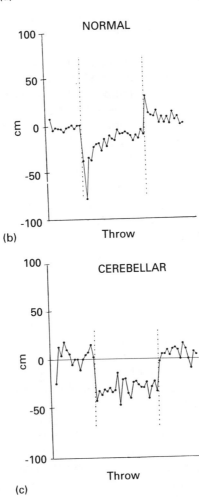

(b)

(c)

off, so that calf muscle activity is opposed by unnecessary antagonist co-contraction (Traub, Rothwell and Marsden, 1980; Nashner, 1983; Diener *et al.* 1984; Diener *et al.* 1990).

Learning

Cerebellar patients seem to have deficits in some forms of motor learning very similar to those described in animal experiments. One experiment investigated how subjects adjust their arm movements to changes in visual feedback (Figure 10.23). Subjects wore prism spectacles which horizontally deviated their gaze, say, 30° to the left. If they then tried to point (rapidly, so that there was no time for visual correction of the movement) at an object in front of them, they mispointed some distance to the right. This is because in order to see an object straight ahead, the eyes are actually deviated to the right, so that at first, they assume the object is also to their right. Continued attempts at pointing show increasing accuracy until subjects can point accurately again at targets before them. This adaptation could have occurred either by a conscious strategy of the subject to point to the left of whatever he observes, or by a recalibration of the automatic process by which we transform visual input into a movement command. Introspectively, it feels as if the latter explanation is correct, and this is borne out by what happens when the prism spectacles are removed. The first attempt at pointing is now to the **left** of the object. This after-effect disappears after several trials, but is an important confirmation that true recalibration of the motor system had occurred. Some cerebellar patients may be able to adapt to an extent when they put on the prism spectacles (although the one illustrated in Figure 10.23 did not), but they do not show any

Figure 10.23 Throwing darts whilst wearing wedge-prism spectacles. These spectacles bend the optic path 15° to the right, so that in order to see objects straight ahead, the subject has to look 15° to the left. Normally, darts are thrown in the direction of gaze, so that subjects would throw 15° to the left in order to try and hit a target in a straight-on position. The subject (a) has recalibrated her motor system so that although she is looking 15° to the left her throw is still straight on. The two graphs (b,c) are plots of the horizontal locations, relative to the target centre of each successive dart hit. Before introduction of the prism, the darts hit close to the centre of the target. Introduction of the prism shifts the hits to the left (down), and with practice (over the next 10–15 trials) they normally return toward the centre of the target as the subject recalibrates the gaze direction/throw direction. After removal of the prism, the hits are normally shifted to the right (up), which shows that the throw direction is still calibrated 15° off the direction of gaze. With practice, the system recalibrates itself to the original value. Graph (c) shows the results from a patient with hypertrophy of the inferior olive. After introduction of the prism, there is no recalibration of the gaze/throw direction, and the throws remain to the left of target centre. After removal of the prisms, the hits land where they did before the introduction of the prisms indicating that there had been no adaptation to the prisms in this subject. (From Thach, Goodkin and Keating, 1992, Figure 3; with permission.)

after-effect when they remove them. The conclusion is that there is a failure to recalibrate their system to changes in visual feedback.

Tremor

Tremor is a rhythmical ('sinusoidal') oscillation of a limb. It is quite separate from the jerky, irregular disturbance of movement which is typical of ataxia.

Cerebellar tremor can occur during movement (kinetic or intention tremor) or may appear after some seconds when the patient holds his arms outstretched against gravity. Because some patients have a more predominant kinetic than postural tremor (or vice versa) it has been suggested that these two types of tremor have different mechanisms. Holmes (1939) thought that postural tremor was a consequence of fatigue or weakness at proximal muscles. He argued that the oscillations of postural (but not kinetic) tremor are always in the line of gravity and hence suggested that they resulted from voluntary efforts to correct limb position. However, it may be just that the voluntary effort required brings out a typical intention tremor. The situation has never been investigated in detail.

Studies in monkeys with cooling probes implanted into the cerebellar nuclei (see above) have suggested that tremor can result from activity in stretch reflex loops. Under normal conditions the timing of activity in these loops is influenced by cerebellar inputs and it prevents oscillations occurring. Some evidence supporting a role of peripheral feedback mechanisms in human cerebellar tremor comes from observations of the effect of mechanical loading on the frequency and amplitude of tremor. For example, if patients are trying to hold their wrist steady in the horizontal plane, adding mass to the hand decreases the frequency of wrist tremor. If they are pressing against a force offered by a torque motor, then increasing the force against which they have to press will increase their tremor frequency. Finally, tremor is much less severe if patients make isometric contractions rather than isotonic contractions. All these manoeuvres change the pattern of afferent input produced by the tremor and hence would be expected to affect the system dominated by an unstable stretch reflex (Homberg *et al.* 1987; Flament and Hore, 1988).

There is one other feature of cerebellar tremor which is a little more difficult to explain. In humans, the tremor is much more pronounced if subjects make movements under visual guidance: in some cases the tremor almost disappears if patients close their eyes. This phenomenon was not observed in the cooling studies performed in monkeys. It may be due to vision having some influence on the 'gain' of the stretch reflex circuitry responsible for tremor (e.g. Sanes, Lewitt and Mauritz, 1988).

Two other forms of tremor are found after lesions to cerebellar input–

output pathway: Rubral tremor and palatal myoclonus (palatal tremor). Rubral tremor is present at rest and worsens on maintaining a posture or during performance of a movement. It occurs after midbrain lesions involving the region of the red nucleus, although there is no evidence that, despite its name, the red nucleus is the **crucial** site of damage for rubral tremor to occur. It seems likely that rubral tremor results from a combined lesion of the cerebello-thalamo-cortical pathway (giving rise to intention and postural components of the tremor) in combination with a deficit of the substantia nigra or its projections (leading to a resting component of tremor: see Chapter 11).

Palatal myoclonus, or palatal tremor is a rare movement disorder in which there are rhythmic movements of the soft palate at around 2 Hz. There are two forms of the condition. **Essential** palatal myoclonus has no known pathological basis and consists solely of movements of the palate accompanied by an irritating ear click. The latter is probably due to contraction of the tensor veli palatini muscle causing the walls of the eustachian tubes to move apart. **Symptomatic** palatal myoclonus may affect other muscles of the body as well as the palate. However, it never involves the tensor veli palatini muscle and so is never accompanied by ear clicks. The strange exception of this muscle is probably related to the fact that, of the muscles of the palate, it is the only one innervated from the trigeminal nerve nucleus rather than the facial or ambiguous nuclei.

In symptomatic essential myoclonus, the inferior olive becomes greatly enlarged. The cells degenerate and swell, a process known as hypertrophic degeneration. It is thought that this may produce abnormal input into the cerebellum. Indeed, even under normal conditions, the cells of the inferior olive have a tendency, because of the particular calcium channel type that they have in their membrane, to fire rhythmically at 10–12 Hz. It has been suggested that the damaged olive may oscillate uncontrollably at a much lower frequency in palatal myoclonus, and this may somehow produce the visible tremor observed. (Deuschl *et al.*, 1990).

Bibliography

Review articles and books

Bloedel, J. R. and Courville, J. (1981) Cerebellar afferent systems, in V. B. Brooks (ed.) *Handbook of Physiology*, sect. 1, vol. 2, part 2, Williams and Wilkins, Baltimore, pp. 735–830.

Brooks, V. B. and Thach, W. T. (1981) Cerebellar control of posture and movement, in V. B. Brooks (ed.) *Handbook of Physiology*, sect. 1, vol. 2, part 2, Williams and Wilkins, Baltimore, pp. 877–945.

Diener, H.-C. and Dichgans, J. (1992) Pathophysiology of cerebellar ataxia. *Movement Dis.*, **7**, 95–109.

Gilman, S. (1992) Cerebellum and motor dysfunction, in A. K. Asbury, G. M. McKhann and W. I. McDonald, *Diseases of the Nervous System. Clinical Neurobiology*. W. B. Saunders, Philadelphia, pp. 319–341.

Gilman, S., Blodel, J. R. and Lechtenberg, R. (1981) *Disorders of the Cerebellum*, S. A. Davies, Philadelphia.

Holmes, G. (1939), The cerebellum of man. *Brain*, **62**, 1–30.

Ito, M. (1984) *The Cerebellum and Neural Control*. Raven Press, New York.

Ito, M. (1989) Long-term depression. *Annu. Rev. Neurosci.*, **12**, 85–102.

Thach, W. T., Goodin, H. P. and Keating, G. J. (1992) Cerebellum and the adaptive coordination of movement. *Annu. Rev. Neurosci.*, **15**, 403–442.

Original papers

Asanuma, C., Thach, W. T. and Jones, E. G. (1983) Cyto-architectonic division of the ventral lateral thalamic region in the monkey. *Brain Res. Rev.*, **5**, 219–235; 237–265; 267–297; 299–322.

Becker, W. J., Kunesch, E. and Freund, H.-J. (1990) Coordination of a multi-joint movement in normal humans and in patients with cerebellar dysfunction. *Can. J. Neurol. Sci.*, **17**, 264–274; 275–285.

Beppu, H., Suda, M. and Tanaka, R. (1984) Analysis of cerebellar motor disorders by visually-guided elbow tracking movements. *Brain*, **107**, 787–809.

Beppu, H., Nagoka, M. and Tanaka, R. (1987) Analysis of cerebellar motor disorders by visually-guided elbow tracking movements. II. Contribution of visual cues on slow ramp pursuit. *Brain*, **110**, 1–18.

Brown, S. H., Hefter, H., Mertens, M. and Freund, H.-J. (1990) Disturbances in human arm movement trajectory due to mild cerebellar dysfunction. *J. Neurol. Neurosurg. Psychiatr.*, **53**, 306–313.

Combs, C. M. (1954) Electro-anatomical study of cerebellar localisation. Stimulation of various afferents. *J. Neurophysiol.*, **17**, 123–143.

Deuschl, G., Mischke, G., Schenck, E., Schulte-Monting, J. and Lucking, C. H. (1990) Symptomatic and essential rhythmic palatal myoclonus. *Brain*, **113**, 1645–1672.

Diener, H.-C., Dichgans, J., Bacher, M. *et al.* (1984) Characteristic alterations of long-loop reflexes in patients with Freidriech's disease and late atrophy of the cerebellar anterior lobe. *J. Neurol. Neurosurg. Psychiatr.*, **47**, 679–685.

Diener, H.-C., Dichgans, J., Guschlbauer, B. *et al.* (1990) Associated postural adjustments with body movement in normal subjects and patients with parkinsonism and cerebellar disease. *Rev. Neurol.*, **146**, 555–563.

Flament, D. and Hore, J. (1988) Comparison of cerebellar intention tremor under isotonic and isometric conditions. *Brain Res.*, **439**, 179–186.

Fortier, P. A., Kalaska, J. F. and Smith, A. M. (1989) Cerebellar neuronal activity related to whole-arm reaching movements in the monkey. *J. Neurophysiol.*, **62**, 198–211.

Ghez, C. and Fahn, S. (1985) The cerebellum, in E. R. Kandel and J. H. Schwartz (eds), *Principles of Neural Science*, Elsevier, New York.

Gilbert, P. F. C. and Thach, W. T. (1977) Purkinje cell activity during motor learning. *Brain Res.*, **128**, 309–328.

Gilman, S. (1969) The mechanism of cerebellar hypotonia: an experimental study in the monkey. *Brain*, **92**, 621–638.

Gorassini, M., Prochazka, A. and Taylor, J. (1992) Cerebellar ataxia without changes in fusimotor (γ) control of proprioception. *Soc. Neurosci. Abs*, **18**, 408.

Groenewegen, H. J., Voogd, J. and Freeman, S. L. (1979) The parasagittal zonal organisation within the olivo-cerebellar projections. II. Climbing fibre distribution in the intermediate and hemispheric parts of cat cerebellum. *J. Comp. Neurol.*, **183**, 551–602.

Hallett, M., Berardelli, A., Matheson, J., Rothwell, J. C. and Marsden, C. D. (1991) Physiological analysis of simple rapid movements in patients with cerebellar deficits. *J. Neurol. Neurosurg. Psychiatr.*, **53**, 124–133.

Homberg, P. E., Hefter, H., Reiner, S. H. and Freund, H.-J. (1987) Differential effects of changes in mechanical limb properties on physiological and pathological tremor. *J. Neurol. Neurosurg. Psychiatr.*, **50**, 568–579.

Hore, J., Wild, B., Diener, H.-C. (1991) Cerebellar dysmetria at the elbow, wrist and finger. *J. Neurophysiol.*, **65**, 563–571.

Inhoff, A. W., Diener, H.-C., Rafal, R. D. and Ivry, R. (1989) The role of cerebellar structures in the execution of serial movements. *Brain*, **112**, 565–581.

Ito, M., Sakurai, M. and Tongroach, P. (1982) Climbing fibre-induced depression of both mossy fibre responsiveness and glutamate sensitivity of cerebellar Purkinje cells. *J. Physiol.*, **324**, 113–134.

Ivry, R. B., Keele, S. W. and Diener, H.-C. (1988) Dissociation of the lateral and medial cerebellum in movement timing and movement execution. *Exp. Brain Res.*, **73**, 167–180.

Jansen, J. and Brodal, A. (1958) Das Kleinhim, in *Mollendorff's Handbuch der Mikroskopischen Anatomie des Menschen*, Springer Verlag, Berlin.

MacKay, W. A. (1988) Cerebellar nuclear activity in relation to simple movements. *Exp. Brain Res.*, **71**, 47–58.

Martinez, F. E., Crill, W. E. and Kennedy, T. (1971) Electrogenesis of cerebellar Purkinje cell responses in cats. *J. Neurophysiol.*, **34**, 348–356.

McDevitt, C. J., Ebner, T. J. and Bloedel, J. R. (1987) Relationships between simultaneously recorded Purkinje cells and nuclear neurones. *Brain Res.*, **425**, 1–13.

Miall, R. C., Weir, D. J. and Stein, J. F. (1987) Visuo-motor tracking during reversible inactivation of the cerebellum. *Exp. Brain Res.*, **65**, 455–464.

Mink, J. W. and Thach W. T. (1991) Basal ganglia motor control. Parts

I, II and III. *J. Neurophysiol.*, **65**, 273–351.

Mugnanini, E. (1983) The lengths of cerebellar parallel fibres in chicken and rhesus monkey. *J. Comp. Neurol.*, **220**, 7–15.

Nashner, L. M. (1983) Analysis of movement control in man using the moveable platform. *Adv. Neurol.*, **39**, 607–619.

Ojakangas, C. L. and Ebner, T. J. (1992) Purkinje cell complex and simple spike changes during a voluntary arm movement learning task in the monkey. *J. Neurophysiol.*, **68**, 2222–2236.

Orioli, P. J. and Strick, P. L. (1989) Cerebellar connections with the motor cortex and the arcuate premotor area: an analysis employing retrograde transneuronal transport of WGA-HRP. *J. Comp. Neurol.*, **288**, 612–626.

Pollock, L. J. and Davis, L. (1927) the influence of the cerebellum upon the reflex activities of the decerebrate animals. *Brain*, **50**, 277–312.

Rispal-Padel, L., Harnors, C. and Troiani, D. (1987) Converging cerebellofugal inputs to the thalamus. *Exp. Brain Res.*, **68**, 47–58; 59–72.

Sanes, J. N., Lewitt, P. A. and Mauritz, K.-H. (1988) Visual and mechanical control of postural and kinetic tremor in cerebellar system disorders. *J. Neurol. Neurosurg. Psychiatr.*, **51**, 934–943.

Sasaki, K., Genba, H. and Mizuno, N. (1982) Cortical field potentials preceding visually-initiated hand movements and cerebellar actions of the monkey. *Exp. Brain Res.*, **46**, 29–36.

Schell, G. R. and Strick, P. L. (1984) The origin of thalamic inputs to the arcuate premotor and supplementary motor areas. *J. Neurosci.*, **4**, 539–560.

Strick, P. L. (1978) Cerebellar involvement in 'volitional' muscle responses to load changes, in J. E. Desmedt (ed.) *Progress in Clinical Neurophysiology*, vol. 8, Karger, Basel, pp. 85–93.

Thach, W. T. (1978) Correlation of neural discharge with pattern and force of muscular activity, joint position and direction of intended movement in motor cortex and cerebellum. *J. Neurophysiol.*, **41**, 654–676.

Thompson, P. D. and Day, B. L. (1993) The anatomy and physiology of cerebellar disease. *Adv. Neurol.*, **61**, 15–31.

Traub, M. M., Rothwell, J. C. and Marsden, C. D. (1980) Anticipatory postural reflexes in Parkinson's disease and other and other akinetic-rigid syndromes and in cerebellar ataxia. *Brain*, **103**, 393–412.

Vilis, T. and Hore, J. (1980) Central neural mechanisms contributing to cerebellar tremor produced by perturbations. *J. Neurophysiol.*, **43**, 279–291.

Wetts, R., Kalaska, J. F., Smith, A. M. (1985) Cerebellar nuclear cell activity during antagonist cocontraction and reciprocal inhibition of forearm muscle. *J. Neurophysiol.*, **54**, 231–244.

Williams, P. L. and Warwick, R. (1975) *Functional Neuro-anatomy of Man*, Churchill Livingstone, Edinburgh.

Wing, A. and Kristofferson, A. (1973) Response delays and the timing of discrete motor responses. *Percept. Psychophys.*, **14**, 5–12.

The basal ganglia **11**

At one time, the term, 'basal ganglia' was used to describe all the large nuclear masses in the interior of the brain, including the thalamus. Gradually, its use has become restricted to five closely related nuclei: caudate, putamen, globus pallidus, subthalamic nucleus and substantia nigra (Figure 11.1). The basal ganglia receive no direct sensory inputs and, like the cerebellum, send no direct motor output to the spinal cord. However, there is no doubt that these structures are involved in the control of movement. All diseases of the basal ganglia in man have some disorder of movement as their primary symptom. These range from an excess of involuntary movements (for instance, chorea) to a poverty and slowness of voluntary movement (for instance, Parkinson's disease). Important as their role may be, there is to date no agreement on the precise function or the mechanism of action of the basal ganglia.

11.1 *Anatomy*

The major part of the basal ganglia is made up of three nuclei: the caudate, putamen and globus pallidus. These three nuclei lie deep in the cerebrum, lateral to the thalamus and separated from it by the internal capsule. The globus pallidus is phylogenetically the oldest of the group and is known as the paleostriatum. Its name (pallidum) derives from the fact that in unstained sections it is paler than caudate and putamen. The globus pallidus is divided into two parts, lateral (or external, the GP$_e$) and medial (internal, or GP$_i$), by the medial medullary lamina. The lateral or external part is larger than the medial or internal part. Although both parts look similar, they have connections with quite different parts of the brain. The phylogenetically newer parts of the striatum (caudate and putamen) are known as the neostriatum, although most books omit the prefix 'neo' and refer to these two nuclei simply as the striatum. In man the caudate and putamen are separated from each other by the internal capsule, whereas in the rat they form

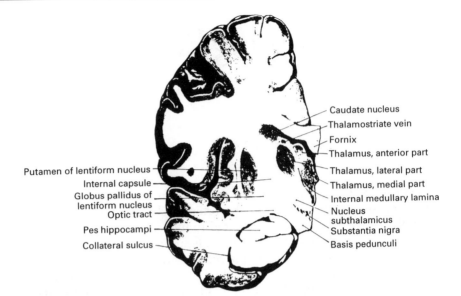

Figure 11.1 labels (clockwise from upper right):
- Caudate nucleus
- Thalamostriate vein
- Fornix
- Thalamus, anterior part
- Thalamus, lateral part
- Thalamus, medial part
- Internal medullary lamina
- Nucleus subthalamicus
- Substantia nigra
- Basis pedunculi
- Collateral sulcus
- Pes hippocampi
- Optic tract
- Globus pallidus of lentiform nucleus
- Internal capsule
- Putamen of lentiform nucleus

Figure 11.1 Anterior aspect of a coronal section through the right cerebral hemisphere showing the main structures of the basal ganglia. (From Williams and Warwick, 1975, Figure 7.156; with permission.)

one single structure. The term **striatum** is descriptive of their appearance in myelin stained sections. A number of fibre bundles known as Wilson's pencils traverse the nuclei, giving them a striped appearance.

The other two nuclei of the basal ganglia, the substantia nigra and subthalamic nucleus, lie in the midbrain. The subthalamic nucleus is a small lens-shaped nucleus situated beneath the thalamus on the dorsomedial surface of the internal capsule at the point where the fibres group together to form the cerebral peduncle. Ventral and caudal to the subthalamic nucleus, and continuous with it at this point, is the substantia nigra. It is the largest nucleus of the midbrain, and is particularly well developed in man. The cells are densely packed with granules of melanin and, in coronal sections of brain, the substantia nigra can be seen with the naked eye as a darkly coloured arched stripe above the cerebral peduncle.

Like the globus pallidus, the substantia nigra is also divided anatomically into two parts which have quite separate connections with other areas of brain. There is no clear demarcation line, but the cells are more densely packed in the dorsal part of the nucleus than in the ventral part, and the area appears darker when examined by eye. The dorsal region is known as the pars compacta, and the ventral region as the pars reticulata (SN_{pr}). As we shall see below, the pars reticulata is often regarded as homologous to the GP$_i$, both structures being major output nuclei of the basal ganglia.

These nuclei of the putamen, caudate, pallidum, nigra and sub-thalamus are the most usually accepted members of the basal ganglia. However, in recent years, additional nuclei have been included and are said to constitute the ventral striopallidal system. The concept is based on similarities between the connections and histochemical features of the neostriatum and both the nucleus accumbens septi and the medium-celled portion of the olfactory tubercle. These two areas are the receiving nuclei (or 'ventral striatum') of the ventral striopallidal system, and receive input principally from limbic areas of cerebral cortex. They have connections with a rostral and ventral extension of the globus pallidus and rostral substantia nigra pars reticulata. The nucleus accumbens is large in rats, but small in humans. (see reviews by Carpenter, 1981; Alexander, DeLong and Strick, 1984 and also *Trends in Neuroscience* Volume 13:10 for more details of basal ganglia anatomy.)

Input–output connections of the basal ganglia

The basal ganglia receive their major input from wide areas of the cerebral cortex, and send most of their output back, via the thalamus, onto the same areas of cortex. Other areas which receive basal ganglia are the superior colliculus, reticular formation and the pedunculopon-tine nucleus. Because of the apparent reduction in volume of projection areas through this corticobasal ganglia–cortical loop, it was at one time thought that the basal ganglia acted to 'funnel' cortical input in some way, so that inputs from many different areas of cortex could interact before the final output was sent back to the cortex. Recent findings have shown that this is not true. Inputs from different cortical areas are kept quite separate as they flow through this loop: there is little, if any, convergence. At present the idea is that there are multiple parallel loops through the basal ganglia from the cortex. The main loops which will concern us here are the motor loop, from motor areas of the cortex and the oculomotor loop from frontal areas of cortex involved in eye move-ment. Within the major loops, there may be several separate 'mini-loops'. Thus, in the motor loop, input from the arm, face and leg are kept separate throughout the basal ganglia.

The concept of multiple parallel loops is at variance with the anatomy of the basal ganglia. For example, even if we imagine that the efferent fibres coursing through the basal ganglia remain topographically or-ganized, the dendritic tree of cells in the globus pallidus, for example, is extremely large and can cover up to 30% of the cross-sectional area of the nucleus. Thus, it would be in a position to receive input from a fairly wide area of the caudate and putamen and provide an opportunity for

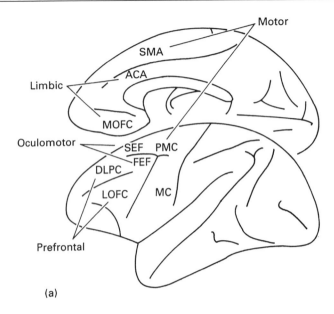

Figure 11.2 (a) Areas of the frontal lobes which receive output from separate basal ganglia circuits. Targets of four different basal ganglia–cortex 'loops' are shown: the motor loop (MC, primary motor cortex; PMC, premotor cortex; SMA, supplementary motor area), the oculomotor loop (SEF, supplementary eye fields; FEF, frontal eye fields), the limbic loop (ACA, anterior cingulate area; MOFC, medial orbitofrontal cortex), and the prefrontal loop (DLPC, dorsolateral prefrontal cortex; LOFC, lateral orbitofrontal cortex).

convergence. Despite this anatomical possibility, electrophysiological studies and neuroanatomical investigations using trans-synaptic tracer techniques show that there is relatively little functional convergence at all.

Details of loop connections

Each of the major loops coursing through the basal ganglia has the same basic organization as that of the motor loop which is shown in Figure 11.2. Cortical areas send glutamatergic projections to the striatum (caudate and putamen). These synapse on the tips of the dendritic spines of the medium spiny neurones which make up 98% of the striatum (Figure 11.3). The remainder of the striatal neurones are large, non-spiny cholinergic interneurones. The medium spiny neurones not only receive the striatal input, they are also the output neurones and send GABAergic axons out to the internal and external segments of the globus pallidus and the substantia nigra pars reticulata. These two projection systems are quite different. Those striatal neurones which end in the external

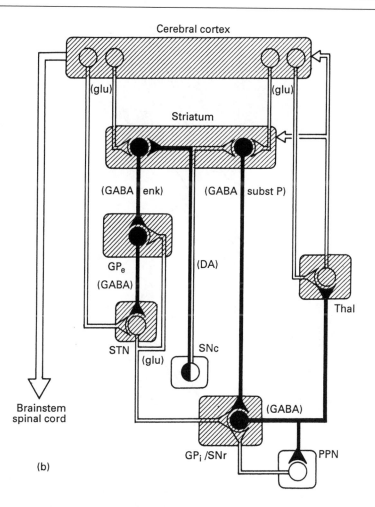

Figure 11.2 Continued. (b) Schematic diagram of the main circuits and neurotransmitters involved in the corticobasal ganglia–cortex circuit. Excitatory connections are shown as open neurones, inhibitory connections are shown as filled neurones. Note the 'direct' pathway from striatum to internal globus pallidus (GP$_i$), and the 'indirect' pathway via the external globus pallidus (GP$_e$) and subthalamic nucleus (STN). PPN, pedunculopontine nucleus, SNc, substantia nigra pars compacta. (From Alexander and Crutcher, 1990a, Figures 1 and 2; with permission.)

pallidal segment co-localize enkephalin and neurotensin as a transmitter with GABA; those neurones which project to the internal pallidal segment co-localize substance P (SP) and dynorphin.

The GABA/SP projection from the striatum to the internal pallidum and substantia nigra pars reticulata forms the so-called 'direct' pathway through the basal ganglia. This is because the internal pallidum and pars

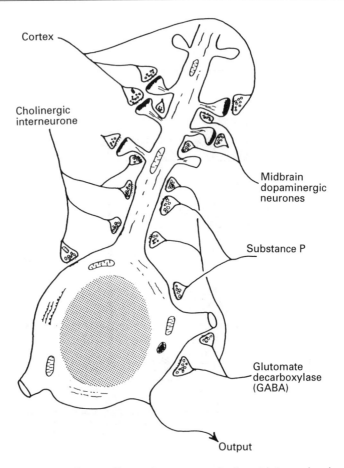

Figure 11.3 Diagram of a medium spiny neurone in the striatum, showing the topography of the synaptic inputs. Inputs from outside the striatum tend to synapse on the more distal parts of the dendritic tree, whilst inputs from local neurones (using substance P, GABA, or acetylcholine as transmitters) synapse more proximally. Note that projections from the cortex synapse at the tips of the spines (asymmetrical, excitatory synapses), whilst dopaminergic inputs from the midbrain substantia nigra end at the base of the spines (symmetrical, inhibitory synapses), and hence are in a position to modulate the transmission of specific cortical inputs. (From Smith and Bolam, 1990, Figure 3; with permission.)

reticulata of the nigra are major output nuclei of the basal ganglia and send GABAergic projections directly to the ventrolateral thalamus and thence to the cortex (the thalamocortical projection uses glutamate as a transmitter). The GABA/enkephalin projection from the striatum to the external pallidum is referred to as the 'indirect' pathway. This is because the external pallidum has very few direct projections to the thalamus. Most of its GABAergic output goes to the subthalamic nucleus (which also receives excitatory glutamatergic input from the cortex), which then projects via an excitatory glutamate pathway to the external pallidum and substantia nigra pars reticulata and thence to the thalamus. The indirect pathway therefore takes in the subthalamic nucleus in its path through the basal ganglia, whereas this is omitted in the 'direct' route.

It should be noted that although the major output target of the external pallidum and substantia nigra pars reticulata is the thalamus, there is also a small projection in the motor loop to the pedunculopontine nucleus of the midbrain. In the oculomotor loop, this projection is to the superior colliculus.

As shown in figure 11.2, the GABAergic synapses are inhibitory, whereas the glutamatergic are excitatory. If we follow the inhibitory and excitatory connections of the direct pathway, we find that excitatory cortical input to the striatum produces disinhibition (i.e. two inhibitory synapses) of the thalamus and hence results in excitation at the cortex. The indirect pathway has the opposite effect. Excitatory cortical input disinhibits the subthalamic nucleus. Increased activity of this nucleus then excites the inhibitory projection to the thalamus from the internal pallidum and pars reticulata of the substantia nigra. Thus, activity in this pathway has an inhibitory effect back on the cortex, and hence might work to brake activity in the direct route. Fig. 11.2 illustrates the two pathways as acting directly on the same output target. However, it is not known whether these pathways actually converge as shown, or whether they remain separate throughout the circuit. If they remain separate then maybe the indirect route should be seen more as a form of 'inhibitory surround' for the direct excitatory pathway.

Role of dopamine

Because of its importance in Parkinson's disease the role of dopamine in basal ganglia function has often warranted particular attention in the past. Within the basal ganglia, the caudate and putamen are the two nuclei which receive dopaminergic inputs, principally from the dopaminergic cells of the substantia nigra pars compacta and a neighbouring cell group in the midbrain known as the A8 region. The dopamine synapses are located at the base of the dendritic spines of the medium spiny neurones of the striatum (Figure 11.3). Since synapses from cortical areas are formed at the tips of these spines, the dopaminergic synapses are in a good position to regulate the effectiveness of cortical inputs at the level of individual spines. For a long time it was not clear whether dopamine had an excitatory or inhibitory function. In fact, it is now thought that the dopaminergic input is inhibitory to the cells which form the indirect (GABA/enkephalin) pathway whereas it is excitatory to the cells of the direct (GABA/SP) pathway. There is even speculation that the excitatory input has the D1 form of the dopamine receptor at the postsynaptic membrane, whereas the inhibitory input may be of the D2 type.

The dopaminergic cells of the substantia nigra pars compacta and its immediate surround receive a major input back from the striatum. However, the cells which provide this projection are quite separate from those which provide both the direct and indirect pathways noted above. This distinction is covered in more detail in the section below on patches/striosomes and islands.

Even though cells of the substantia nigra pars compacta have dendrites which descend into the pars reticulata they are thought not to receive any direct striatal input. Terminals of striatal efferent fibres probably synapse only with the cells of the pars reticulata.

Motor circuit (Figure 11.4)

A large proportion of the total input to the basal ganglia comes from the sensorimotor areas of cortex. These include the precentral motor, premotor, supplementary motor, and postcentral sensory areas (including area 5) of cortex. Their fibres terminate almost exclusively (in primates) on the putamen, which is the receiving nucleus of the 'motor loop'. The inputs form a topographic body map, following the arrangement of the cortical mantle above. Dorsal parts of the putamen receive from medial (or leg) areas of cortex, while ventral parts receive from lateral (or face) areas of cortex. Within the gross area of arm representation, cortical projections from somatosensory and motor areas may overlap.

The cortical terminals form discrete patches in the putamen which probably correspond with the patchy physiological representation found in microrecording and stimulation studies (see below). This patchy organization does not correspond to the striosome/matrix organization described below.

Outputs from the striatum group together in the bundles known as Wilson's pencils, and project to the caudoventral parts of both segments of the globus pallidus, and caudolateral substantia nigra pars reticulata. The projections are topographically organized, preserving the dorsoventral orientation within the putamen. The GP_e sends output to the subthalamic nucleus and thence to GP_i (and back to GP_e), forming the 'indirect' pathway through the motor loop. The final stage in the output of both 'direct' and 'indirect' pathways is from the internal segment of the globus pallidus and substantia nigra pars reticulata to the ventrolateral thalamus. Thalamic efferents of the GP_i travel in two separate fibre bundles through the descending axons of the internal capsule. The bundles are known as the ansa lenticularis, which arises from the outer

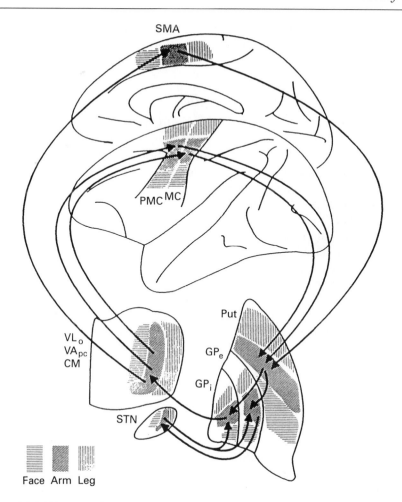

SMA

PMC MC

Put

VL$_o$
VA$_{pc}$
CM

GP$_e$

GP$_i$

STN

Face Arm Leg

Figure 11.4 Diagram of the motor loop through the basal ganglia, showing that the somatotopic subdivisions of the input remain separate throughout the circuit. Note that although the thalamic output of the circuit is shown as projecting to all three of the major motor areas (SMA, supplementary motor area; PMC, premotor area; MC, primary motor area), the major site of termination is in the SMA. (From Alexander and Crutcher, 1990a, Figure 4; with permission.)

part of the GP$_i$, and the lenticular fasciculus, from the inner part of the GP$_i$. After crossing the internal capsule, they join and meet ascending thalamic afferents from substantia nigra pars reticulata to form a single fibre bundle called the thalamic fasciculus. Pallidal fibres in the 'motor loop' end on distal dendrites of cells in the VL$_o$. Fibres from the 'motor' part of the substantia nigra pars reticulata are believed to end in VL$_m$. The major output of these thalamic nuclei is back to cortical motor areas, in particular the supplementary motor area (Chapter 8). However there are some (less dense) terminals in the dorsal premotor cortex and even the primary motor cortex.

Other outputs of the basal ganglia

Although the cerebral cortex appears to be the main output target for the basal ganglia, its fibres project to several other regions. The output to the superior colliculus has been mentioned above in relation to the 'oculomotor loop'. Outputs also end in the centromedian nucleus of the thalamus, which together with all other midline thalamic nuclei sends some fibres back to the striatum. These are collaterals of axons which also project to diffuse areas of cortex. There are also projections from GP_i to the habenular nucleus and the pedunculopontine nucleus. The latter is the largest and in the cat can account for 8% of basal ganglia output. The axons travel in the pallidotegmental tract and terminate in a small mass of cells partly embedded in the superior cerebellar peduncle. Unfortunately, little is known about the connections of the pedunculopontine nucleus. It receives some direct afferents from cerebral cortex, and has been implicated in the control of walking in the cat.

Patches, striosomes and islands

Superimposed on the parallel pathways through the basal ganglia is another, less understood, organizational entity. Despite the fact that the medium spiny neurones of the striatum (i.e. the caudate and putamen) look similar throughout the nuclei, it now appears that there are at least two distinct subpopulations, each with their own input/output connections. These subpopulations were originally detected using different techniques in different species and were given different names. Workers found out that there were 'islands' of more densely packed cells within a less dense matrix; there were groups of cells ('striosomes') which showed less dense staining for acetyl cholinesterase; finally, there were 'patches' of neurones rich in opiate receptors. It is now thought that all these techniques identify the same groups of cells in the striatum. For convenience these groups are sometimes called striosomes; the remainder of the striatal neurones are termed matrix. There are many other biochemical differences between these two neuronal subpopulations (summarized in the article by Graybiel, 1990). (As discussed in the description of the motor loop above, the cortical input seems to cluster into groups. These clusters are not striosomes: the putamen is virtually all matrix. Instead, this may be a further subdivision within the matrix region, which has provisionally been termed matrisome.)

Several of the inputs and outputs of the striatum respect the striosome/matrix division.

(1) Most somatosensory and motor areas of the cortex project to matrix; indeed, the putamen, which is the main recipient of the 'motor' loop, is virtually all matrix. In contrast many striosomes receive input from limbic structures.

(2) The connections to and from the dopamine systems of the midbrain also conform to the striosome/matrix pattern. Only striosomal neurones send projections to the dopamine containing cells of the substantia nigra pars compacta. However, both striosome and matrix receive dopaminergic input (from the pars reticulata and the closely adjacent A8 dopamine region, respectively).

The segregation between striosomes and matrix leads to the interesting consequence that limbic input to the basal ganglia might regulate transmission through the motor (and other) loops of the matrix compartment. First, limbic areas project mainly to the striosomes which then have a major input to the dopaminergic-containing cells to the midbrain. These then project back upon the matrix (and striosomes), where their synapses, at the base of the dendritic spines are in an ideal position to modulate the effectiveness of cortical inputs located at the tips of the spines (Figure 11.5). The implication is that attentional emotional factors from the limbic system might change the effectiveness of transmission through the basal ganglia motor loop.

11.2 *Electrophysiological recordings from behaving animals*

Resting discharge

Neurones in different parts of the basal ganglia have different levels of resting discharge. This ranges from slow or absent in the cells of the caudate, putamen and substantia nigra pars compacta, to high frequencies of 60–100 Hz in cells of the globus pallidus and substantia nigra pars reticulata. Neurones of the subthalamic nucleus have a maintained but slow rate of spontaneous firing. Since the output of the caudate and putamen is inhibitory onto the cells of the globus pallidus, it may be that the high firing frequencies of globus pallidus neurones result partly from lack of inhibition from slowly firing cells of the striatum. Further subdivision of cell firing patterns has been seen in the globus pallidus. Cells in the GP_i have a sustained high frequency firing; GP_e cells fire in high frequency bursts and are of two subtypes, one with relatively long

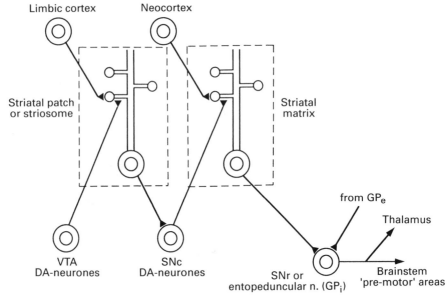

Figure 11.5 Striosome and matrix organization of the striatum. This diagram shows that a major input to the striosomes comes from limbic areas of cortex, whereas the main input to the matrix is from sensorimotor cortex. These inputs are shown as terminating on the tips of the dendritic spines of the medium spiny neurones. One of the output targets of the striosomal compartment is the dopamine cells of the substantia nigra pars compacta (SNc). These then project back to the matrix compartment, where the terminals on the base of the dendritic spines of the striatal neurones can modulate input from sensorimotor cortex. Output from the sensorimotor striatum is shown projecting to the substantia nigra pars reticulata (SNr) or the internal globus pallidus (GPi, or entopeduncular nucleus). The projection to external pallidus is not shown. VTA, ventral tegmental area. (From Smith and Bolam, 1990, Figure 6; with permission.)

bursts and the other with relatively short bursts of firing (DeLong and Georgopoulos, 1981.)

Relationship of discharge to movement

As electrophysiologists study neurones which are more and more distant from the peripheral motor apparatus, it becomes more difficult to define the precise relationship of their firing patterns to movements being performed. Some neurones of the basal ganglia show phasic changes in their discharge frequency in relation with movements of the body. Such cells can be studied to reveal the relationship of discharge to various parameters of movement such as speed, amplitude or force. Other neurones are apparently silent and unrelated to movement performance. Their discharge is not related to movement *per se* but is dependent on both movement and the context in which the movement is made. They are said to be 'context-dependent' cells.

Movement-related cells: somatotopy
Many cells in putamen, globus pallidus, substantia nigra pars reticulata and subthalamic nucleus show phasic changes in their firing frequencies in association with movements of the contralateral side of the body. As expected on the basis of the anatomical projections, there is a somatotopic organization of these cells: in the putamen, globus pallidus and subthalamic nucleus, cells related to the movement of the legs lie dorsal to those related to movements of the face. Arm-related neurones are found between these two populations. Neurones of the SN_{pr} are related only to orofacial and saccadic eye movements (Figure 11.6). Within the putamen, this dorsoventral somatotopic arrangement parallels the mediolateral organization of the somatomotor cortex lying above.

The rostrocaudal extent of the putaminal movement related neurones is much larger than that of the motor cortex. This is because the putamen receives input from premotor, motor and sensory areas of cortex, all three of which may contribute to the discharge of putaminal neurones. Movement-related neurones of the putamen are found in clusters. Cells lying between these clusters and the cells of substantia

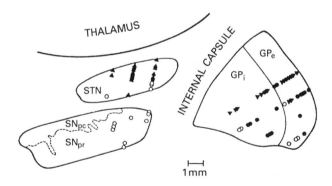

Figure 11.6 Composite figure made from the results of several experiments showing somatotopic grouping of movement-related neurones in the internal and external segments of globus pallidus (GP_i, GP_e), the substantia nigra pars reticulata (SN_{pr}) and subthalamic nucleus (STN). Representative coronal sections of each structure are shown, although each is taken at slightly different anteroposterior levels of the brain. Filled triangles, leg; filled circles, arm; open circles, face. The representation of each body part extends over most of the anterior–posterior extent of each structure. Note the clustering of orofacial neurones in SN_{pr} and absence of any neurones related to limb movements. Cells of SN_{pc} did not show any phasic discharges during movement of any part of the body. (From DeLong and Georgopoulos, 1979, Figure 1; with permission.)

nigra pars compacta and the caudate nucleus are not directly related to movement (e.g. DeLong, Crutcher and Georgopoulos, 1985).

Movement-related cells: firing pattern
There are two aspects of the neuronal firing pattern in the basal ganglia which are accepted by all authors. The first is that basal ganglia neurones tend to discharge rather late with respect to the onset of movement. In reaction time tasks to simple visual or somatosensory stimuli the majority of cells in both the putamen and globus pallidus tend to fire **after** the onset of movement compared with the 50% or so of motor cortex cells which fire before movement. Indeed, some authors think that basal ganglia discharge may be signalling the **end** of a movement rather than the start. There are two implications: first, the late basal ganglia discharge may reflect input from motor cortical areas involved in the movement; second, basal ganglia probably play little role in the initiation of movement under these reaction time conditions. They are therefore more likely to have a role in the execution of movement which is already under way. This is consistent with other findings in both the lesions studies and in human disease. Indeed, in simple reaction time tasks, initiation of movement is relatively little affected; the main deficit is slowness in the performance of tasks.

The second important feature of basal ganglia firing is that the majority of movement-related neurones discharge in relation to the direction of the movement rather than the force necessary to achieve it. In a task involving, say, elbow flexion movements against either an opposing or assisting load, the pattern of EMG activity can be dissociated from the direction of movement (see Chapter 9). With an opposing load, the movement is made using agonist activity, whereas with an assisting load, the movement is made using the braking action of the antagonist. In the putamen, a minority of the neurones have a discharge which reflects the pattern of EMG activity, while most of them have a discharge related to the direction of movement (Figure 11.7). This is true also for globus pallidus and subthalamic nucleus.

There are three main models which take these accepted facts of late discharge and directional-sensitivity, combine them with other aspects of neuronal firing, and then attempt a general explanation of how this cell activity might contribute to the overall control of movements. Since they are quite different theories, they are best discussed separately.

Hypothesis 1: The basal ganglia work by disinhibiting areas of the motor system and thus 'allow' movement to occur.

Since much of the work which supports this hypothesis comes from the study of eye movements it is necessary first to outline the connections of the 'oculomotor' loop through the basal ganglia. The major cortical

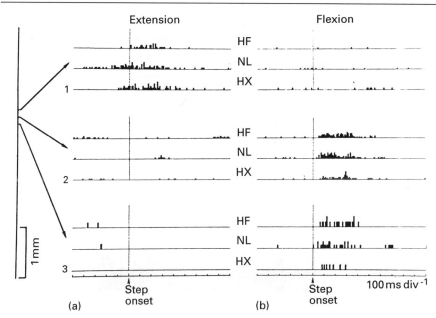

Figure 11.7 Relationship between firing frequency and direction of elbow movement in three different neurones encountered during the course of a single microelectrode penetration (left) through the monkey putamen. The task was to extend (a) or flex (b) the elbow rapidly in response to movement of a visual target. Movements were made either with no load (NL), with a constant load (150 g) opposing flexion (HF) or with a load opposing extension (HX). Histograms of average unit activity for each cell in the six different tasks are shown. Vertical dashed line indicates time of onset of movement. The first neurone (1) was active during extension movements, whereas the other two neurones (2 and 3) were active during flexion movements. Thus, activity was related to direction in these cells, but not the force of the movement. (From Crutcher and DeLong, 1984, Figure 3; with permission.)

output areas are the frontal eye fields and supplementary eye fields of the dorsolateral pre-frontal cortex and regions in the posterior parietal cortex. These regions send axons to the caudate nucleus and thence to the substantia nigra pars reticulata and internal globus pallidus. In addition to the output back via thalamus to cortex, there is a substantial output from the substantia nigra pars reticulata to the superior colliculus in the brainstem. This is an area closely involved in the control of the oculomotor nuclei themselves.

The main fact essential to the disinhibitory hypothesis is a continuous high discharge rate of the neurones of the substantia nigra pars reticulata and the internal globus pallidus in animals at rest. This output is inhibitory, and it is suggested that it provides a tonic 'braking' action on cells of the motor thalamus. Since the striatal output to the substantia nigra pars reticulata and the internal pallidum is also inhibitory, then

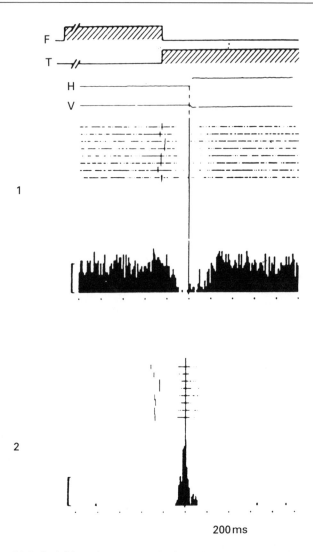

Figure 11.8 Activities of a neurone in the substantia nigra pars reticulata (1), and one of its presumed target cells in the superior colliculus (2) during a visually guided saccadic eye movement in a monkey. The traces in the upper part of the figure illustrate the target and eye movements. The top trace shows the duration of the fixation point display (F), the second trace the time of appearance of the target (T), and the third trace shows when the horizontal eye movement began (H). The trace labelled V shows the vertical component of the eye movement. The accompanying neuronal discharges in the nigra and colliculus are shown both as dot raster displays and time histograms. The nigral cell discharges at a high rate during fixation, but is suppressed just before the onset of the saccade. At the same time, the collicular cell fires. After completion of the saccade, the basal firing rates in each structure are resumed. (From Hikosaka and Wurtz, 1983, part IV, Figure 7; with permission.)

activity in the striatum will cause disinhibition of thalamic cells. That the circuitry can actually work this way in practice can be shown by injecting the excitatory transmitter glutamate into the striatum of a rat. Immediately after injection, striatal neurones increase their firing rate and this is accompanied by a decrease in the tonic firing cells of the substantia nigra pars reticulata. Finally, in the two projection targets of the pars reticulata, the ventromedial thalamus and superior colliculus, there is a large increase in discharge.

The cells of the oculomotor loop illustrate this pattern of firing behaviour very well. For example, in Figure 11.8 a neurone in the substantia nigra pars reticulata decreased its discharge rate about 100 ms before the onset of a saccadic eye movement to a visual target. In similar trials, this was accompanied by increased discharge of neurones in the superior colliculus.

Although the eye movement control system appears to make good use of disinhibition of basal ganglia target structures more detailed studies show that eye movement is not an **obligatory** result of decreased basal ganglia output. Thus, there are occasions when eye movements occur without obvious changes in basal ganglia firing; conversely,

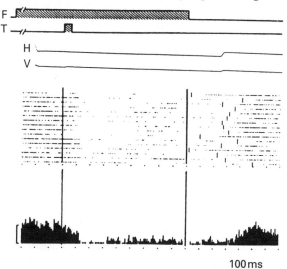

100 ms

Figure 11.9 Experiment showing that a reduction in nigral cell discharge does not cause an obligatory eye movement. The monkey had to perform a memory-guided saccade. It maintained its gaze on the fixation point (F) whilst a target (T) was displayed transiently in the peripheral field. When the fixation point was removed, the monkey made a horizontal saccade (H) to the point where the target had been displayed. The raster and the histogram at the bottom show the discharge in a cell in the substantia nigra pars reticulata. Note that the cell discharge decreases after presentation of the target, and is maintained until the eye actually makes the saccade. (From Hikosaka and Wurtz, 1983, part III, Figure 7; with permission.)

pauses in nigral neuronal discharge occur without causing the eyes to move. An example of the latter is shown in Figure 11.9. In this experiment a monkey fixated a visual stimulus while a target light was presented briefly in the peripheral field. The monkey was trained not to make a saccade to the target while the fixation stimulus remained on. However, the animal had to remember the location of the target. Some seconds after the target disappeared the fixation light was removed and the monkey made a saccadic eye movement to the remembered position of the target. Under these circumstances, some cells decrease their discharge after target presentation and remain at a low rate until the remembered saccade is made. It is as though the substantia nigra pars reticulata output was facilitating the colliculus until such time as the saccade was released.

In summary, there is evidence that certain types of movement may occur after release of tonic inhibition from basal ganglia structures (e.g. Hikosaka and Wurtz, 1988; Chevalier and Deniau, 1990).

Hypothesis 2: The basal ganglia are involved in turning off 'automatic' postural activity so that voluntary movement can occur. They also prevent unwanted muscle activity during focal tasks. This is the hypothesis of Mink and Thatch (1991). It differs from the hypothesis outlined above in two respects: (i) it proposes an important role of the basal ganglia in control of postural, rather than prime mover activity, and (ii) it emphasizes the **prevention** of movement, and, as such, it implies that the important change in output during movement is an increase and not a decrease in activity.

There are two strands to this argument. First, the authors point out that even though cells might be directionally selective in one task, this may not be consistent. Given a different task, cells might have quite a different directional preference. They argued that direction could not be the most important basic parameter for which these cells were coding. They next asked whether the task itself was more important. Monkeys made wrist movements in different directions, but sometimes they would move slowly, sometimes fast. Sometimes they followed a target display and on other occasions they made self-paced movements from memory. Putaminal neurones fired in relation to one or more of these task combinations, but in no case could Mink and Thatch establish with certainty that there was a **preferred** movement type in which the **majority** of pallidal neurones were interested. The conclusion of this strand of the argument was that pallidal firing is not related consistently to any of the types of movement that they studied.

The problem of what exactly this pallidal discharge was coding for made Mink and Thatch go back and review the results of lesion studies on the performance of movement. In addition, they made their own chemical lesions of the pallidum to confirm this data. Injection of small amounts of muscimol or kainic acid into the pallidum inactivates neurones without affecting activity in fibres of passage which traverse the nuclei. These substances produced effects on the monkey's movements similar to those seen in basal ganglia diseases in man. There was (i) co-contraction of antagonist muscles, (ii) the limbs took up a flexed position, (iii) movements were made more slowly than normal, and (iv) the animals had difficulties in turning off muscle activity (e.g., in letting go of an object).

Their interpretation of these lesion studies was that, normally, pallidal activity was involved in switching off unwanted muscle activity. Any role in producing the prime movement itself was small. This would fit with the known inhibitory action of the pallidal output. In addition, it would explain why cell discharge was not obviously and consistently related to any aspect of movement performance. The discharge would be related not to the observable movement made by the monkey but by definition to the unobservable activity which might have occurred had these cells not fired. Failure to switch off inappropriate activity might lead to co-contraction and a flexed posture. Movements would not easily be released and would tend to be slow. It should be emphasized again that unlike the theory of hypothesis 1, this proposal involves increased basal ganglia output during movement, rather than a decrease.

Hypothesis 3: As noted above, the discharge of neurones in the basal ganglia is related to many different aspects of movement. Because of this, Alexander and Crutcher (1990b) have recently suggested that the basal ganglia do not perform one single movement related function but that they are involved in several different aspects of movement at the same time. In other words, they propose, in today's computerese a parallel-processing model of basal ganglia function.

Their experiments emphasize that the discharge of basal ganglia neurones is not only related to the execution phase of movement (i.e., the size, speed or direction of the movement when it is made), but also reflect preparatory activity which occurs **prior** to movement. Figure 11.10a shows the typical arrangement for investigating preparatory activity in these experiments. It is similar to the procedure used to examine preparatory activity in the premotor and supplementary motor areas described in Chapter 9. The monkey can move a cursor to the left or right on a screen by flexing or extending his wrist. An initial practice trial begins with the cursor in the centre of the screen. Some seconds later, a target appears on either the left or the right and the monkey must move the cursor to capture it. After this, he returns the cursor to the centre to await the test trial. This is signalled by the appearance of both targets (on the left and the right) simultaneously, and the task is to move in the same direction as in the practice trial. In

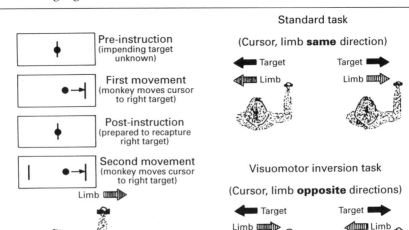

Figure 11.10 (a) Illustration showing the arrangement for studying preparatory and movement-related activity in an elbow flexion/extension task. The monkey faced a display, holding its arm in the neutral position, and made a movement towards a target light when it appeared on either the left or right of the screen. The arm was then returned to the initial position. After a few seconds, two target lights appeared on both the left and right of the screen simultaneously. The monkey had to move in the same direction as it had moved in the first trial. (b) Diagram of arrangement to dissociate the direction of movement of the target from the direction of movement of the arm. In the standard task, the arm moved in the same direction as the target, whereas in the inversion task, the monkey had to move his arm in the opposite direction to the target. (From Alexander and Crutcher, 1990b, Figure 1; with permission.)

order to achieve this, the monkey must store in memory the direction of the practice trial until onset of the test movement. To make things just a little more complex, like the experiment of Figure 11.7, the movement can be assisted or loaded by a torque motor to make the direction of the next movement independent of the muscle force needed to achieve it.

One third of movement-related cells in the putamen showed preparatory activity in this task and the majority of these were related to the direction, rather than the muscle activity needed for the next movement. Similar patterns of firing can be found in both the cells of the primary motor and supplementary motor areas, although there were far more in the latter than the former (see also Chapter 8).

There is one final twist to this story. The experiment described above defines two types of preparatory information: the direction of the second movement **and** the direction of the target which must be captured. Is the preparatory activity related to the movement as such, or is it just 'reminding' the monkey which of the two forthcoming targets it will be important to move towards? In order to separate out movement- and target-related influences, Alexander and Crutcher could reverse the relationship between the movement of the monkey's wrist and the cursor movement on the screen. The same target direction could then be made to signal two directions of movement (Figure 11.10b).

By using this dissociation, it was found that most directionally-selec- tive preparatory cells were actually related to the direction of the visual target rather than the actual elbow movement to be made (Figure 11.11). Conversely, when they examined the firing pattern of cells which discharged preferentially during the movement (rather than before onset of movement) then the cells more often signalled the direction of limb movement rather than that of the target.

To sum up, the movement-related cells of the basal ganglia may fire during the movement or even in the preparatory period beforehand. Their firing may be related either to the direction to the target displace- ment, direction of limb movement, or the muscle activity needed to produce the movement. In technical language, this neuronal discharge reflects either (i) target-level variables, (ii) trajectory/kinematic vari- ables or (iii) dynamics/muscle-level variables. During the preparatory period the firing of most neurones is related to target displacement whilst during performance of movement most neuronal firing is related to the direction of movement. It should also be noted that just because the cells have been denoted movement related, their firing may also be influenced by somatosensory input (see below).

The wide variety of movement-related neuronal firing, all occurring in the same structure (and also reflected in primary motor and supple- mentary motor area discharge) suggested to Alexander and Crutcher that several different neuronal operations were happening at the same time within the basal ganglia. They suggested that neurones with similar firing patterns might be linked together throughout the basal ganglia circuitry and form multiple parallel information processing loops within a major anatomical 'motor' loop. Their idea is that the brain may not work in a serial fashion. In other words, the does not (i) identify the position of a target then (ii) work out which direction to move in order to achieve it and finally (iii) decide what level of muscle activity is needed to move the limbs. Instead, the brain may be performing all three functions at once.

Hypothesis 4: The basal ganglia are involved in the execution of sequences of movement. They play a role in (i) performance of most 'automatic' movements and (ii) in changes from one subunit of a sequence to the next. This is an idea which stems from clinical observations of patients with Parkinson's disease who perform complex sequences of move- ments much worse than expected from the examination of their

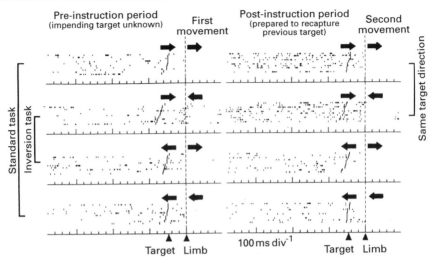

Figure 11.11 Rasters showing the discharge of a neurone in the putamen with target dependent preparatory activity in the task illustrated in Figure 11.10b. The responses during both the first and the second movements are shown. The monkey had no indication of the direction of the first movement and therefore showed no pre-instruction activity in the first movement. However, since it had to move in the same direction in the second movement, the cells could show both post-instruction and movement-related activity. In the first set of rasters, the monkey had to move the arm in the same direction as the target (to the right). Prior to the second movement, there was a sustained increase in firing in the post-instruction period. In the second set of rasters, the monkey had to move the arm in the opposite direction to the target (to the left). Even though the movement was different, there was still the same increase of cell activity in the post-instruction period. The third and fourth set of rasters show the results when the target moved in the opposite direction. Note that the cell discharge was reduced in the post-instruction period before the second movement in both cases even though the direction of arm movement was opposite. (From Alexander and Crutcher, 1990b, Figure 6; with permission.)

performance in each subunit of the task on its own (see below). Brotchie, Ianseck and Horne (1991) again take as their starting point the fact that there is no obvious and consistent relation between basal ganglia discharge and parameters of simple movement. Because of this, they made the task a little more complex. Their monkeys had to perform a sequence of wrist movements consisting of jumps to the right or left each separated by a holding phase.

They emphasized two findings in their experiments. The first was that pallidal discharges were usually most prominent in trials in which (i) the movements were predictable, and (ii) which were performed by monkeys with the least error. For example, discharge might be weak on the first movements of a new block of trials but would increase on subsequent repetitions of the movement (Figure 11.12a). The authors argued that basal ganglia activity was greatest during movements made with the least conscious intervention. Tasks which were difficult or unpredictable, i.e. those which the authors imagined would require the most conscious intervention, had the smallest changes in basal ganglia activity. Thus, their first idea was that as movements became more and more 'automatic', control is passed from a conscious, presumably cortical level to basal ganglia structures.

The second finding that they emphasized was that in a sequence of two movements, some cells had a small burst of firing just before the onset of the second movement. This burst was related both to the timing of the movement and also, in some cases, to its direction (see Figure 11.12b). If the monkey could not predict the time or direction of the second movement, this burst did not occur. In the authors' terms, it may have been an internal signal indicating a transition between subunits in a sequence. In a way, this is another example of the basal ganglia controlling the automatic aspects of a predictable movement.

(In some respects, the burst of firing during sequences of movement is similar to the predictive activity of basal ganglia neurones described above by Alexander and Crutcher. However, the latter authors emphasized the prolonged change in activity, starting immediately after the warning signal and lasting until the onset of movement itself. In Brotchie *et al.*'s experiment, the first movement is equivalent to a warning signal, but in this case, the predictive activity is burst-like, and anticipates the second movement by only 100–200 ms. Whether these represent different types of behaviour entirely, or subsets of a single response type, modified by the experimental conditions is not clear.)

Somatosensory inputs to basal ganglia

Many of the movement-related cells in putamen, globus pallidus and subthalamic nucleus also respond to somatosensory stimuli. However, there are a small number of cells which respond either to somatosensory input or to movement. The most effective somatosensory inputs come from muscle or joint particularly from the region related to discharge during active movement. There is little effective input from superficial cutaneous receptors. In general, somatosensory responses recorded when the animal is at rest are much weaker than movement-related discharges. Many of the inputs to putamen, globus pallidus and subthalamic nucleus are directional: that is, they give opposite responses to opposite directions of joint rotation. As in the motor cortex, there are a number of directionally selective movement-related cells which also receive directionally specific sensory input (see Figure 11.6).

Within the whole class of somatosensory-related cells there is a certain somatotopy of face, arm and leg representations, although this does not compare with the discrete representation seen in the primary sensory areas of cortex. In the basal ganglia, there is intermixing between, say, the representation of the shoulder, elbow and wrist within the general area for the arm, although areas of arm and leg input remain separate.

   Timing of discharge to sensory inputs suggests that the main source of afferent information comes from the cerebral cortex. Putaminal neurones discharge 30–50 ms after onset of a torque pulse disturbance to the wrist. Globus pallidus neurones discharge slightly later with a latency of 40–60 ms. This compares with about 30 ms in motor cortex.

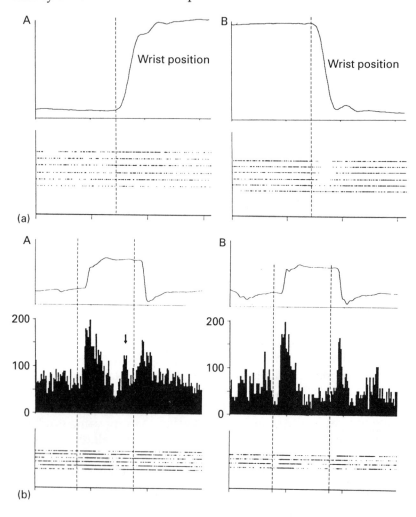

Firing pattern of dopaminergic cells in the substantia nigra pars reticulata

On anatomical grounds (see above) it has been proposed that dopamine input to the striatum may form a pathway whereby limbic (i.e., motivational/emotional) inputs modulate transmission in the motor loop. The firing pattern of pars compacta cells seems to fit with this idea. Monkeys were trained in two tasks. In the first, they reached out in their own time and obtained food from inside a covered box. In the second, the door of the box was closed and the signal to reach out was opening of the door. Dopaminergic cells of the pars compacta had a low (less than 8 Hz) rate of resting discharge, which contrasted with the rapid firing (80–100 Hz) of their neighbours in the pars reticulata. In the first task, this discharge remained virtually constant except for a burst of firing when the monkey touched the food. In the second task, opening of the door produced the burst of firing, whereas the later contact with food failed to produce any effect. Other experiments showed that dopaminergic neurones respond well to the noise of door opening or even to unrelated but, novel, stimuli in the environment. Thus, firing of dopaminergic cells does not seem to be directly related to movement as such. Instead, pars compacta discharge seems to reflect motivational arousal. As such, they may be able to set the effectiveness of transmission in the basal ganglia motor loop without themselves triggering each individual component of the movement (Romo and Schultz, 1990).

Figure 11.12 (a) Rasters of two pallidal neurones whose discharge demonstrated a change in movement-related activity during repeated performance of a wrist flexion/extension task. Raster A illustrates a neurone with an excitatory response; raster B a neurone with an inhibitory response. The trials are aligned on movement onset (dotted line), and the rasters are ordered from bottom upwards. They show that the cell had a minimal response on the first trial (bottom row of dots), but that in subsequent trials there was a phasic change in the discharge. (b) Predictive activity of a pallidal neurone in a sequential movement. In A, the monkey had to make a sequential movement consisting of wrist flexion to a target, hold for 1 s and then return to the start position. The trials were performed in blocks so that after the first trial, the monkey could predict the direction of the movements required. In this instance, the cell showed two major bursts of movement-related activity, but in addition there was a third burst (arrow) of activity just before the onset of the second movement. In B, the task was the same except that it was randomized with other directions of movement so that the monkey could not predict what the direction of the second movement would be. In this case the two movement-related bursts of activity were still present, but the predictive burst prior to the second movement was absent. Top trace shows the wrist movement, the second trace a histogram of the cell firing pattern, and the third trace shows rasters of neural discharge in single trials. (From Brotchie, Ianseck and Horne, 1991, Figures 3 and 8; with permission.)

11.3 *Effects of lesions of the basal ganglia*

Because the basal ganglia lie deep within the hemispheres, it is difficult to produced localized lesions with conventional anatomical or electrolytic techniques without producing damage to surrounding structures such as the hypothalamus, internal capsule and overlying cortex or white matter. Such conventional techniques could not, for example, distinguish between the pars compacta and pars reticulata of the substantia nigra even though the two parts have quite separate effects on the motor control. In addition, destructive lesions affect both the cells within an area and the fibres passing through that area on their way to other structures. Thus the effects of a lesion may result from loss of input to a part of the brain quite separate from that where the lesion is placed.

For these reasons it is sometimes difficult to interpret older work on the effect of basal ganglia lesions. As a general rule, it was found that animals with lesions of the globus pallidus or substantia nigra moved less and more slowly than usual. After large lesions they assumed a flexed posture. Lesions of the subthalamic nucleus had the opposite effect: hyperkinesia characterized by an uncontrollable excess of movement.

Recent experiments have shown that unilateral pallidal lesions in trained monkeys produce deficits very similar to those described in patients with Parkinson's disease. The animals were taught to make rapid flexion and extension movements at the wrist or elbow, either in their own time or in response to a visual trigger signal. The globus pallidus was then inactivated (both segments) by one of two methods. In one method, cooling probes were implanted unilaterally for reversible cooling of the pallidal tissue. Neurones were inactivated at temperatures of about 28°C. Although this produced little obvious effect on the overall behaviour of the monkey, there was a dramatic change in performance of the arm movement task. The most impressive effects were a decrease in the speed and extent of movement, sometimes associated with a smaller increase in reaction time to the visual stimulus (Figure 11.13). During experiments in which the monkey was allowed no visual feedback of arm position, the elbow joint tended to assume a flexed position.

Similar effects were seen with the second method of pallidal inactivation, injection of kainic acid. This is a neurotoxic substance which first inactivates and then destroys cell bodies with relative sparing of fibres of passage. In these experiments it was observed that despite the slowing of movement, there was no change in the **order** of muscle activation in complex tasks involving postural fixation of joints as well as prime movement about one single joint.

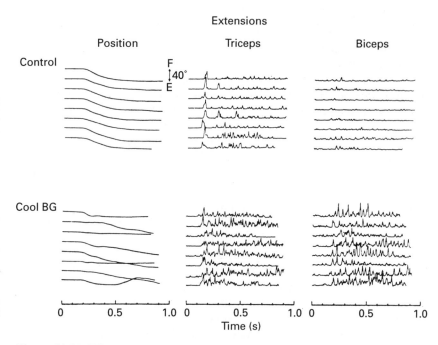

Figure 11.13 Effect of basal ganglia cooling (primarily the external segment of the globus pallidus) on extension movements at the elbow made in response to a visual 'go!' Signal at $t = 0$ ms. Elbow position and rectified EMG activity from triceps and biceps are shown for consecutive single trials. In the control state, the movements are fast and repeatable, being performed mainly with a phasic burst of activity in triceps, followed by a lower level of sustained activity. After cooling (Cool BG), the responses are far more variable; some are slow (4th, 5th and 7th movements), some are smaller than required (1st), some do not start properly (2nd, 3rd and 6th), and one rebounded back towards the original position shortly after starting (8th). All the movements were accompanied by co-contracting EMG activity in biceps and triceps. (From Hore and Vilis, 1980, Figure 5; with permission.)

In some experiments, it was possible to stimulate the globus pallidus during a rapid arm movement. Stimulation during the time when pallidal neurones normally change their activity produced a change in the duration of the movement as well as in the underlying EMG pattern. This was an important finding since it shows that even though pallidal discharge usually begins **after** a movement has started **changes** in pallidal output can still influence the movement as it progresses. This has led to the idea that in fast, simple, movements, pallidal discharge may be important in controlling the end, rather than the start of the movement (Hore and Vilis, 1980; Anderson and Horak, 1985; Anderson and Turner, 1991; Mink and Thatch, 1991).

MPTP

N-methyl-4-phenyl-1,2,3,6-tetrahydropyridine (MPTP) has become one of the most famous chemicals of the past decade. It was discovered when a group of drug abusers on the west coast of the United States were found to be suffering from acute, severe and permanent parkinsonian symptoms following self administration of a batch of 'synthetic narcotics'. It was later discovered that the batch was contaminated by MPTP and that when this substance was administered to primates it caused a profound parkinsonian state remarkably similar to that of human Parkinson's disease. The animals developed akinesia, rigidity and tremor and sat immobile in a flexed position. In quantitive studies, both the reaction time and the movement time were increased. The condition responded well to L-dopa.

MPTP is a neurotoxin which has a remarkably high specificity for the dopaminergic cells of the substantia nigra pars compacta. Other dopamine or noradrenergic cell groups may also be affected, but this is usually at higher dose levels. The precise mechanism of the effect is not yet known, but it is obviously important in view of the possible relationship it may have with the cause of Parkinson's disease itself. A remarkable feature of MPTP is its species specificity: dramatic effects at low doses are limited to primates. Consistent selective neuronal destruction is not seen in subprimates.

There are several studies on the effects of MPTP administration on the pattern of neuronal discharge in the basal ganglia (Tremblay, Filion and Bedard, 1989; Filion and Tremblay, 1991). The most obvious change is that the 'resting' discharge of neurones in the internal pallidum is increased, whereas that of the external pallidal neurones is decreased. This can be understood on the basis of the model of basal ganglia involvement in Parkinson's disease discussed in the next section and shown in Figure 11.2.

In addition, there is an increased responsiveness and lack of selectivity in the response of pallidal cells to peripheral input. Rather than responding only to, say, manipulation about a specific joint, neurones respond well to inputs from many joints, sometimes even on both sides of the body. Dopamine therefore seems to play a role in focusing activity in basal ganglia circuitry. If dopaminergic activity in normal animals has a strong input from limbic structures, this may be one way in which attention can change activity in the motor loop.

Finally, it should be mentioned that in the past, the chemical 6-hydroxydopamine which is toxic to dopamine cells was used to produce selective lesions in the nigrostriatal dopaminergic pathway by placing small injections into ascending dopamine systems. This technique is still used, and may produce more reliable and permanent damage to the dopaminergic pathways than MPTP.

Summary

(1) The flow of information through the basal ganglia is through parallel loops.

(2) The major output of the basal ganglia from the internal pallidal segment and the substantia nigra pars reticulata is inhibitory. Since the neurones of both structures have a constant high basal firing rate it is assumed that the basal ganglia exert tonic inhibition on cortical motor areas. Removal of this inhibition may assist movement, but on its own may be insufficient to produce movement. Increased inhibition may reduced unwanted movement.

(3) Inter-loop connections exist but are sparse.

(4) Dopamine may modulate transmission through the motor loop by its synapses on the base of dendritic spines of striatal medium spiny neurones. Input to the dopaminergic system is mainly from the striosomal compartment of the striatum which receives substantial input from limbic areas of cortex. Dopaminergic output affects all parts of the striatum. Thus, limbic input may have indirect access to the motor pathway.

11.4 *Models of basal ganglia disease in humans*

The circuitry shown in Figures 11.2 and 11.14 can be used as a first approximation to understand the symptoms of hyper- and hypokinesia which are common to several basal ganglia diseases in humans. Indeed, it has even been used to **predict** a new possible treatment for Parkinson's disease (Albin, Young and Penney, 1989; Bergman, Wichmann and DeLong, 1990).

Hyperkinesias

Ballism and chorea are two common hyperkinetic movement disorders associated with the basal ganglia lesions. Ballism occurs after lesions of the subthalamic nucleus and is characterized by uncontrollable, rapid movements of the contralateral arm and (less frequently) leg. In the model of Figure 11.2, we see that removal of subthalamic nucleus results in loss of excitatory input to the internal pallidum and substantia nigra pars reticulata. This will reduce the inhibitory output of the basal ganglia, and hence release extra, unwanted movement.

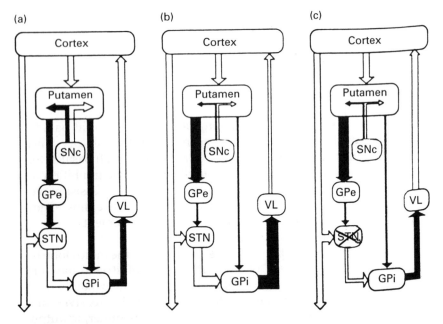

Figure 11.14 (A) Normal connectivity in the cortico-basal ganglia-cortex circuit. Open arrows are excitatory connections, filled arrows are inhibitory connections. Note the direct pathway from putamen to internal globus pallidus (GP$_i$) and the indirect pathway via the external pallidus (GP$_e$) and subthalamic nucleus (STN). VL, ventrolateral nuclear region of the thalamus. (B) The situation in MPTP-induced parkinsonism. Degeneration of the dopaminergic projection from the substantia nigra pars compacta (SNc) leads to underactivity in the direct pathway (thin arrow from striatum to GP$_i$), and overactivity in the output of the indirect pathway from STN to GP$_i$ (thick open arrow). The result is excessive activity in the output neurones of the GP$_i$ (thick filled arrow), and a reduction in the final excitatory output from thalamus to cortex. (C) How this situation can be normalized by lesioning the STN. Less excitatory drive in the indirect pathway to GP$_i$ leads to a reduction in the pallidal inhibition of ventrolateral thalamus. (From Bergman, Wichmann and DeLong, 1990, Figure 1; with permission.)

Chorea is a major symptom of Huntington's disease. The excessive, involuntary movements are usually said to be less abrupt and wild than those of ballism, but this is not always the case. In some instances, the movements of the chorea and ballism are identical. In the early stages of Huntington's disease, when chorea is most prominent, post mortem brain studies have shown a selective loss in the GABA/enkephalin projection from the striatum to the external pallidal segment. On the model, this will result in less inhibition of the external pallidal cells and hence to increased inhibition of the subthalamic nucleus. If we imagine excessive inhibition of the subthalamic nucleus to be similar to a sub-thalamic lesion, then it is easy to see how excessive movement may occur. The model also helps explain why injection of bicuculline, a GABA-receptor antagonist, into the external pallidal segment can produce chorea in the monkey. The injection blocks striatal inhibition of neurones in the external pallidum. These neurones will therefore increase their firing and produce inhibition of the subthalamic nucleus as proposed for Huntington's disease.

Hypokinesia (Figure 11.14)

In Parkinson's disease, there is selective degeneration of the dopaminergic innervation of the striatum from the substantia nigra pars compacta. From the model we can see that this should have two effects: removal of dopaminergic **excitation** to the projection to the internal pallidum, and removal of dopaminergic **inhibition** of the projection to the external pallidum. Since the striopallidal pathway is inhibitory to both external and internal segments, this leads to excessive inhibition of the external pallidum and a reduction of inhibition of the internal pallidum. Following the indirect pathway further, the decreased external pallidal activity results in overactivity in the subthalamic nucleus and excessive facilitation of the internal pallidal segment. Thus the internal pallidum is overactive because of reduced inhibition throughout the direct route and extra facilitation through the indirect route. Such extra inhibitory output from the basal ganglia is thought to prevent movement and lead to hypokinesia and bradykinesia.

As noted above, MPTP poisoning has the same consequence, and in this case it has been possible to verify electrophysiologically that such changes in neuronal firing in the external and internal pallidum actually do occur as predicted by this model (Figure 11.15). In addition, the postulated role of the subthalamic nucleus in contributing to the extra activity of the internal pallidal segment led to the discovery that lesions of the subthalamic nucleus in MPTP-induced parkinsonism can alleviate most of the symptoms of that condition, including both rigidity and bradykinesia. Such a discovery is important because of the possibility of developing new surgical or pharmacological treatments of Parkinson's disease involving inactivation of the subthalamic nucleus.

Although the model is remarkably effective in some respects, it should always be clear that it is only a model and as such cannot explain everything. For example, pallidal lesions alone never give rise to involuntary movements. Indeed, large pallidal lesions in animals produce hypokinesia. Despite this, pallidal lesions in the past have been used as a treatment for Parkinson's disease. The balance between hypo- and hyperkinesia may therefore not be as simple as the model may lead us to believe.

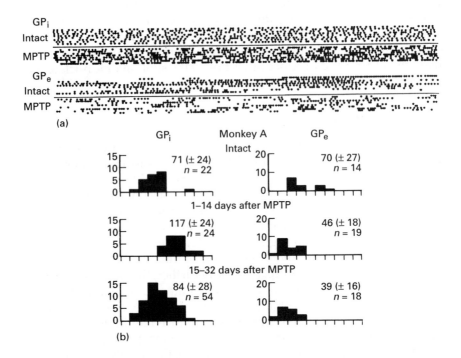

Figure 11.15 (a) Two traces show rasters of neuronal activity in the internal and external segments of the globus pallidus (GP$_i$ and GP$_e$) at rest in intact and MPTP treated monkeys. The firing rate of GP$_i$ cells increases, whilst that of GP$_e$ cells decreases. (b) The distributions of mean firing rates in one monkey at different times before and after MPTP treatment. The numbers represent the mean and standard deviation of the distribution and the number of neurones studied. Rates in the GP$_i$ increase, whereas those in the GP$_e$ decrease. (From Filion and Tremblay, 1991, Figures 1 and 3; with permission.)

11.5 *Pathophysiology of diseases of the basal ganglia in humans*

Diseases of the basal ganglia encompass a wide variety of different movement disorders. Common to all these conditions are either an excess of (abnormal) involuntary movements or a lack of spontaneous movement. In addition, there may be changes in muscle tone and defects of postural reflexes. (General reviews are provided by DeLong and Georgopoulos, 1981; Marsden, 1982.)

Some brief definitions may be useful. **Hyperkinesia** refers to an excess of movement; **akinesia** or **hypokinesia** to a lack of spontaneous

Figure 11.16 Typical posture of a patient with Parkinson's disease: flexed trunk, neck and arms, with a shuffling gait (Courtesy of Dr N. Quinn).

movement. Akinesia may also be used to refer to a lack of normal associated movements such as swinging of the arms when walking. In contrast to these terms which describe the amount of movement, **bradykinesia** defines the speed of movement: bradykinetic movements are slow (whether or not there is an excess, or a lack of them).

Parkinson's disease (Figure 11.16)

Parkinson's disease is the most common disorder of the basal ganglia. Histological examination of the post mortem brains shows, in most cases, a degeneration of neurones in the substantia nigra and locus coeruleus. There is a concomitant dramatic decrease in the dopamine content of the striatum due to degeneration of the nigrostriatal tract. This decrease is more pronounced in the putamen than in the caudate. Interestingly, for symptoms to appear, it has been estimated that there must be at least an 80% decrease in the dopamine content of the striatum. If more than 20% of the nigrostriatal terminals are left intact, then the nervous system seems capable of compensating for the deficiency. The dopaminergic projection from the ventral tegmental area to the accumbens (the mesolimbic dopamine system) is not greatly affected in Parkinson's disease. Degeneration is seen in other areas of the brain (for instance, dorsal motor nucleus of the vagus, substantia innominata, hypothalamus, sympathetic ganglia and many others), but it is not so obvious as in the nigra. Because of this it is believed that the abnormalities of movement, which are the main symptom of the disease, are due mainly to striatal dopamine deficiency. The success of L-dopa replacement therapy in the treatment of Parkinson's disease supports this view.

The classic symptoms of Parkinson's disease are akinesia, tremor and rigidity. To this might be added bradykinesia and postural instability. Patients have a poverty of associated and spontaneous movements (akinesia): their face has a mask-like expression, their arms do not swing while walking and they sit or stand motionless, with their body flexed at the waist, elbows and knees. All movements are slow (bradykinesia), giving the impression of a very deliberate and careful action. Despite the difficulties with voluntary movement, there may be a conspicuous tremor, especially of the hands which only disappears during sleep. This is known as a **rest** tremor and has a frequency of 4–6 Hz. Usually, it appears early in the disease, beginning in the hands and later affecting the lips and tongue and then the feet. In the hands, this tremor has a very characteristic motion, consisting of opposition and adduction of the thumb towards the index finger, usually described as a 'pill-rolling' tremor. In the early stages of the disease, this tremor may disappear when the patient moves his hands. However, as time passes the tremor persists during voluntary movement. Such **action** tremor has a higher frequency (6–8 Hz) than the rest tremor.

In relaxed patients, passive manipulation of the joints meets with more resistance than normal (rigidity). The increased tone is evenly distributed between extensor and flexor muscles (unlike the increased tone in spasticity), and may be accompanied by a phenomenon known as 'cog-wheeling'. This is due to superimposition of tremor onto the increased muscle tone, and gives the examiner the feeling of moving the joint through a ratchet-like device. Finally, patients have postural difficulties, especially in the later stages of the disease. A firm, unexpected push to the shoulders may easily overbalance such individuals. Stepping reactions are lost, and there is a failure to throw the limbs out to protect the body during falling.

The advent of positron emission tomography (PET) has allowed direct visualization of dopamine function in the parkinsonian brain. Subjects are given an intravenous injection of a labelled form of L-dopa, 6-L-[^{18}F] fluorodopa which (after decarboxylation to the dopamine form) is taken up actively by synaptic nerve terminals of dopaminergic fibres. It is therefore thought to be a suitable tracer for presynaptic dopamine, the degree of uptake probably being proportional to the number of dopamine terminals and to the concentration of the decarboxylase enzyme.

In Parkinson's disease striatal fluorodopa uptake is decreased, particularly contralateral to the clinically most affected side of the body. In the early stages of the disease, the body of the putamen shows the greatest loss, with caudate and anterior putamen being relatively spared. At this time, dopamine deficiency in the basal ganglia motor loop is presumably responsible for the clinical symptoms observed. Indeed, the degree of both akinesia and rigidity correlate well with the levels of fluorodopa uptake. The amount of tremor, however, does not correlate so well.

The success of L-dopa therapy coupled with the widespread and relatively non-specific dopaminergic innervation of the striatum has led to a new approach to the possible treatment of Parkinson's disease. This involves implantation of dopamine rich mesencephalic brain tissue from 6–7-week-old aborted human fetuses into the striatum of affected patients. Extensive work in animals has shown that such tissue can survive and grow in adult brain. The graft releases dopamine into the host tissue and can cause postsynaptic dopamine receptor binding sites to return to normal density in previously denervated tissue. It seems possible that the same may be true in humans. A small number of patients with Parkinson's disease have been successfully transplanted with human fetal mesencephalic grafts. Serial PET studies have shown increasing fluorodopa uptake by the graft in the months following implantation. This has been accompanied by clear signs of clinical improvement, which in some cases has led to a reduction or even total withdrawal of previous antiparkinsonian medication (Lindvall *et al.* 1992).

Physiological studies in Parkinson's disease

Rigidity

The possible mechanisms of Parkinsonian rigidity have been discussed in Chapter 6. Briefly, the increased tone felt on passive manipulation of the joints is due to a combination of the patients' difficulty in relaxing their muscles, and also to abnormal stretch reflex activation of the muscles. There is no evidence for increased excitability in the spinal monosynaptic pathway which might explain the increased stretch reflexes. However, long-latency stretch reflexes often are enhanced and excessive activity of these reflexes may be partly responsible for the rigidity of Parkinson's disease (Figure 11.17). There is no universal agreement on which anatomical pathways mediate these long-latency reflexes in humans. However, there is one attractive theory which explains the influence of the basal ganglia on such responses: (i) the long-latency reflexes may use a transcortical reflex pathway (via thalamus,

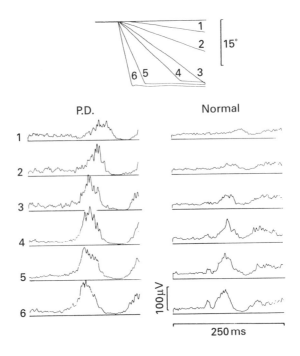

Figure 11.17 Average stretch reflex EMG responses in the flexor pollicis longus following ramp stretches of different rates applied to the thumb (top traces, 1–6). On the left are responses from a severely rigid patient with Parkinson's disease (P.D.); on the right are the responses from a normal subject. In the patient, only the long-latency component of the reflex (latency of 50 ms) can be seen; in the normal, at high-stretch velocities, a small, short-latency component can be distinguished. Note how the sensitivity to stretch is greatly increased in the patient, as is the maximum size of the reflex response. (From Rothwell *et al.*, 1983b, Figure 3; with permission.)

sensory cortex, motor cortex and spinal cord); (ii) electrical stimulation of the supplementary motor area (Chapter 9) is known to suppress activity in monkey motor cortex evoked by muscle stretch; (iii) much of the basal ganglia output is directed towards the SMA. Thus it is possible that defects of basal ganglia function produce rigidity via their action on the SMA and long-latency reflexes.

This hypothesis is not proven, however. Other workers favour an entirely different explanation. They suggest that slowly conducting group II afferents from secondary muscle spindle afferent terminals contribute strongly to the long-latency stretch reflex. The long latency of the response would then be due to slowly conducting afferent pathways, rather than to the long distance of the supraspinal pathway.

They suggest that the basal ganglia may have some influence on the excitability of this pathway. (Rothwell *et al.* 1983b; Meara and Cody 1992 give further details and references on rigidity and stretch reflexes.)

Tremor

Parkinsonian rest tremor has a frequency of 4–6 Hz, and is produced by alternating activity in antagonist muscles. Motor units are recruited in their normal order during bursts of tremor, but may fire twice, at high frequency (50 Hz or so) during each burst. In those patients in whom tremor persists during voluntary movement (action tremor), the frequency increases to 6–8 Hz, and the EMG activity sometimes becomes synchronous in antagonists. Because the rest and action tremor are believed by some workers to have different mechanisms, they will be discussed separately.

Rest tremor can be produced in experimental animals by lesions in the ventromedial tegmentum of the midbrain. The critical lesion must interrupt four pathways involving the substantia nigra and red nucleus: nigrostriatal, cerebellorubral, cerebellothalamic (which travel through the red nucleus) and rubro-olivary tracts (Figure 11.18). Interruption of all these pathways is important: neither electrolytic lesions of the substantia nigra nor 6-hydroxy-dopamine lesions of the ascending dopamine pathways from substantia nigra produce tremor. Similarly, isolated lesions of the superior cerebellar peduncle (or dentate nucleus, from which many of the fibres arise), which contains the ascending cerebellar efferents, does not produce rest tremor. Instead, ataxia and intention tremor (see Chapter 10) are seen.

The conclusion is that in animals, rest tremor can only be produced by damage to the nigrostriatal dopamine pathway combined with damage to cerebellar outflow. Both the thalamus and sensorimotor cortex seem to have an important role in producing the tremor. Microelectrode recordings from both structures reveal groups of neurones which fire bursts of impulses at the same frequency and which accompany the EMG bursts in the limbs. These bursts are seen even after peripheral deafferentation, indicating that they do not need peripheral feedback to occur. Lesions in either thalamus or cortex abolish the rest tremor. The particular region of the thalamus which is lesioned is equivalent to the nucleus ventralis intermedius in man, an area which projects to the motor cortex. Thus, it is not clear whether motor cortical activity arises because it is driven by thalamic activity, or whether the bursting, rhythmic discharge of neurones is a consequence of some interaction between thalamic and cortical circuits. Thalamic neurones show a propensity for rhythmic firing at 5 Hz during sleep and drowsiness, but this usually disappears in the alert state. It may be that a

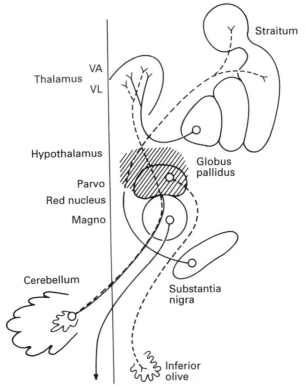

Figure 11.18 Diagram to show the main nervous pathways which have to be lesioned in the monkey to produce a parkinsonian rest tremor. Note loss of the ascending nigrostriatal, cerebellorubral and cerebellothalamic pathways, and of the descending rubro-olivary pathway. (From Pechadre, Larochelle and Poirier, 1976.)

ventral tegmental lesion, which removes several sources of thalamic input reinforces this natural tendency.

Several observations on parkinsonian tremor in humans are consistent with these data from other animals.

(1) Bursts of neuronal discharge in phase with rest tremor can be recorded from the ventralis intermedius nucleus of the thalamus during neurosurgery. Electrical stimulation or lesion of this area immediately abolishes resting tremor.

(2) In the few cases that it was attempted, dorsal rhizotomy to remove feedback from the trembling limb did not abolish tremor.

(3) Damage to the motor cortex, or a stroke in the internal capsule producing hemiplegia, abolishes rest tremor. Thus, in humans, as in the experimental animal, resting tremor seems to be highly dependent upon spontaneous bursting discharge in the thalamus and cortex.

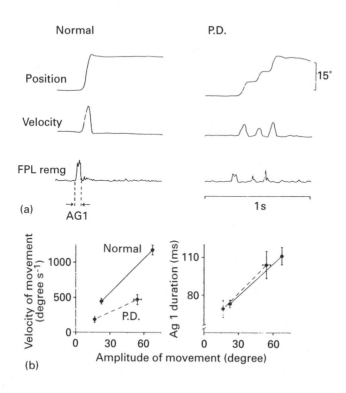

Figure 11.19 Deficits in rapid self-paced thumb (a) and wrist (b) movements in bradykinetic patients with Parkinson's disease. The upper panel shows a representative rapid thumb flexion movement of 20° in a normal subject and a patient with Parkinson's disease. The thumb was firmly clamped at the interphalangeal joint so that no postural activity was needed to steady it. The normal subject produces the movement in a single step, using a large burst of EMG activity in the flexor pollicis longus (FPL) muscle. The patient (P.D.) makes the movement in a series of three small steps, with correspondingly small bursts of EMG activity. The bottom panel shows an analysis of similar rapid flexion movements made at the wrist in a group of eight normal and eight patients with Parkinson's disease. Only the initial step of the patients' movements was analysed. The graph on the left shows that movements of a given extent are much slower in patients than normals. However, the duration of the first burst of agonist EMG activity (AG1) was the same (right). This indicates that the basic instructions for the movement were intact. (Unpublished observations of Berardelli, Rothwell, Dick *et al.*)

From time to time, the question is raised of a possible role for stretch reflex mechanisms in producing tremor. Long-latency stretch reflexes are enhanced in Parkinson's disease, so that it is conceivable that this could lead to instability in the reflex system and result in a self-sustaining oscillation. Peripheral input can undoubtedly influence rest tremor, but only when relatively large inputs are given. For example, forcible extension of the wrist can sometimes 'reset' rest tremor, interrupting the tremor cycle and causing it to restart in phase with the stretch. Similarly, single large electrical stimuli to a peripheral nerve can also 'reset' the phase of ongoing rest tremor. However, there is no indication that under normal conditions the tremor is entirely driven by peripheral input. One good piece of evidence to support this comes from microneurographic recordings from muscle spindle afferents during rest tremor. The spindles fire two bursts of impulses per tremor cycle, once when the receptor bearing muscle is stretched (by antagonist contraction), and once when the receptor bearing muscle contracts. A burst of muscle spindle activity during muscle contraction indicates α–γ coactivation similar to that seen in a voluntary movement. If the muscle contraction had been driven solely by the stretch reflex, the spindles would have been unloaded and would fire only one burst of impulses per tremor cycle.

Although the animal model of resting tremor reproduces many of the aspects of parkinsonian rest tremor in man, there are two features of human Parkinson's disease that remain unaccounted. First, it is not necessary to produce combined damage to both nigral and cerebellar pathways in order to produce resting tremor in man. It is usually thought that damage limited to nigral output as in Parkinson's disease or in MPTP-induced parkinsonism, is sufficient to cause tremor. Nevertheless, there are indications that this may not be entirely true. As pointed out above, PET studies show that fluorodopa uptake correlates well in individual patients with clinical symptoms of akinesia and rigidity, but does not correlate with measurements of the severity of tremor. Some other factor other than striatal dopamine deficiency must be important in producing tremor, but as yet it is unknown. The second problem is that it is odd that lesions of the thalamic nucleus ventralis intermedius in man abolish tremor so effectively. This nucleus receives no direct basal ganglia input, yet is the most effective site for relief of parkinsonian, as well as other forms of tremor. One possible explanation is a lesion in that area interrupts fibres of passage from the basal ganglia to other thalamic nuclei, but at present this theory is speculative.

Action tremor of Parkinson's disease has not been so extensively studied. Some investigators believe that it has a different mechanism to that involved in rest tremor. The reasons for this are: (i) its higher frequency (6–8 Hz); (ii) the frequent appearance of co-contracting EMG activity in antagonist muscles, rather than alternating activity as in rest tremor; and (iii) it is sometimes possible to record tremor which consists of **both** an 8 Hz and a 5 Hz component simultaneously.

A phenomenon related to action tremor is 'cog-wheeling'. This is the feeling of a ratchet-like resistance to movement which can be detected during passive manipulation of a patient's joints. It is produced by bursts of EMG activity in the appropriate muscles which have the same frequency as action tremor. It is believed to be due to superimposition of action tremor onto the increased muscle tone (Lamarre and Joffroy, 1979; Marsden, 1984, 1990).

Akinesia
Akinesia is a term which is often used in a broad (but strictly incorrect) sense to refer to several types of movement deficit in Parkinson's disease. These are:

• lack of spontaneous movement, especially noticeable as an immobile and mask-like expression of the face;
• lack of associated movements (such as swinging the arms when walking);
• an increased reaction time to external stimuli;
• slowness of movement;
• a tendency to make smaller movements than required (for instance, small steps when walking, or small movements of a pen while writing – micrographia);
• fatiguability of repetitive movements, seen, for example, in tapping the hands for any length of time;
• difficulty in performing two different movements at the same time (for instance, rising from a chair to shake hands is performed by normal individuals in one continuous action, whereas the patient with Parkinson's disease will first rise from his chair and only then will he extend his hand in greeting).

Such changes are very similar to those seen in monkeys after lesions of the substantia nigra or globus pallidus. Different books may refer to all or one of these phenomena as manifestations of akinesia. However, since the symptoms may be dissociated from one another in certain patients, it is likely that, in fact, they are not all different manifestations of the same deficit. Akinesia is more strictly defined as a lack of spontaneous or associated movement. Bradykinesia refers to the slowness of voluntary movements.

Poverty of movement is a deficit which is not only seen in patients with Parkinson's disease. Lack of spontaneous movement in conscious patients is seen in a variety of other conditions, involving damage to other parts of the brain, such as the frontal lobes. Pure akinesia is extremely difficult to measure since it represents the absence of movement. Counting the number of movements made under strictly controlled conditions (for example, through a two-way mirror) is the most direct way of measuring akinesia. Another method is to count the rate of spontaneous eye-blinking, which is reduced in patients with Parkinson's disease.

There have been many studies of reaction time in patients with Parkinson's disease. Although many different types of effect have been described, it should be noted that in general, reaction time deficits in Parkinson's disease are relatively small, especially when compared with much larger changes seen in the speed of their movement. This is consistent with electrophysiological data showing that the firing of basal ganglia neurones changes relatively late in simple reaction time tasks, implying that their role in initiating movements is limited. Reaction time studies in man are useful for a different reason: providing information on processes involved in preparation for a movement.

Most studies to date have found that compared with normal subjects, the simple reaction time of patients with Parkinson's disease is increased, whilst their choice reaction time is much less affected. The experiments are usually of the following design: in a simple reaction time task, the subject knows in advance what movement he will have to make (e.g. move to the left), and the 'go' signal, which is often an audio tone, simply tells the subject **when** to move. In a choice reaction time task, the subject does not know which movement he may have to make. For example in a two choice task he may have to move either to the left or the right. When the 'go' signal is given, it not only tells the subject **when** to move, but also **where** to move (e.g. a low tone indicates moves left, a high tone move right).

Choice reaction times, because of the extra processing required when the 'go' signal is given are always longer than simple reaction times. Simple reaction times are short because subjects can prepare some aspects of the movement in advance. The preferential impairment of patients in simple reaction time tasks can therefore be explained by a failure to make use of advance information. In other words, even though the patients know what movement they will have to make next, they do not use this information as effectively as normals to speed up their reaction. In neural terms, it is hypothesized that the lack of dopamine input interferes with the activity of the preparatory neurones in the basal ganglia. This activity may normally facilitate appropriate circuits for the forthcoming movement, and hence reduce the time taken to release the movement by other structures. The situation might be analogous to the discharge pattern of some substantia nigra pars reticulata cells in memory-guided saccades. They reduce their firing rate after brief presentation of the target and this is maintained until the release of the saccade by the later 'go' signal. The decreased pars reticulata output to the colliculus may increase collicular excitability (being insufficient on its own to evoke a saccade) and decrease the time taken for the final voluntary command to make the saccade.

Another line of evidence supports the suggestion that preparation for movement may be affected in Parkinson's disease. The Bereitschaftspotential or premovement potential is decreased in Parkinson's disease. As described in Chapter 8, this is a slow EEG potential which can be recorded from the scalp prior to self-initiated voluntary movement. It consists of three components, the NS1 (lasting from about 1.5–0.6 s prior to the movement, the NS2 (lasting from 0.6 s to just before movement onset) and the MP (a small potential appearing just prior to movement). All these potentials seem to be generated in both primary motor and supplementary motor areas. The early parts of the potential (NS1) are produced by activity in both hemispheres whereas the later parts have a larger contribution from the hemisphere contralateral to the moving side. In Parkinson's disease, the NS1 component is reduced in amplitude, and this has been taken as evidence for lack of basal ganglia outflow to one of the main targets of the motor loop, the supplementary motor area (Dick, Rothwell Day *et al.*, 1989).

Bradykinesia, the slowing of movements, is more easily measured. A simple experiment is to ask patients to flex their wrist as rapidly as possible through several different distances. Normal subjects perform such tasks in a very repeatable and stereotyped manner. The EMG consists of a burst of activity in the agonist, flexor muscles and is followed by a burst of activity in the antagonist, extensor muscles. The agonist burst provides the impulsive force for the movement, and the antagonist helps to brake the movement so that it does not overshoot the end point. In large movements, the agonist burst is large and because of this the peak velocity is greater than in small movements. The duration of the EMG bursts and their relative latencies are remarkably constant from one individual to another. Because agonist and antagonist bursts are often complete before the limb has reached its final end point, and because the EMG pattern can be seen even in totally deafferented individuals, the 'ballistic burst pattern' is thought to represent a single central programme of muscle activation.

When patients with Parkinson's disease attempt these tasks, two striking deficits are seen (Figure 11.20). Their movements are much slower than normal and the size of the movement is smaller than normal. Because of this latter effect, patients often make large movements in a series of smaller steps. In the EMG this is reflected as a decrease in size of the first burst of agonist EMG activity which results in a smaller impulsive force for the movement than normal. If the movement is made in a number of small steps, then there is an accompanying number of small EMG bursts in the agonist. The important feature of the EMG pattern is that although it is reduced in amplitude,

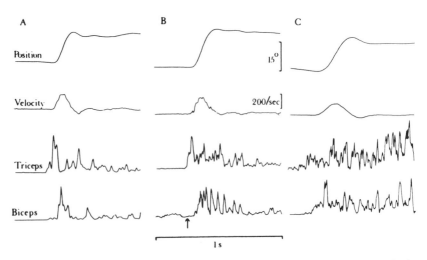

Figure 11.20 Single, rapid, extension movements of the elbow through 20° in three subjects with dystonia of different severity and extent. (a) Segmental dystonia affecting the arm and shoulder; (b) Segmental dystonia of the arm, shoulder and neck; (c) generalized dystonia. In (a), the movement and the EMG pattern is normal, with the typical initial burst of EMG in the agonist (triceps), followed by a burst in the antagonist (biceps). In (b) and (c), the EMG pattern becomes more disorganized. The bursts increase in duration, and there is pronounced co-contraction between biceps and triceps. The movements are slower than normal. (From Rothwell *et al.*, 1983a, Figure 4; with permission.)

the duration of the bursts and the latency of the antagonist burst are quite normal. Because of this it is thought that the parts of the basal ganglia affected in Parkinson's disease do not play a direct role in producing the basic 'ballistic burst pattern'. The deficit is in scaling the amplitude of this pattern to match that required in the movement.

Although deficits in simple movements made about a single joint are easy to measure, it is often found that the impairments on this task are not correlated well with the clinical measures of bradykinesia. However, clinical evaluation is usually carried out by asking patients to perform complex tasks which often involve free movement of the whole arm in a sequence of different actions (like touching each finger in turn with the thumb). In fact, when more complex actions are examined quantitatively, additional deficits show up in the performance of parkinsonian patients.

One such task is the 'squeeze' and 'flex' task devised by Benecke *et al.* (1986, 1987). Subjects sit with their arm in a lightweight manipulandum, allowing free movement at the elbow. At the end of the manipulandum they grasp a stiff force transducer between the thumb and finger. Various combinations of task can then be performed: elbow flexion through a given angle, squeeze of the force transducer to a given force, simultaneous performance of flex and squeeze, or a sequential movement consisting of a squeeze followed by a flex. Patients with Parkinson's disease perform each simple movement (flex or squeeze) more slowly than normals. Unexpectedly, when they perform both movements together, or sequentially, their performance deteriorates even further. Each individual component of the movement is made even more slowly than usual, and, in the sequential task, there is a much longer interval between the start of the squeeze and the start of the flex than normal. This extra slowing, which is never seen in normal subjects is thought to reflect the clinical problems that the patients have when trying to do more than one thing at once. It indicates that the basal ganglia have a crucial role to play in the performance of complex movements, both in running two different tasks simultaneously and in timing the onset of tasks in a sequence. Perhaps the former results from a deficit in the flow of information through the multiple parallel loops within the main motor loop of the basal ganglia. Perhaps the latter is due to absence of predictive cell discharge normally seen in some pallidal neurones during sequential movements.

A striking feature of the akinesia of patients with Parkinson's disease is a way in which visual input can affect movement. Patients may find it easier to walk along a pattern of stripes painted on the ground on a plain surface. In contrast, many patients 'freeze' when trying to walk through a doorway. Even if the door is held open, they halt on the threshold and find it impossible to pass through. Their feet appear to be stuck on the ground. Frequently, the only way to get the patient moving again is to use another visual input. Placing one's foot before the patient so that he has to step over it to carry on, usually proves to be effective.

The effect of visual input on movement can be analysed in a laboratory task. Normal subjects and patients with Parkinson's disease can learn to track a moving visual target on a computer screen by moving a joystick backwards and forwards. If the target moves in a predictable way, subjects learn the task and begin to move the joystick in advance of their perception of target movement. In this way, they can keep up with the target much better than if they were lagging its motion by one reaction time. Parkinsonian subjects are almost as good as normal in learning to predict target motion. However, they fare much worse if the target is blanked from the screen for part of the trial. They may even fail to perform the required movement even though they know what to do. They are much more dependent on visual feedback to retrieve their movement from memory than normal subjects. In the same way as they fail adequately to use advance information in simple reaction time tasks, during tracking, they have difficulties accessing stored information about the movement unless visual feedback is allowed.

These deficits are sometimes discussed in terms of Goldberg's distinction between internally and externally guided movements (see Chapter 9). Patients with Parkinson's disease have a problem in performing internally-guided movements, whilst externally-guided (i.e. visually-guided in this instance) movements are relatively normal. If the supplementary motor area is indeed an important control area for internally-guided movement, then the problem in Parkinson's disease may reflect inadequate basal ganglia activation of supplementary motor area through the 'motor' loop. (Schwab, Chafetz and Walker, 1954; Flowers, 1976, 1978; Day *et al.*, 1984; Evarts, Teravainen and Calne, 1981; Marsden, 1990).

Postural difficulties
The disorders of posture seen in Parkinson's disease are beautifully described by Martin in his classic book on the basal ganglia and posture (1967). Patients walk with a flexed posture, bent at the knees, elbows and trunk. They lose their righting reactions: seated on a tilt table, they fail to keep their trunk vertical when the table is tilted from side to side. They also lose stepping reactions, usually seen following an unexpected push to the shoulders, and protective reactions of the arms when falling. The result is that not only do patients with Parkinson's disease have frequent falls, but they also damage themselves more than normal when they do fall. These postural reactions are triggered by proprioceptive and/or labyrinthine input.

Another type of postural reaction is abnormal in Parkinson's disease. These are the anticipatory postural reactions described in Chapter 8. They are evoked at short latency in postural muscles by proprioceptive input from a distant part of the body. For example, short forward pulls to the arm can evoke reflex activation of the soleus muscle with a latency of 80–90 ms. These reflexes are used to prevent the anticipated body sway that the pull to the arm is likely to generate. They are not as powerful as the stepping and protective reactions referred to above and might be envisaged as providing a 'fine tuning' mechanism for postural stability. Anticipatory postural reflexes are small or absent in patients with Parkinson's disease who have a history of postural instability (Traub, Rothwell and Marsden, 1980).

The flexed posture typical of Parkinson's disease also is seen in animals with lesions of the basal ganglia. However, in animals the lesions must be large and bilateral for this to happen. In Parkinson's disease, too, the postural instabilities are prominent only in the later stages of the illness, when the disease may spread beyond the substantia nigra. Martin (1967) believed that postural difficulties were the principal symptom of the disease. They would be evident, he thought, not only in examining whole body posture but also in failure of postural fixation at proximal joints when making movements with a distal extremity. As pointed out above, this idea is now believed to be wrong. Distal movements may be slow even in circumstances when postural fixation is not needed. Akinesia and bradykinesia are not produced by defects in postural control. It is more likely that they contribute to postural instability by affecting operation of postural reflex systems.

Chorea

Chorea is a common symptom of basal ganglia disease, and refers to involuntary, purposeless and irregular movements of the body affecting proximal or distal joints. Choreic movements are relatively slow and of small amplitude compared with hemiballismus (see below). Almost all muscle groups may be affected, but the chorea will seem to move randomly from one part of the body to another. Facial and limb movements are equally affected. Grimacing, twitching or writhing of the limbs and head all occur unpredictably.

Chorea may result from many different causes, one of the most common of which, and that on which most studies have been performed, is Huntington's disease. This is a dominantly inherited disease with symptoms involving not only chorea but also a progressive decrease in higher cortical function. The symptoms do not appear until about the age of 30–40 years, by which time the next generation has usually been born; 50% of these children will have the disease. Involuntary movements and psychiatric changes often begin together, and are seen at first only as increased restlessness and eccentricity. However, chorea gradually becomes more florid until the movements are almost incessant. Patients become incontinent, unable to feed themselves and severely demented. Death follows within 10 years of the onset of symptoms. In the final stages of the disease, the involuntary movements become less frequent and the muscle tone greatly increased.

In the early stages of the disease there is a relatively selective loss of the GABA/enkephalin neurones in the striatum which project to the GP_e (Albin *et al.*, 1992). Later, all the medium spiny neurones of the striatum are affected and the pathology becomes widespread, with the principal affected areas of brain being the caudate, putamen, globus pallidus and cerebral cortex. In the cerebral cortex, frontal and precentral regions are affected with cell loss mainly in the deeper layers. It is believed that chorea is the result of basal ganglia deficit, whereas dementia is caused by cortical cell loss.

Paradoxically, despite the excess of involuntary movements, voluntary movements of patients with Huntington's disease are, like those of

Parkinson's disease, slower than normals. However, they are much more variable than those of patients with Parkinson's disease possibly because of contamination by involuntary activity. The question arises as to whether it is possible, on the basis of our present models of the basal ganglia, to explain this strange combination of hyper- and brady-kinesia. One hypothesis is that the functional consequences of changes in activity of the direct and indirect pathways through the basal ganglia are not equivalent. As pointed out above, underactivity in the projection from subthalamic nucleus to the internal pallidal segment (produced by excessive inhibition from the external pallidal segment) may play a crucial role in hyperkinesia. Perhaps underactivity in the direct path-way to the internal pallidal segment is important in producing bradykinesia. Thus, in Parkinson's disease, lack of dopamine results in underactivity in the direct striatal projection to internal pallidal seg-ment, and, via an opposite action on the indirect pathway, to excess excitatory output from the subthalamic nucleus. This would produce both bradykinesia and hypokinesia. In the midstages of Huntington's disease, when all cells of the striatum may be affected, reduced output from the subthalamic nucleus would result in hyperkinesia, and a lack of activity in the striopallidal projection to the internal pallidal segment to bradykinesia. However the theory is speculative and is difficult to reconcile with the discovery that all parkinsonian symptoms (akinesia, bradykinesia and tremor) of the MPTP treated monkey are alleviated by a lesion of the subthalamic nucleus.

Other studies of the physiology of Huntington's disease have shown that cortical somatosensory evoked potentials (the response of the cortex to peripheral nerve stimuli) are absent or reduced, as are long latency stretch reflexes in the small hand muscles. Short latency, spinal stretch reflexes are present. If long latency stretch reflexes use a trans-cortical pathway, then their absence, and that of the somatosensory evoked potential, may be due to two possible reasons. First, in Huntington's disease there is loss of cells in the cortex and in the thalamocortical projection neurones. Thus, the cortical response to sen-sory input would be reduced. Second, there may be a basal ganglia output to the reticular nucleus of the thalamus, which is thought to be able to exert a gating action on transmission through the sensory relay nuclei. If there was excessive gating in Huntington's disease, this would also reduce cortical responses to sensory input. Interestingly, long latency stretch reflexes are present in forearm and biceps muscles, even though they are absent in the hand, suggesting that the reflexes in these muscles may not be transcortical in origin. (Thompson *et al.*, 1988 give more details of the pathophysiology of Huntington's disease.)

Athetosis describes distal writhing movements of the fingers, usually accompanied by similar slow movements of the tongue. A characteristic feature is the posture of the outstretched arm: adducted at the shoulder, partially flexed at the elbow and wrist with hyperextension of the fingers. The movements are superimposed upon this posture. Athetosis is uncommon, but often seen in children with cerebral palsy. Birth trauma or malformation is believed to produce damage to the putamen which is responsible for the athetosis.

Torsion dystonia

Torsion dystonia is a term used to describe twisted (torsion) sustained postures of the limbs, neck or trunk. In contrast to the forever changing, fleeting muscle contractions of chorea, torsion dystonia describes a relatively fixed posture, maintained by abnormal muscle activity (dystonia). Torsion dystonia is rare and, in most instances, its cause is unknown. If all parts of the body are affected the condition is known as generalized torsion dystonia or dystonia musculorum deformans. This usually begins in childhood, perhaps affecting one leg while walking, and spreads gradually to affect all four limbs, the trunk and neck. The sustained muscle contractions contort the body into grotesque postures, making it impossible for the patient to walk, clothe or feed himself. If symptoms begin in adult life, the progression is likely to be much less severe. The most frequently affected part is the neck, which becomes twisted to one or other side of the body (torticollis). Sometimes the spasms are limited to one hand or forearm, and only occur when the patient performs some delicate manual task such as writing or typing or playing a musical instrument (writers', typists' or musicians' cramp).

Even in severe cases of generalized torsion dystonia, investigation of the brain usually reveals no gross abnormality. However, there are examples where symptoms of torsion dystonia have been linked with basal ganglia damage. Wilson's disease, a condition resulting from an abnormality of copper metabolism, often presents with abnormal movements and dystonic postures and, in some cases, this is associated with damage to the putamen. In addition, since the advent of the CT brain scanner, several cases have been reported of individuals with localized lesions of the putamen and globus pallidus, and contralateral dystonia on the opposite side of the body. No specific neurochemical abnormalities have yet been revealed, so that it is not yet possible to fit dystonia into the circuitry used to explain other diseases of the basal ganglia.

Little is known about the physiology of torsion dystonia. The muscle spasms are characterized by co-contraction of antagonist muscles, rather than the more usual reciprocal pattern seen in many normal voluntary movements. Because of this it has been suggested that there may be an associated disorder of reciprocal inhibition in these patients (see Chapter 6). Unlike the situation in Parkinson's disease, the 'ballistic burst pattern' of EMG activity in rapid limb movements may be severely disrupted in torsion dystonia. Burst duration, latency and the reciprocal relationship between antagonists all may be affected (Figure 11.20) (Rothwell *et al.*,1983a).

Hemiballismus

As in the monkey, hemisballismus in man results from lesions of the subthalamic nucleus, usually due to vascular infarction. Recovery occurs within a few weeks. The abnormal movements consist of gross, rapid, flinging movements of the contralateral arm (and sometimes leg), particularly at proximal joints.

Bibliography

Review articles and books

Crossman, A. R. and Sambrook, M. A. (1989) *Neural mechanisms in disorders of movement*, John Libbey, London.
Ciba Foundation (1984) Functions of the basal ganglia, *Ciba Foundation Symposium*, **107,** Pitman, London.
Marsden, C. D. (1982) The mysterious motor function of the basal ganglia, *Neurology*, **32,** 514–539.
Trends in Neuroscience (1990), volume 13, no. 10. Special edition on the basal ganglia.

Original papers

Albin, R. L., Young, A. B. and Penney, J. B. (1989) The functional anatomy of basal ganglia disorders, *Trends Neurosci.*, **12,** 366–375.
Albin, R. L., Reiner, A., Anderson, K. D. *et al.* (1992) Preferential loss of striato-external pallidal projection neurones in presymptomatic Huntington's disease, *Ann. Neurol.*, **31,** 425–430.
Alexander, G. E., DeLong, M. R. and Strick, P. L. (1986) Parallel organisation of functionally segregated circuits linking basal ganglia and cortex, *Annu. Rev. Neurosci.*, **9,** 357–381.
Alexander, G. E. and Crutcher, M. D. (1990a) Functional architecture of basal ganglia circuits: neural substrates of parallel processing, *Trends Neurosci.*, **13,** 266–271.
Alexander, G. E. and Crutcher, M. D. (1990b) Preparation for movement: neural representations of intended direction in three motor areas of the monkey, *J. Neurophysiol.*, **64,** 133–150; 150–163; 164–178.
Anderson, M. E. and Horak, F. B. (1985) Influence of the globus pallidus

on arm movement in monkeys, Parts 1, 2, and 3, *J. Neurophysiol.*, **52**, 290–304; 305–322; **54**, 433–448.

Anderson, M. E. and Turner, R. S. (1991) A quantitative analysis of pallidal discharge during targeted reaching movement in the monkey, *Exp. Brain Res.*, **86**, 623–632.

Benecke, R., Rothwell, J. C., Dick, J. P. R. *et al.* (1986) Performance of simultaneous movements in patients with Parkinson's disease, *Brain* **109**, 739–757.

Benecke, R., Rothwell, J. C., Dick, J. P. R. *et al.* (1987) Disturbance of sequential movements in patients with Parkinson's disease, *Brain*, **110**, 361–379.

Bergman, H., Wichmann, T. and DeLong, M. R. (1990) Reversal of experimental parkinsonism by lesions of the subthalamic nucleus, *Science*, **249**, 1436–1438.

Brotchie, P., Ianseck, R. and Horne, M. K. (1991) Motor function of the monkey globus pallidus, Papers 1 and 2, *Brain*, **114**, 1667–1702.

Carpenter, M. B. (1981) Anatomy of the corpus striatum and brainstem integrating systems, in V. B. Brooks (ed.), *Handbook of Physiology*, sect. 1, vol. 2, part 2, Williams and Wilkins, Baltimore, pp. 947–995.

Chevalier, G. and Deniau, J. M. (1990) Disinhibition as a basic process in the expression of striatal function, *Trends Neurosci.* **13**, 277–280.

Crutcher, M. D. and DeLong, M. R. (1984) Single cell studies of the primate putamen, Parts 1 and 2, *Exp. Brain Res.*, **53**, 233–258.

Day, B. L., Dick, J. P. R. and Marsden, C. D. (1984) Patients with Parkinson's disease can employ a predictive motor strategy, *J. Neurol. Neurosurg. Psychiatr.*, **47**, 1299–1306.

DeLong, M. R. and Georgopoulos, A. P. (1979) Motor function of basal ganglia as revealed by studies of single cell activity in the behaving primate, *Adv. Neurol.*, **24**, 131–140.

DeLong, M. R. and Georgopoulos, A. P. (1981) Motor functions of the basal ganglia, in V. B. Brooks (ed.), *Handbook of Physiology*, sect. 1, vol. 2, part 2, Williams and Wilkins, Baltimore, pp. 1017–1061.

DeLong, M. R., Crutcher, M. D. and Georgopoulos, A. P. (1985) Primate globus pallidus and subthalamic nucleus: functional organisation, *J. Neurophysiol.*, **53**, 530–543.

Dick, J. P. R., Rothwell, J. C., Day, B. L. *et al.* (1989) The Bereitschaftspotential is abnormal in Parkinson's disease, *Brain*, **112**, 233–244.

Evarts, E. V., Teravainen N. H. and Calne, D. B. (1981) Reaction time in Parkinson's disease, *Brain*, **104**, 167–186.

Filion, M. and Tremblay, L. (1991) Abnormal spontaneous activity of globus pallidus neurones in monkeys with MPTP induced parkinsonism, *Brain Res.*, **547**, 142–151; 152–161.

Flowers, K. A. (1976) Visual 'closed loop' and 'open loop' characteristics of voluntary movement in patients with parkinsonism and intention tremor, *Brain*, **99**, 269–310.

Flowers, K. A. (1978) Lack of prediction in the motor behaviour of parkinsonism, *Brain*, **101**, 35–52.

Graybiel, A. M. (1990) Neurotransmitters and neuromodulators in the basal ganglia, *Trends Neurosci.*, **13,** 244–253.

Hikosaka, O. and Wurtz, R. H. (1983) Visual and oculomotor functions of monkey substantia nigra pars reticulata, parts 1–4, *J. Neurophysiol.*, **49,** 1232–1301.

Hore, J. and Vilis, T. (1980) Arm movement performance during reversible basal ganglia lesions in the monkey, *Exp. Brain Res.*, **39,** 217–228.

Lamarre, Y. and Joffroy, A. J. (1979) Experimental tremor in monkey: activity of thalamic and precentral cortical neurones in the absence of peripheral feedback, *Adv. Neurol.*, **24,** 109–122.

Lindvall, O., Widner, H., Rehncrona, S. *et al.* (1992) Transplantation of foetal dopamine neurons in Parkinson's disease: one year clinical and neurophysiological observations in two patients with putaminal implants, *Ann. Neurol.*, **31,** 155–165.

Marsden, C. D. (1984) Origins of normal and pathological tremor, in L. J. Findley and R. Capildeo (eds), *Movement Disorders: Tremor*, Macmillan, London, pp. 37–84.

Marsden, C. D. (1990) Neurophysiology, in G. Stern (ed.) *Parkinson's Disease*, Chapman & Hall, London, pp. 57–98.

Martin, J. P. (1967) *The Basal Ganglia and Posture*, Lippincott, Philadelphia.

Meara, R. J. and Cody, F. W. J. (1992) Relationship between electromyographic activity and clinically assessed rigidity studied at the wrist joint in Parkinson's disease, *Brain*, **115,** 1167–1180.

Mink, J. W. and Thach, W. T. (1991) Basal ganglia motor control, Parts 1, 2, and 3, *J. Neurophysiol.*, **65,** 273–351.

Pechadre, J. C., Larochelle, L. and Poirier, L. J. (1976) Parkinsonian akinesia, rigidity and tremor in the monkey, *J. Neurol. Sci.*, **28,** 147–157.

Romo, R. and Schultz, W. (1990) Dopamine neurones of the monkey midbrain: contigences of responses to active touch to self-initiated arm movement, *J. Neurophysiol.*, **63,** 592–606; 607–624.

Rothwell, J. C., Obeso, J. A., Day, B. L. *et al.* (1983a) Pathophysiology of dystonias, *Adv. Neurol.*, **39,** 851–863.

Rothwell, J. C., Obeso, J. A., Traub, M. M. *et al.* (1983b) The behaviour of the long-latency stretch reflex in patients with Parkinson's disease, *J. Neurol. Neurosurg. Psychiatr.*, **46,** 35–44.

Schwab, R. S., Chafetz, M. E. and Walker, S. (1954) Control of two simultaneous motor acts in normals and in parkinsonism, *Arch. Neurol. Psychiatr.*, **72,** 591–598.

Smith, A. D. and Bolam, J. P. (1990) The neural network of the basal ganglia as revealed by the study of synaptic connections of identified neurones, *Trends Neurosci.*, **13,** 259–265.

Thompson, P. D., Berardelli, A., Rothwell, J. C. *et al.* (1988) The coexistence of bradykinesia and chorea in Huntington's disease and its implications for theories of basal ganglia control of movement, *Brain*, **111,** 223–244.

Traub, M. M., Rothwell, J. C. and Marsden, C. D. (1980) Anticipatory postural reflexes in Parkinson's disease and other akinetic-rigid syndromes and in cerebellar ataxia, *Brain*, **103,** 393–412.

Tremblay, L., Filion, M. and Bedard, P. J. (1989) Responses of pallidal neurones to striatal stimulation in monkeys with MPTP induced parkinsonism, *Brain Res.*, **498,** 17–33.

Williams, P. L. and Warwick, R. (1975) *Functional Neuro-anatomy of Man*, Churchill Livingstone, Edinburgh.

Index